Rishabha Malviya, Sonali Sundram, Babita Gupta
Sustainability in Healthcare

De Gruyter Series on Smart Computing Applications

Edited by
Prasenjit Chatterjee, Dilbagh Panchal, Dragan Pamucar and Sharfaraz Ashemkhani Zolfani

Volume 10

Rishabha Malviya, Sonali Sundram,
Babita Gupta

Sustainability in Healthcare

mHealth, AI, and Robotics

DE GRUYTER

Authors
Dr. Rishabha Malviya
Galgotias University
Greater Noida
India
rishabhamalviya19@gmail.com

Prof. Sonali Sundram
Galgotias University
Greater Noida
India
sonaliaim13@gmail.com

Babita Gupta
Galgotias University
Greater Noida
India
guptababita2001@gmail.com

ISBN 978-3-11-143635-7
e-ISBN (PDF) 978-3-11-143643-2
e-ISBN (EPUB) 978-3-11-143658-6
ISSN 2700-6239

Library of Congress Control Number: 2024935429

Bibliographic information published by the Deutsche Nationalbibliothek
The Deutsche Nationalbibliothek lists this publication in the Deutsche Nationalbibliografie; detailed bibliographic data are available on the internet at http://dnb.dnb.de.

Cover image: monsitj/iStock/Getty Images Plus
Typesetting: Integra Software Services Pvt. Ltd.
Printing and binding: CPI books GmbH, Leck

MIX
Papier | Fördert gute Waldnutzung
FSC® C083411

www.degruyter.com

Preface

The ongoing search for sustainability is a fundamental requirement that connects innovation, technology, and ethical practices in the complex field of healthcare. Healthcare is changing, and with it, so is the need to develop systems that support human and environmental health by restoration as well as sustaining.

The book *Sustainability in Healthcare: mHealth, AI, and Robotics* highlights the significant intersections between sustainability, technology, and human welfare. To create a sustainable healthcare ecosystem, renowned experts and thought leaders come together to prepare chapters showing revolutionary approaches that leverage the potential of mHealth, artificial intelligence, robots, and cutting-edge techniques.

This book explores sustainable hospital services, robotics integration, and healthcare design. It highlights the importance of artificial intelligence, additive manufacturing, mobile health applications, biomedical engineering, waste management, sustainable supply chains, and telemedicine in pandemic preparedness. The book discusses ethical issues, practical solutions, and imaginative ideals, incorporating sustainability dimensions like social justice, environmental conservation, and fiscal responsibility into healthcare.

The editors have put together an extensive compilation of information, valuable suggestions, and a dedication to developing a paradigm for healthcare that supports long-term viability and adaptability. By highlighting the developments in mHealth AI and robotics, editors believe that this book will motivate healthcare professionals, researchers, and inventors, on a revolutionary path toward a sustainable healthcare future.

https://doi.org/10.1515/9783111436432-202

Foreword

It gives me immense pleasure to present the book *Sustainability in Healthcare: mHealth, AI and Robotics* an insightful chapter collection that has been thoughtfully put together by the editors. The pursuit of sustainability has become a motivating factor in the healthcare sector, pointing the way toward a future where compassion and innovation coexist and the health of patients and the environment are of utmost importance. Dr. Malviya's commitment to this cause is evident throughout this extensive book.

This collection is an example of Dr. Malviya's unwavering dedication to designing a book that prioritizes long-term wellness while also attending to patients' urgent needs. Each chapter contains a mosaic of insights expertly crafted by recognized experts, scholars, and trailblazers in their respective professions.

Editors highlights an exceptional vision for a sustainable healthcare paradigm through skillful curation of a wide range of topics, from the integration of robotics to the ethical considerations surrounding artificial intelligence, from the crucial role of additive manufacturing to the profound impact of telemedicine during pandemics.

The editor of this book, Dr. Rishbaha Malviya, provides a plethora of information, creative approaches, and game-changing concepts that promote a future in which healthcare and sustainability are integrated. Those who aspire to a more sustainable, ethical, and cutting-edge healthcare landscape appreciate him for his commitment, knowledge, and determination.

Best Wishes.

Dr. Dhruv Galgotia
CEO,
Galgotias University, Greater Noida

https://doi.org/10.1515/9783111436432-203

Contents

About the book

Sustainability in Healthcare: mHealth, AI and Robotics is an in-depth discussion of sustainable methods in the healthcare industry. It looks into critical issues, providing insights and ideas for improving sustainability, integrating technology, and addressing environmental, social, and economic concerns. The first and second chapters focus on building sustainable hospital healthcare services, with an emphasis on rural and community healthcare improvement. Chapter 2 investigates the use of robots in healthcare for sustainability, emphasizing their potential to help achieve the Sustainable Development Goals (SDGs). Chapter 3 describes healthcare sustainability and its dimensions, examining a conceptual framework, variables influencing sustainability, and long-term innovation. Chapter 4 focuses on the implementation of AI in healthcare, focusing on ethical concerns as well as its role in accomplishing SDGs and climate action. Chapter 5 examines the possibilities of additive manufacturing in sustainable healthcare, focusing on its role in environmental sustainability, waste reduction, and medical improvements such as tissue engineering and drug delivery. Chapter 6 examines the role of mobile health in patient-centered treatment, analyzing its economic, environmental, social, and behavioral viability. Chapter 7 addresses the use of biomedical engineering in healthcare developments, focusing on diversity and research issues. Chapter 8 discusses healthcare waste management and comprehensive disposal and treatment techniques. Chapter 9 investigates the economic, social, and environmental implications of sustainability in healthcare supply chain management. Chapter 10 highlights the role of telemedicine in healthcare sustainability management during pandemics.

https://doi.org/10.1515/9783111436432-205

Chapter 1
Developing sustainable hospital healthcare services

Abstract: The concept of sustainability is a complicated and controversial topic, frequently presenting challenges in terms of its definition and implementation in practical activities, particularly within the complex healthcare environment. In modern healthcare systems, the development of sustainable hospital healthcare services is of great significance. It is because of a growing awareness of the related issues that affect environmental degradation, economic limits, and societal well-being. The chapter provides a thorough investigation into the crucial transition toward sustainability in the healthcare industry. The study highlights the multidimensional influence of sustainability in healthcare, highlighting its central position in the industry. First, the analysis explores the environmental aspect, providing a detailed explanation of the significant ecological impact of hospitals resulting from their energy usage, resource consumption, and waste production. The environmental effect causes climate change and health problems. Second, the economic aspect is discussed, showing how sustainable healthcare practices may save money. A hospital's financial stability may be greatly strengthened by achieving greater efficiency in energy use, waste management, and procurement. Lastly, the societal dimension is examined, emphasizing the societal benefits of sustainable healthcare, including improved public health outcomes and community well-being. The study analyzes the multidimensional influence of sustainability in healthcare, demonstrating its central position in the industry and also demonstrates the critical requirement for a transition toward sustainable healthcare services in healthcare facilities.

1.1 Introduction

Healthcare systems aim to enhance public health, but come with social and environmental impacts that may negatively impact individuals and the environment. A sustainable hospital structure and administration are essential for promoting wellness and health among patients. A sustainable hospital structure refers to a construction that possesses the ability for convenient maintenance and demonstrates functionality from environmental, social, and economic perspectives. This is accomplished to effectively address the various objectives and requirements of all stakeholders involved. The academic health community has been studying and writing about the negative effects of pollution on human health for a long time. They have looked at specific toxicants like asbestos, lead, mercury, and pesticides, along with various combinations like outdoor and indoor air pollutants. According to estimates from the World Health

https://doi.org/10.1515/9783111436432-001

Organization and the Institute for Health Statistics, pollution caused 9 million premature deaths globally in 2015, accounting for 16% of all deaths [1, 2]. Due to limited pollutant estimates, efforts have begun to extend them and quantify the current and future illness burden from climate change [3]. Healthcare institutions around the nation are undergoing continuous transformations in order to optimize the delivery of healthcare services with reduced funding. To attain economies of scale, current developments include bigger and more specialized hospitals. It is essential to consider this aspect both as a primary necessity and as a component of quality since healthcare facilities must possess the ability to maintain high standards even in dynamic situations. Sustainable construction can be readily managed. It is also useful from an environmental, social, and economic perspective. Public healthcare has an important effect on national economies and is controversial both nationally and locally. Recent data demonstrates how the expenses of the present healthcare system are rising; in Italy, for example, they grew from 5.5% of GDP in 2000 to 7.5% of GDP in 2012 [1, 4].

When considering the Italian context, certain challenges arise about the physical infrastructure, in addition to concerns over administration expenses and excessive expenditures. The majority of these structures are currently outdated and deteriorating, hence affecting their ability to provide an effective service to the general public. Issues are not limited to just the existing facilities; they also pertain to newly constructed hospitals [2, 4]. It is less widely known that the medical industry contributes significantly to pollution and emissions of greenhouse gases (GHGs), which contribute to global warming. A recent investigation has indicated that the healthcare industries in the United States, Australia, England, and Canada collectively release an annual total of 748 million metric tons of carbon dioxide equivalents [5]. If the healthcare industries of these nations were considered separate entities, their GHG emissions would be ranked eighth globally [5]. Furthermore, it is important to note that the healthcare industry plays a significant role in the use of natural resources. This consumption contributes to the existing imbalances in ecosystems, which are emblematic of the Anthropocene era. The connection between environmental concerns and health issues has received attention in the field of planetary health [6].

Hospitals are an essential, yet are costly component of any healthcare system. High expenses are related to both the original investment and the sustained management of a hospital [7]. The establishment of a sustainable healthcare sector and a reasonable return on investment in terms of effective and useful medical facilities are both required [8]. Due to many factors, including advancing technology, aging populations, more patient engagement, and creative care delivery methods, a majority of the current healthcare systems are changing quickly [9, 10]. A strong return on these significant public investments is required. However, hospitals are frequently out-of-date and unable to satisfy the increased expectations, which has resulted in an increase in the amount of money that the government spends on healthcare facilities [11]. Typically, the structures and fixed expenditures comprise a large amount of hospital expenses [12]. According to McKee et al. [13], many decision-makers in the healthcare industry believe that "bigger hospitals are not always better [14]." Bigger hospitals are commonly perceived as more efficient than smaller ones, even when both are operating at their maximum capacity, mostly due to the concept of economics of scale [15, 16], when evaluated in terms of finances. Although the presence of big hospitals might potentially enhance operational efficiency inside their premises, it is important to acknowledge that evaluating their sustainability requires consideration of additional factors. The healthcare industry is often recognized as a highly demanding sector. The healthcare business experiences significant growth on a worldwide scale, comprising a wide range of healthcare facilities such as hospitals, clinics, nursing homes, and institutes that aim to deliver primary, secondary, and tertiary levels of patient care [17].

1.2 The operational aspects of the healthcare industry and its impact on the environment

The healthcare sector constitutes around 10% of the worldwide economic output, with significant variation across different nations [18]. Similar to several other service industries, the relative ecological impact of the healthcare industry at a worldwide scale is comparatively less when compared to other industrial sectors. However, it has been calculated that healthcare is responsible for 4.4% of global GHG emissions, with certain nations contributing a significantly greater amount. In general, the healthcare industry is characterized by its significant energy consumption due to the continuous operating requirements, essential mechanical services such as lighting, heating, ventilation, and cooling, and the necessity for high-performance surroundings to facilitate the provision of services [19]. Research examining healthcare GHG emissions in the 36 nations belonging to the Organizations for Economic Cooperation and Development, or OECD, together with India and China, has been conducted. These countries collectively represent 54% of the global population and contribute to 78% of worldwide output [20]. The United States had the greatest healthcare emissions per individual, at about 25 times that of India, which had the lowest. According to one estimate, the healthcare sector in the USA accounts for 8.5% of all emissions, more than all other nations, and about 25% of global healthcare emissions.

Furthermore, GHG emissions have been rising in the US recently [4]. The healthcare sector in the United States has significant interconnections with industrial operations that are responsible for a substantial portion of the country's air, water, and soil pollution, the emissions originating from both direct and indirect sources within the healthcare industry, as well as their potential adverse impacts on public health. The study utilized economic input and output life cycle evaluation modeling to evaluate the adverse public health and environmental consequences. In the year 2013, the healthcare industry contributed known proportions of air pollution emissions and their associated effects at the national level. These effects encompassed acid rain (12%), emissions of GHGs (10%), pollution formation (10%), criteria-airborne contami-

Table 1.1: An overview of greenhouse gas emissions originating from the healthcare sector, categorized according to their respective scopes.

The scope of emissions	Related actions
Scope 1: Direct, health-related industries	Medicinal center operations and vehicles (used to provide patient transport) that rely on fossil fuel burning, and the use of medicinal gases and refrigerants
Scope 2: Indirectly, through the cost of providing healthcare facilities with the energy needed to run their utilities	Electricity generation, air conditioning, heating, steam, and icy water
Scope 3: Indirect, connected to facilities and other financial resources, tools, devices, goods, pharmaceuticals, and so on	Pharmaceuticals, clothing, data centers, and uses further down the line (home care, waste management, etc.)

nants (9%), and depletion of stratospheric ozone (1%), as well as cancer-causing and noncarcinogenic airborne pollutants (1–2%).

Common categorizations of the health sector's environmental impact [4] (Table 1.1):

Scope 1 covers the primary emissions originating directly from healthcare facilities, such as those arising from the utilization of automobiles for patient transportation, the release of anesthetic gases, and the combustion of fuels for heating and various operational purposes. Health institutions have the capacity to design policies intended for reducing these GHG emissions. Scope 2 emissions include indirect emissions, such as those that arise from the generation of energy that is acquired to power healthcare facilities. Scope 3 emissions encompass indirect sources, including the purchasing of medications, buying of medical supplies, and the emissions resulting from wastes and home-based care. These emissions present a possibility for prominent healthcare organizations to apply pressure on suppliers, encouraging them to enhance the sustainability of their operations through the implementation of assessments and procurement practices that promote responsible enterprises. According to a recent analysis of US healthcare emissions, Scope 1 emissions made up about 9% of the total, Scope 2 emissions made up about 11%, and Scope 3 emissions made up about 80% of the total [4]. Scope 2 emissions, however, are very sensitive to whether renewable energy sources or fossil fuels are used to power the utility system in the region of operation. The generation of power also has a significant effect on Scope 3 emissions from the health sector. If the entire healthcare industry in the United States used renewable energy to power its electric infrastructure, it would decrease emissions by 39%. Several countries, such as the United States, Canada, Australia, and certain Asian nations, experienced a rise in emissions, but not at the same rate as the growth in healthcare consumption. The rate of growth in healthcare emissions in both Turkey and China exceeds that of expenditure on healthcare [21].

The following are instances of carbon emissions originating from healthcare-related sources: Some medical procedures generate a lot of GHG emissions. Desflurane, a common anesthetic gas, causes ten times the emissions of less polluting anesthesia [22] and contributes to the high HVAC energy consumption of operating rooms (because of extremely higher exchange rates of air that minimize the possibility of infection). The energy consumption of an operating room is three to six times that of a regular hospital room. Succinylcholine had the smallest environmental impact of the 20 injectable anesthetics studied at 11 kg CO_2 equivalents (kgCO_2eq), whereas dexmedetomidine had the most at 3,000 kgCO_2eq [23]. It was extremely difficult to estimate the carbon impact of pharmaceuticals, which may be responsible for 25%, 18%, and 10% of the healthcare sector's environmental impacts in Canada, Australia, and the United States, respectively [24], during the study because it required creating life cycle inventory of GHG emissions for 130 different substances used in the production of these drugs. There is still a significant amount of effort that has to be undertaken to establish comprehensive inventories for healthcare institutions, enabling them to provide precise and reliable data about their environmental impact. A comprehensive analysis of existing research about the carbon footprint of healthcare organizations revealed that hardly six of them included all three aspects [25]. However, various evaluations have been conducted to analyze the carbon emissions and other environmental consequences associated with numerous products within the supply chain. These products include metered dose inhalers, anesthesia equipment, central venous catheters, dental burs, disposable surgical kits, instruments and plastics used in medical applications, hospital gowns, radiology equipment, and incontinence products [26, 27]. Estimations have been conducted for a wide range of medical operations, encompassing childbirth, hysterectomy, dialysis, plastic surgery, cataract extraction, diverse minimally invasive surgical interventions, critical care, and postoperative treatment, emergency services, as well as telemedicine services.

1.3 Developing sustainability of rural healthcare hospitals

Small rural towns worldwide commonly share demographic characteristics with their neighbors and are routinely shown to have a variety of issues related to the global phenomena of rural decline along with urbanization drift. It is widely recognized that rural people exhibit poor health conditions and face the challenge of an aging demographic. In Australia, several state and national policies and initiatives have been implemented to address such conditions. However, few people in rural areas are convinced that their hometown is the same as a local or national town of apparently comparable size. Rural and remote populations experience elevated morbidity and mortality rates in comparison to their urban counterparts, alongside constrained accessibility to healthcare services. The limited access is hindered by factors such as insufficient service availability, increased costs, shortages in healthcare personnel, transportation difficulties, and the deterioration of rural infrastructure [28–31]. It is widely recognized that the complexities of the healthcare supply are a barrier to improving the health of people. Numerous program assessment standards, vertical funding and organization of health services, limited cross-sector collaboration, and short-term and insufficient funding are among the issues that have been mentioned. Complex regulations and procedures are implemented in a varied manner throughout the nation, resulting in a lack of comprehension of neighborhood context and culture [32–35]. Additionally, investigations into potential solutions have vertical attention and do not take a comprehensive view of how health services are delivered. The majority of initiatives to enhance healthcare in rural hospitals have been receptive, time-constrained, ill-coordinated, and centered on certain occupational groups or disease types. This has led to inconsistent service delivery levels that are not properly correlated to need [36]. Figure 1.1 shows the hospital sustainability in rural and urban regions. The connection between residence and socioeconomic status has been investigated. Mayhew's 1861 explanation of the correlation between crime and other factors, such as the availability of education, is considered the starting point for the discipline of social geographies [37]. The work is still relevant today and serves as the foundation for a report "The Social Determinants of Health" that was published by the World Health Organization. It says that a person's health and life chances have a strong connection to their environment. This information should change how health services are given and how health policy is used in Australia. However, Vinson points out that a community's internal relationships and unique qualities can change (for better or worse) even the best policy or program meant to improve the health and welfare of people [37].

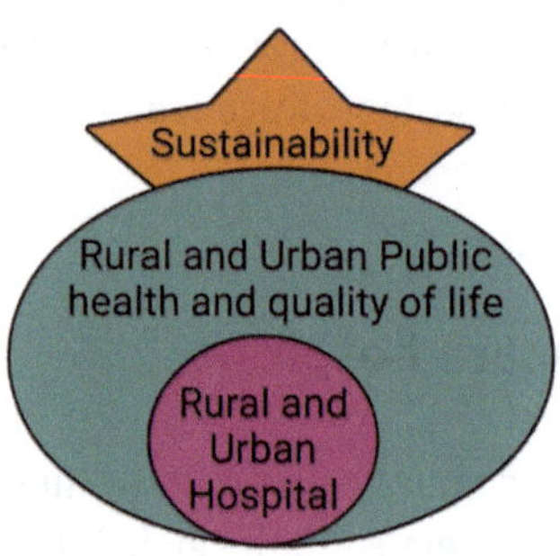

Figure 1.1: The hospital sustainability in rural and urban regions.

1.3.1 Approach for improving community healthcare

However, being recognized as a healthcare approach, the development of communities has a background in social practice in Western nations rather than health practice [38]. In Australia, community development strategies are often the responsibility of the local government or charitable organizations rather than state or federal health authorities. The development of communities presents a viable approach to creating novel frameworks for delivering health services. Such an approach encompasses a comprehensive perspective on health, encourages collaboration across sectors, assesses and evaluates innovative practices, and integrates local context and culture. It is imperative, however, that these initiatives adhere to rigorous methodological and ethical standards. Two elements of a community development strategy – identification and needs analysis – can help identify issues faced by a specific population [36]. The target population is characterized by one or more identifying attributes, such as age, gender, or geographical region. The process of needs identification includes the gathering of information related to the conditions, challenges, and available resources of the target population. The process includes assessing the significance of the identified requirements and determining the most suitable approaches to address them. The significance of conducting a requirements analysis consists of its ability to inform service planning by identifying gaps and obstructions in service provision, determining the target population of service users, documenting persistent disadvantages, and offering opportunities for engaging in advocacy efforts. The platform serves as a medium for facilitating community development activities [39].

The most effective approach in community healthcare development is to help individuals determine what they need and then collaborate with them to assist in coming up with creative solutions to those problems. While the ideal procedure is apparent, determining which demands are more important based on values is more complicated. Sometimes, the description of the target group or the definition of the issue might suggest which requirements are valued more highly than others. For example, Stevens observed inequality in the needs assessment procedures within the British National Health System, namely between the requirement for healthcare services and the requirement for a sense of well-being and overall health [39, 40].

1.4 The concept of hospital sustainability

In 2002, Harlem Brundtland, the Director-General of the World Health Organization, claimed that there exists a fundamental interconnection between sustainable development and health. Brundtland emphasized that the achievement of a healthy lifestyle is both a consequence and a potent and invaluable method of realizing sustainable development. The acceptance of health as a valuable resource and a catalyst for economic progress and reducing poverty is essential [41]. The interrelationship of health and sustainable development has been consistently emphasized in significant political sources. The enhancement of services within the hospital environment through sustainable development is a sector whereby the improvement of at least one dimension is achieved without causing negative impacts on the other dimensions. In an optimal circumstance, the objective is to find solutions that maximize all dimensions [42]. Energy and material efficiency in sustainably constructed facilities reduces their impact on the environment. Before evaluating the importance and acceptance of sustainable design in different regions of the world, it is important to understand the underlying principles that govern it [43]. The sustainable building structure is built on the trinity of environmental, social, and economic considerations. Reducing the consumption of nonrenewable resources, toxicity, waste, and energy consumption, all contribute to environmental sustainability. Examining the whole cost of the project's execution is one way to provide an economic perspective. Improving the structure's accessibility will have a positive effect on society [44]. Prior research will be used to identify significant terms and indications for sustainable hospital design, resulting in guidelines for healthcare authorities to apply sustainable construction principles (Table 1.2).

Table 1.2: Sustainability criteria for healthcare facility development.

Dimensions of environmental sustainability	Categories	Indications
Ecological sustainability	Site	Sustainable urban design Gardening and landscaping Ventilation from the outside in public buildings The structure was designed such that it would make the most of the site's natural resources (such as sunlight, wind, and the path of the Sun through the sky)
	The building's plan and orientation	Linking building to other site buildings Building places to adapt to direct sunshine and pollution Green design (green roofs, green walls) Reducing SO_2, NOx, PM, CO
	Water	Reuse rainfall and gray water for irrigation and minimize drinking water usage Building water-saving systems
	Energy	Use daytime natural light Intelligent devices track energy usage per consumer demand Intelligent systems monitor the user's energy consumption
	Pollution	Reduce CO_2 emissions. Reduce harmful pollutant emissions.

Table 1.2 (continued)

Dimensions of environmental sustainability	Categories	Indications
Social sustainability	Inhabitant and patient health and well-being	Reduce finish product toxicity using sustainable materials Heat sensitivity Indoor environment and design quality Space flexibility, adaptation, and expansion Furniture and equipment
	Accessibility	Easy public transit access Easy access to facilities Area allocation Promoting cycling and walking
	Awareness and education related to sustainability	Educating building residents on the principles of sustainability
Economical sustainability	The life cycle expenses	The initial expense Operational and maintenance expenses Building durability
	Get energy	Renewable energy generation Reducing and treating all pollutants
	Initiatives to reduce expenses (savings)	Integration of automation technology into the building management system
	Minimize expenses of transportation	The use of alternative fuels, electric cars, and the usage of bicycles, walking, virtual clinics, and telemedicine are all examples of sustainable mobility strategies that hospitals might use

1.4.1 Healthcare and health system sustainability

Sustainable healthcare is defined as "the provision of high-quality healthcare and the enhancement of the public's health without decreasing natural resources or causing serious environmental damage," according to the Sustainable Development Unit (SDU), which was established in 2008 by England's National Healthcare System. Without reducing the quality of care, sustainable healthcare involves the use of healthcare products, procedures, and services that have greater environmental sustainability [45, 46]. According to the World Health Organization, an environmentally sustainable health system is characterized by its ability to enhance, preserve, or rehabilitate health, while simultaneously minimizing adverse effects on the environment and maximizing opportunities to restore and enhance it. This approach ultimately benefits the health and well-being of both current and future generations [47].

The term "health system" encompasses several elements that have a role in facilitating health and social care, including the National Health Service, social services, and public health [48]. According to the SDU, a sustainable health system is defined as one that operates within the existing environmental and social resources to safeguard and enhance health outcomes for both present and future generations. It involves participating in activities to decrease carbon emissions, minimize waste and pollution, optimize the utilization of limited resources, develop the capacity to withstand the impacts of climate change, and enhance the capabilities and resources of communities [49]. Since hospitals care for individuals in the community, they have a natural connection to issues of social responsibility and environmental sustainability. The provision of healthcare to patients has constantly constituted a fundamental aspect of hospital infrastructure, connecting perfectly with the societal aspects, encompassed under the concept of sustainability. The concept of sustainability is essential for addressing green hospital issues involving numerous elements, including the efficient utilization of natural resources, establishment of sustainable energy, introduction of conservation of energy system, and the reduction of carbon dioxide (CO_2) emissions during planning, implementation, and rehabilitation [50].

The development of the green hospital movement may be connected to a time a few years ago when the US Green Building Council (USGBC) introduced the Leadership Certification in Energy and Environmental Design (LEED) criteria for constructing facilities. Even though they are more expensive to build initially, studies have shown that environmentally friendly hospitals end up saving money on their energy bills. The use of green design principles in hospitals has been found to have a positive impact on patient outcomes and personnel retention. The hospital facilities that have recently been erected or refurbished have obtained LEED certification [51].

1.4.2 Hospital sustainability

Sustainable design is widely regarded as a "best practice" due to its ability to effectively mitigate the life cycle and maintenance expenses associated with a building, while also minimizing the adverse ecological consequences of construction. Failure to incorporate sustainable solutions in healthcare environments can result in health issues stemming from unsustainable building practices. The issues encompass carbon emissions, which have the potential to cause cancer or other diseases, sickening building syndrome, respiratory ailments, anxiety, fluctuations in mood, sadness, and allergies. Moreover, the phenomenon gives rise to a certain degree of irony, as individuals seek solace and recovery in an environment that gradually endangers their health due to the presence of harmful emissions and poisonous compounds [52]. For environmental and health reasons, it is necessary to examine the full life cycle of structure components and goods, as well as the construction industry as a whole, about its physical, ecological, and human settings at the regional, global, and local levels [53]. Healthcare has measurable goals and objectives that support sustainability and the capability of the healthcare system to thrive in its social, economic, and environmental environments, since the World Summit on Sustainable Development in Johannesburg in 2002 recognized health as one of five big priorities for the future [54]. The achievement of the hospital's primary company operations, coupled with requirements for medical excellence and health advancement, is examined for the sustainability of the economy, society, and the environment as part of hospital-wide sustainable development [55]. In alternative terms, the concept of substantial sustainability within the context of hospitals can be understood as an expansion of quality requirements to encompass dimensions like as ecological, social, and economic considerations. Therefore, the method applies to the principles of quality and techniques employed by contemporary hospital facilities. The central focus of the Sustainability Triangle is on healthcare and health promotion, which are considered necessary services. The system also incorporates sustainability criteria as extra quality standards for making decisions related to vital services, as seen in Figure 1.2.

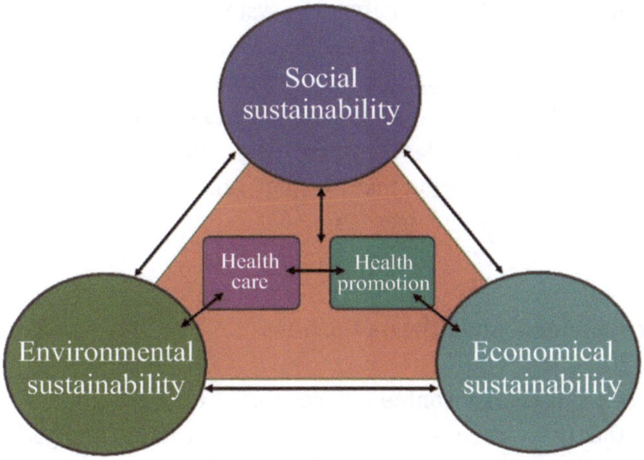

Figure 1.2: The hospital sustainability triangle.

1.5 Sustainability in hospital healthcare facilities

The increasing global importance of sustainability has prompted the healthcare industry to consider adopting the "greening movement" in its operations due to philosophical, sociological, and economic factors. Hospital healthcare development has received more attention recently because of shifting demographics, shifting illness patterns, and emerging medical technologies. Concerns about their flexibility to adjust to changing conditions, preferences, and medical technology are frequently addressed. However, hospitals are also required to include sustainability ideas in their design and operations to minimize the development industry's negative effects on the environment [56–58]. However, even though these structures constitute an important component of economic activity in Western countries, hospitals have exhibited delayed response in resolving the problem of sustainability [57]. To achieve sustainability in hospital healthcare facilities, it is necessary to accept methods that are both environmentally responsible and economically effective, all the while protecting the health of the surrounding population. The design of green buildings, energy efficiency, the reduction of waste, the use of renewable energy sources, ethical purchasing, and participation with the local community are essential components. In addition to lowering their influence on the surrounding environment, sustainable practices also improve the quality of care provided to patients and reduce overall operational expenses. Limited research exists on sustainable healthcare unit advancement, with most studies focused on the disposal of waste along with energy efficiency [58]. Currently, there is a lack of a comprehensive evaluation of the environmental aspects associated with the construction of hospital healthcare facilities.

In order to enhance the assessment of environmental sustainability in healthcare facilities during the initial design stage, various certification systems have been developed. Among the most significant examples are BREEAM, LEED, and the Green Certification Guide for Healthcare (The internationally recognized certification systems LEED and BREEAM serve as indicators of a building's quality of environmental design.) While these tools are popular among development professionals because of their simplicity of use, their subjective evaluation methodology raises questions about their effectiveness in creating sustainable structures [59].

1.5.1 Evaluation method for sustainable healthcare facilities

Evaluating healthcare facilities for their environmental, economic, and social sustainability is essential to ensuring that they accomplish their respective sustainability goals. In addition to maintaining a healthy environment for patients and employees, sustainable healthcare facilities aim to leave as small an ecological footprint as possible, enhance the quality of care provided to patients, and lower the overall cost of operations. The study on sustainable development in healthcare has gained significant popularity in the last two decades. The term "sustainability" might be challenging to precisely define; however, it is crucial to acknowledge that healthcare and environmental degradation are of significant importance, which supports the concerns of all relevant parties [43]. The effect of the environmental change on the provision of healthcare services is just now being explained by the healthcare sector. The introduction of the Healthcare Building Sustainability Assessment (HBSA) methodology is facilitating the use of various sustainable healthcare development approaches in projects worldwide, hence enhancing their global market competitiveness. The healthcare structure is a complicated and contradictory architectural category, functioning as a connected structure of several systems. It is a concentrated collection of resources, including personnel, machinery, and materials [60]. This is why developers and other interested parties need to work together when developing a healthcare facility. Sustainability challenges in the built environment continue to be the focus of environmental, social, and economic architectural thought, all of which are developing new levels of understanding and integrated methods. The subject of "how to generate concepts for sustainability from architectural perspectives" is crucial [61].

The utilization of HBSA methodologies can effectively address this inquiry by including a growing number of criteria that immediately correlate with the geographic design of architecture. If the application of sustainable environments falls below being fully maximized, the significance of sustainability as a concept would be considered inconsequential. Therefore, it is imperative to enhance the participation of developers in the effective application of such technologies. In the context of healthcare facilities, effective implementation of organizing space may facilitate the fulfillment of several criteria defined by the existing Healthcare Building Sustainability Assessment (HBSA) techniques. It includes enhancing user comfort and incorporating sustainable technology advancements. Healthcare facilities serve as exemplary instances for addressing the concerns mentioned above, as their personnel are not just engaged in patient care, but rather in assisting individuals [62].

1.5.2 Methods for evaluating sustainability in the Flemish area

In order to enhance the understanding of the sustainability of hospitals and address the current developments regarding building accreditation, the BREEAM, a new construction program was implemented in Belgium. Furthermore, due to the absence of a specialized BREEAM New Construction guidebook for healthcare, building professionals in Flanders depend on the BREEAM Bespoke procedure. The criteria relevant to the building and its units are established by the Building Research Establishment under such a scheme [63]. Due to the intricate nature of hospitals, which include several departments, it is imperative to establish appropriate criteria as required. The Flemish Infrastructures Fund for Person-related Matters (VIPA) has recently authorized the creation of the instrument, Duurzaamheidsmeter zorg (Sustainability meter), to address the lack of a specialized certification system adapted for the healthcare facilities in the Belgian setting. Stevanovic et al. [64] discovered the resemblance between this tool and BREEAM in their investigation, demonstrating that it is mostly based on the new building scheme with the addition of social and cultural factors. By incorporating the aforementioned criteria, VIPA attempted to address two sustainability pillars, namely environmental and social, as opposed to BREEAM, which focuses solely on ecological concerns. Stevanovic et al. [64] found that the instrument, based on the new building scheme, is equivalent to BREEAM, and includes social and cultural factors. VIPA introduced standards for environmental and social sustainability, whereas BREEAM focuses on ecological concerns. VIPA aimed to adopt a comprehensive strategy by adopting criteria for the economic pillar. The Duurzaamheidsmeter zorg management criteria now include corporate social responsibility to encourage institutions to take a more active role in sustainability [65].

The qualitative instrument "Duurzaamheidsmeter zorg" for the Flemish required establishing needs, to assist construction practitioners in analyzing the sustainability of hospital projects. Various urban planners as well as architects who used the tool noticed various drawbacks and inadequacies. A quantitative methodology in sustainability evaluation is needed for greater reliability. Analyzing the sustainability of hospitals is a significant problem due to their complex nature, which encompasses several building types, including residential, office, and service buildings, all within a single facility. Sustainable healthcare in Flemish is evaluated in terms of environmental impact, financial viability, and social acceptability [65]. Carbon footprint analysis, biodiversity assessments, and resource efficiency measures are used to evaluate environmental sustainability, whereas life cycle cost analyses and return on investment evaluations are used to evaluate economic sustainability. Patient satisfaction, employee happiness, and community integration are prioritized in order to ensure social sustainability. Furthermore, responsibility and compliance may be ensured by using performance indicators and healthcare legislation, specific to the Flemish region [65]. These assessment strategies support a sustainable healthcare ecosystem that helps patients and the environment in the Flemish area through robust data collecting, benchmarking, stakeholder involvement, ongoing improvement, and transparent reporting. Examining the present state of healthcare in the Flemish region as well as the hospital models anticipated for the future is the first step in addressing sustainability from a life cycle perspective. Subsequently, it is important to do a preliminary life cycle assessment and life cycle cost study on one or several hospitals situated in the Flemish region. The analysis will facilitate the identification of areas of significant environmental impact and technological complexities that arise when employing a quantitative life cycle strategy [66].

1.6 Conclusion

In conclusion, sustainable hospital healthcare services are essential to global environmental responsibility, economic viability, and social well-being. The comprehensive exploration has highlighted the multifaceted nature of sustainability in healthcare, emphasizing its ecological, financial, and social dimensions. The environmental dimension demonstrates hospitals' significant environmental effects and the critical necessity for environmentally sensitive procedures. Hospitals can substantially reduce their ecological footprint through green building design, energy efficiency measures, waste reduction strategies, and responsible procurement, thereby contributing to a healthier environment and mitigating climate change. Sustainability in healthcare provides real financial advantages. Hospitals that adopt sustainable practices have the potential to achieve significant financial benefits by decreasing the use of energy, implementing effective waste management strategies, and making responsible procurement strategies. The financial advantages not only contribute to the financial sustainability of a hospital but also enable the allocation of resources toward the improvement of patient care and infrastructure. In addition, the societal aspect demonstrates the significant effect that sustainable healthcare has on both the general population's health and the well-being of the community. Sustainable hospitals serve as prominent examples of community involvement, collaborating with many stakeholders to efficiently address both healthcare requirements and environmental issues. It promotes healthier communities by increasing responsibility and empowerment outside hospitals. The investigation provides strong evidence of the transformational potential of sustainability in the healthcare industry through the case studies and best practices that are provided. The development of sustainable hospital healthcare services is the shared responsibility of healthcare professionals, administrators, and legislators. Healthcare organizations may establish a long-lasting beneficial influence by adopting sustainability as a fundamental principle. By adopting sustainability as a fundamental principle, healthcare establishments have the potential to establish a sustainable beneficial influence, thereby contributing to a global environment where healthcare is not solely efficient, but also maintains ethical standards, demonstrates environmental responsibility, and maintains its financial sustainability. In conclusion, the establishment of sustainable healthcare services within hospitals is a pressing matter that necessitates the dedication of healthcare practitioners, administrators, and politicians. It involves not only organizing effectiveness in healthcare but also in assuming ethical responsibilities, practicing responsible environmental management, and ensuring financial viability. The scientific investigation shows the interplay of these aspects and emphasizes the need for collaboration in order to advance healthcare toward a state of sustainability and comprehensive well-being.

References

[1] Landrigan PJ, Fuller R, Acosta NJR, Adeyi O, Arnold R, et al. The Lancet Commission on pollution and health. Lancet. 2018;391(10119):462–512.

[2] Shaffer RM, Sellers SP, Baker MG, De Buen Kalman R, Frostad J, Suter MK, Anenberg SC, Balbus J, Basu N, Bellinger DC, Birnbaum L. Improving and expanding estimates of the global burden of disease due to environmental health risk factors. Environmental Health Perspectives. 2019;127 (10):105001.

[3] Watts N, Amann M, Arnell N, Ayeb-Karlsson S, Beagley J, Belesova K, Boykoff M, Byass P, Cai W, Campbell-Lendrum D, Capstick S. The 2020 report of the Lancet Countdown on health and climate change: Responding to converging crises. The Lancet. 2021;397(10269):129–70.

[4] Eckelman MJ, Huang K, Lagasse R, Senay E, Dubrow R, Sherman JD. Health care pollution and public health damage in the United States: An update: Study examines health care pollution and public health damage in the United States. Health Affairs. 2020;39(12):2071–79.

[5] Sherman JD, MacNeill A, Thiel C. Reducing pollution from the health care industry. Jama. 2019;322 (11):1043–44.

[6] Whitmee S, Haines A, Beyrer C, Boltz F, Capon AG, de Souza Dias BF, Ezeh A, Frumkin H, Gong P, Head P, Horton R. Safeguarding human health in the Anthropocene epoch: Report of The Rockefeller Foundation–Lancet Commission on planetary health. The Lancet. 2015;386 (10007):1973–2028.

[7] Rechel B, Wright S, Barlow J, McKee M. Hospital capacity planning: From measuring stocks to modeling flows. Bulletin of the World Health Organization. 2010;88:632–36.

[8] Weisz U, Haas W, Pelikan JM, Schmied H. Sustainable hospitals: A socio-ecological approach. GAIA-Ecological Perspectives for Science and Society. 2011;20(3):191–98.

[9] Bromley E. Building patient-centeredness: Hospital design as an interpretive act. Social Science & Medicine. 2012;75(6):1057–66.

[10] Scott WR, Ruef M, Mendel P, Caronna C. Institutional change and healthcare organizations: From professional dominance to managed care. Contemporary Sociology. 2001 Dec;47(2):300–302.

[11] Rechel B, Wright S, Edwards N, Dowdeswell B, McKee M. Conclusions and critical success factors. In: Investing in Hospitals of the Future (pp. 249–64). 2009. European Observatory on Health Systems and Policies. http://www.euro.who.int/pubrequest

[12] Abel-Smith B, Mossialos E. Cost containment and health care reform: A study of the European Union. Health Policy. 1994 May 1;28(2):89–132.

[13] McKee M, Edwards N, Atun R. Public-private partnerships for hospitals. Bulletin of the World Health Organization. 2006;84:890–96.

[14] McKee M, Healy J, Edwards N, Harrison A. Pressures for change. European Observatory on Health Care System Series, Hospitals in Changing Europe (pp. 36–58). 2002. ISBN 0 335 20928 9.

[15] Posnett J. Are bigger hospitals better. In: McKee M, Healy J (eds.), Hospitals in a Changing Europe (pp. 100–18). Open University Press Buckingham: Philadelphia; 2002.

[16] Pantzartzis E, Edum-Fotwe FT, Price AD. Sustainable healthcare facilities: Reconciling bed capacity and local needs. International Journal of Sustainable Built Environment. 2017 Jun 1;6(1):54–68.

[17] Talib F, Rahman Z. An interpretive structural modeling for sustainable healthcare quality dimensions in hospital services. International Journal of Qualitative Research in Services. 2015;2(1):28–46.

[18] Tennison I, Roschnik S, Ashby B, Boyd R, Hamilton I, Oreszczyn T, Owen A, Romanello M, Ruyssevelt P, Sherman JD, Smith AZ. Health care's response to climate change: A carbon footprint assessment of the NHS in England. The Lancet Planetary Health. 2021 Feb 1;5(2):e84–92.

[19] Castro MD, Mateus R, Bragança L. A critical analysis of building sustainability assessment methods for healthcare buildings. Environment, Development and Sustainability. 2015;17:1381–412.

[20] Pichler PP, Jaccard IS, Weisz U, Weisz H. International comparison of health care carbon footprints. Environmental Research Letters. 2019;14(6):064004.

[21] Karliner J, Slotterback S, Boyd R, Ashby B, Steele K, Wang J. Health care's climate footprint: The health sector contribution and opportunities for action. European Journal of Public Health. 2020 Sep;30:165–84 cf3.

[22] MacNeill AJ, Lillywhite R, Brown CJ. The impact of surgery on global climate: A carbon footprinting study of operating theatres in three health systems. The Lancet Planetary Health. 2017;1(9):e381–88.

[23] Parvatker AG, Tunceroglu H, Sherman JD, Coish P, Anastas P, Zimmerman JB, Eckelman MJ. Cradle-to -gate greenhouse gas emissions for twenty anesthetic active pharmaceutical ingredients based on process scale-up and process design calculations. ACS Sustainable Chemistry & Engineering. 2019;7 (7):6580–91.

[24] De Soete W, Jiménez-González C, Dahlin P, Dewulf J. Challenges and recommendations for environmental sustainability assessments of pharmaceutical products in the healthcare sector. Green Chemistry. 2017;19(15):3493–509.

[25] Alshqaqeeq F, Esmaeili MA, Overcash M, Twomey J. Quantifying hospital services by carbon footprint: A systematic literature review of patient care alternatives. Resources, Conservation and Recycling. 2020;154:104560.

[26] Jeswani HK, Azapagic A. Life cycle environmental impacts of inhalers. Journal of Cleaner Production. 2019;237:117733.

[27] Willskytt S, Tillman AM. Resource efficiency of consumables–Life cycle assessment of incontinence products. Resources, Conservation and Recycling. 2019;144:13–23.

[28] Campbell A, Manoff T, Caffery J. Rurality and mental health: An Australian primary care study. Rural & Remote Health. 2006;6(3):1–9.

[29] Allan J, Ball P, Alston M. Developing sustainable models of rural health care: A community development approach. Rural & Remote Health. 2007;7(4):1–3.

[30] Alston M. From local to global: Making social policy more effective for rural community capacity building. Australian Social Work. 2002;55(3):214–26.

[31] Kinnear S, Kabir Z, Mann J, Bricknell L. The need to measure and manage the cumulative impacts of resource development on public health: An Australian perspective. Current Topics in Public Health. 2013 May;15:125–44.

[32] George AS, Mehra V, Scott K, Sriram V. Community participation in health systems research: A systematic review assessing the state of research, the nature of interventions involved and the features of engagement with communities. PLoS One. 2015 Oct 23;10(10):2–4.

[33] Braveman P, Gottlieb L. The social determinants of health: It's time to consider the causes of the causes. Public Health Reports. 2014 Jan;129(1_suppl2).

[34] Rudolph L, Caplan J, Mitchell C, Ben-Moshe K, Dillon L. Health in all policies: Improving health through intersectoral collaboration. NAM Perspectives. 2013 Sep 18; 2–5.

[35] Mahmood MA, Moss J, Karmaliani R. Community context of health system development: Implications for health sector reform in Pakistan. EMHJ-Eastern Mediterranean Health Journal. 2003;9(3):464–71.

[36] Chenoweth L, McAuliffe D. The Road to Social Work and Human Service Practice with Online Study Too Ls 12 Months. Cengage AU: Australia; 2017 Sep 11.

[37] Vinson T. Community Adversity and Resilience: The Distribution of Social Disadvantage in Victoria and New South Wales and the Mediating Role of Social Cohesion. Melbourne: Jesuit Social Services; 2004.

[38] Ching CR, Hibar DP, Gurholt TP, Nunes A, Thomopoulos SI, Abé C, Agartz I, Brouwer RM, Cannon DM, de Zwarte SM, Eyler LT. What we learn about bipolar disorder from large-scale neuroimaging: Findings and future directions from the ENIGMA Bipolar Disorder Working Group. Human Brain Mapping. 2022 Jan;43(1):56–82.

[39] Ife JW. Community Development: Creating Community Alternatives-vision, Analysis, and Practice. Longman Australia: Melbourne; 1995.

[40] Stevens A, Raftery J, Simpson S, Mant J. Health Care Needs Assessment, Vol 2: The Epidemiologically Based Needs Assessment Reviews. London: Radcliffe; 2005.

[41] Brundtland GH. World summit on sustainable development: The importance of health in economic development makes it a priority. BMJ. 2002;325(7361):399–400.

[42] Aydın D, Yaldız E, Buyuksahin S. Sustainable hospital design for sustainable development. In: Proceedings of the 8th International Conference on Urban Planning, Architecture, Civil and Environment Engineering, Dubai, UAE 2017 Dec (pp. 21–22).

[43] Ullah Z, Thaheem MJ, Waheed A, Maqsoom A. How sustainability in healthcare sector challenges guidelines and code development: A framework for design of sustainable hospital buildings. In: The

10th International Conference on Engineering, Project, and Production Management 2020 (pp. 213–25).
[44] Zadeh RS, Xuan X, Shepley MM. Sustainable healthcare design: Existing challenges and future directions for an environmental, economic, and social approach to sustainability. Facilities. 2016 Apr 4;34(5/6):264–88.
[45] Pencheon D. Developing a sustainable health care system: The United Kingdom experience. Medical Journal Australia. 2018 Apr 16;208(7):284–85.
[46] Nicol E. Sustainability in healthcare: Efficiency, effectiveness, economics and the environment. Future Healthcare Journal. 2018 Jun 1;5(2):81.
[47] Sherman JD, Thiel C, MacNeill A, Eckelman MJ, Dubrow R, Hopf H, Lagasse R, Bialowitz J, Costello A, Forbes M, Stancliffe R. The green print: Advancement of environmental sustainability in healthcare. Resources, Conservation and Recycling. 2020 Oct 1;161:104882.
[48] Alsawaf ES, Albadry AM. Principles for the sustainable design of hospital buildings. International Journal of Sustainable Development and Planning. 2022 Oct 1;17(6):1797–1808.
[49] Pencheon D. Making health care more sustainable: The case of the English NHS. Public Health. 2015 Oct 1;129(10):1335–43.
[50] Setyowati E, Harani AR, Falah YN. Green building design concepts of healthcare facilities on the orthopedic hospital in the tropics. Procedia-Social and Behavioral Sciences. 2013;101:189–99.
[51] Prugsiganont S, Jensen PA. Identification of space management problems in public hospitals: The case of Maharaj Chiang Mai Hospital. Facilities. 2018 May 8;37(7/8):435–454.
[52] Mostepaniuk A, Akalin T, Parish M. Practices pursuing the sustainability of a healthcare organization: A systematic review. Sustainability 2023;15:2353. doi:10.3390/su15032353.
[53] Kumari S, Kumar R. Green hospital-A necessity and not option. Journal of Management Research and Analysis. 2020;7(2):46–51.
[54] Sahamir SR, Zakaria R. Green assessment criteria for public hospital building development in Malaysia. Procedia Environmental Sciences. 2014;20:106–15.
[55] Borges de Oliveira K, de Oliveira OJ. Making hospitals sustainable: Towards greener, fairer and more prosperous services. Sustainability. 2022 Aug 8;14(15):9730.
[56] Martinsen DS, Vrangbæk K. The Europeanization of health care governance: Implementing the market imperatives of Europe. Public Administration. 2008 Mar;86(1):169–84.
[57] Prasad S. Changing Hospital Architecture (288 pages). London, United Kingdom: RIBA Publications; 2008.
[58] Capolongo S, Bottero MC, Lettieri E, Buffoli M, Bellagarda A, Birocchi M, Cavagliato E, Dervishaj A, Di Noia M, Gherardi G, Gola M. Healthcare sustainability challenge. In: Capolongo S, Bottero M, Lettieri E, Buffoli M, Bellagarda A, Birocchi M, Cavagliato E, Dervishaj A, Noia M, Gherardi G, Gola M, Mantua F, Miljatovic S, Nickolova M, Rostagno M, Speranza S, Volpatti L (eds.), Improving Sustainability during Hospital Design and Operation: A Multidisciplinary Evaluation Tool (pp. 1–9). Springer International Publishing; 2015.
[59] Aspinal S, Sertyesilisik B, Sourani A, Tunstall A. How accurately does BREEAM measure sustainability?. Creative Education. 2013;3(07):1.
[60] McGain F, Naylor C. Environmental sustainability in hospitals – a systematic review and research agenda. Journal of Health Services Research & Policy. 2014 Oct;19(4):245–52.
[61] Allacker K, Khan A. Introduction to architecture and sustainability: Critical perspectives for integrated design. In: Khan AZ, Allacker K (eds.), Architecture and Sustainability: Critical Perspective for Integrated Design (1st ed., pp. 13–36). Brussels: ACCO; 2015.
[62] Clark PA, Malone MP. What patients want: Designing and delivering health services that respect personhood. In: Improving Healthcare with Better Building Design (pp. 1, 2, 15–35). Springer International Publishing; 2006.

[63] Buffoli M, Capolongo S, Bottero M, Cavagliato E, Speranza S, Volpatti L. Sustainable Healthcare: How to assess and improve healthcare structures' sustainability. Annali Di Igiene: Medicina Preventiva e Di Comunità. 2013 Sep 1;25(5):411–18.

[64] Gerritsen M. Vlaanderen en Nederland: één taal, twee culturen. Neerlandia/Nederlands van nu. 2014;1:26–29.

[65] Stevanovic M, Allacker K, Vermeulen S. Hospital building sustainability: The experience in using qualitative tools and steps towards the life cycle approach. Procedia Environmental Sciences. 2017 Jan 1;38:445–51.

[66] Stevanovic M, Allacker K, Vermeulen S. A review on requirements and existing qualitative tools for designing sustainable large-scale healthcare facilities: A case study in the context of Flanders. In: Proceedings of International Conference CISBAT 2015 Future Buildings and Districts Sustainability from Nano to Urban Scale (pp. 511–16), 2015.

Chapter 2
Promoting the adoption of robots in healthcare for sustainability

Abstract: The use of robots in healthcare environments has demonstrated significant progress, presenting a potential opportunity for addressing sustainability issues within the healthcare sector. The use of robotic technology in the healthcare sector, with a specific focus on its sustainability implications –this statement emphasizes the potential of robotic technology for improving the sustainability of healthcare systems via the reduction of resource use, enhancement of operational efficiency, and improvement of patient care. The chapter also examines potential challenges to using robots in healthcare and emphasizes the need for collaborative initiatives among many stakeholders to effectively incorporate robots into healthcare systems. It demonstrates the crucial significance of robots in the progression of healthcare sustainability, providing an understanding of a potential healthcare future that is both ecologically friendly and operationally effective. In conclusion, the chapter discusses the significance of collaboration and research in promoting the implementation of robots in healthcare systems to promote sustainability. The advancement of robotic technology is dependent upon the collaboration of technology developers, healthcare providers, and research institutes. Robotic applications are always being improved to become more efficient and sustainable through research and innovation. In conclusion, the chapter provides a thorough examination of how robotic technology might be used to improve sustainability in the healthcare industry. By addressing multiple dimensions, including resource optimization, enhanced precision, telemedicine, waste reduction, extended reach, supply chain efficiency, data utilization, energy conservation, and social considerations, it provides a comprehensive picture of the transformative potential of robots in healthcare. The healthcare industry needs sustainability to deal with growing problems like a lack of resources, rising healthcare costs, and more patients who require good care. The use of robotic technology has emerged as a technique that has the potential to bring about considerable improvements in sustainability within the healthcare sector. The present chapter explores many complex functions of robots within the healthcare sector, as well as their influence on the concept of sustainability.

2.1 Introduction

The healthcare industry associated with robots has had substantial growth, particularly within the industrial sector, over many years. However, more focus has been assigned to the healthcare industry in comparison to other industries, possibly due to the inherent challenges related to delivering personalized care, which is a fundamental aspect of

https://doi.org/10.1515/9783111436432-002

the healthcare service. The growing epidemic of the coronavirus has significantly altered the trajectory of the development of robots that assist in patient care, which was previously restricted by multiple barriers. Rapid assessment and use of robotics are currently being implemented to reduce the transmission of viruses and germs. This is achieved via the implementation of disinfection protocols, supply chain management, and telehealth services. Since the condition is so infectious, there is also a growing need for robotics to support nurses and other healthcare professionals in improving efficiency [1], decreasing the amount of patient-caregiver interaction and making up for the increased number of healthcare worker deaths. Robotic applications have the potential to provide sustainable and cost-effective healthcare without affecting the quality of treatment. While robot technology may be used in many different fields, its impact on medical treatment is particularly considerable. As an example, robots have the potential to enhance screening efficiency by using thermal sensors to measure temperatures. Furthermore, several nations have depended on the collection of nasopharyngeal and oropharyngeal swabs as a means for conducting diagnostic tests for COVID-19. This may take a long time since it requires collecting, handling, transporting, and testing [2]. The healthcare sector is receiving substantial attention because of the rapid advancements and difficulties associated with technology, the high level of financial investment required, and the limitations in the storage of digital information on the effects of accessibility on customer experiences [3].

The implementation of an automatic procedure has the potential to lower the possibility of infection and minimize mistakes made by people. Additionally, it may enable frontline employees to allocate their attention toward higher-priority responsibilities. The most significant social challenges of the coming decades are related to changes in demographics and the difficulties that arise for society as a result, including the need to contribute to healthcare deficiencies, an increase in the requirement for sustainable care, lack of qualified workers, and demand on social security systems [4]. The use of robots for the care of old people includes a wide range of potential applications. Recently, there has been a modification to the conversation around the use of social employees, with an increased focus on their role as a social support system rather than a traditional care service [5]. However, while examining discussions about the absence of qualified professionals, particularly in the field of geriatric care, it is important to not just focus on social assistance but rather adopt a more comprehensive approach.

The purpose of the investigation is to evaluate, with the use of social, medicinal, and supportive robotic technologies, the level of acceptability of healthcare robotics for older people. The study assesses the broad level of acceptability of robots and identifies the key aspects that impact individuals' attitudes to using them. The primary objective of this work is to provide significant and informative contributions. It also explores the possible commercial opportunities for these products and predicts the development of further sustainable healthcare services resulting from their commercialization [6]. According to recent research [7], it is projected that the market for eldercare will have an annual growth rate of 6.8%. Similarly, a separate study conducted by Research and Markets [8] has calculated a compound annual growth rate of almost 15% for robots. It suggests the presence of a substantial market potential that may be utilized via the adoption of robotics in healthcare sustainability. The statement highlights the identification of many study categories related to the use of robots within the healthcare sector [9]. Examples of surgical robotic systems [10, 11], laparoscopic surgery, tele-rounding robots [12], robot-assisted treatment [13–15], caregivers and patient assistance [16–18], as well as robotic applications in dental- and bioprostheses [19], are significant occurrences in the field of research. With further technological development, the potential for even more medical robot uses develops. Humanoid robots, which possess the capability of carrying out activities from a distant location, have significant importance as well. These robots possess human-like qualities that help healthcare givers and patients, particularly in contaminated environments. These remotely controlled robots are required to contain certain qualities to facilitate human-robot interaction. Tele-operated semiautonomous robots can assist healthcare workers during epidemics, reducing the time spent in contaminated areas and wearing personal protective equipment in high temperatures and humidity, especially in West Africa. In addition, several healthcare robots designed for other reasons were recognized [20]. The autonomous robot named Cody can independently wash human limbs. A study assessing the patient's response to this technology revealed that the level of acceptability is significantly influenced by the perception of the robot's aim [21]. In addition to their many applications, some robots have been used in the field of rehabilitation [22]. These robots are capable of supporting activities such as patient lifting, as well as performing blood sample procedures (Table 2.1) [23, 24]. The chapter provides an extensive overview of the existing and future uses of robotics in healthcare facilities. It suggests the careful selection of robots specifically designed for patient care.

2.2 Robotics in the healthcare sector

Robots play a significant role in the healthcare industry by performing many integrated activities, including monitoring patients, clinical treatments, prostheses, rehabilitation, and e-health. These duties include the use of mechanical and electronic components, such as detecting spatial dimensions, motion, or force, as well as using electronic sensors and other related techniques [25]. The field of robotics has potential economic, social, and healthcare benefits, especially for patient populations with specific requirements such as people with disabilities, stroke survivors, and those with cognitive disorders. In the field of healthcare, robotics has the potential to be used in several applications such as intelligent medical capsules, surgical procedures, prostheses, analysis and management of motor coordination, robot-assisted social and cognitive therapy, and monitoring devices for patients [25].

The industry for medical robots has recently expanded, and they are now widely employed in healthcare around the globe since they can perform quickly and with low risk, for instance during cardiac and prostate procedures, rehabilitation, and smart prosthetics. The field of social robotics has the potential to effectively monitor and provide motivation to patients. The exponential increase in the use of robots in the healthcare industry can be attributed to the escalating need for technological innovation, which is expected to drive future expansion in the healthcare sector. The use of less invasive surgical robots is becoming prevalent on a global scale, along with growing populations and demonstrating effectiveness in addressing various medical conditions such as orthopedics and neurological problems. The use of robots has been shown to have promise in nations such as China, India, and Brazil [26]. The healthcare industry is being transformed by robotics in patient care and medical operations. Robotic device systems give surgeons unmatched precision and minimally invasive capabilities, improving patient outcomes and recovery times. Telemedicine and telerobotics provide remote consultations, diagnosis, and surgeries, bridging geographical gaps and improving healthcare in underprivileged areas. Exoskeletons and assistive devices provide tailored and consistent therapy to mobility-impaired patients. Social and service robots help older people with everyday work and companionship. Robots speed drug discovery and diagnosis and reduce human error in labs. In modernized surgical procedures, computer-integrated systems exhibit superior precision compared to surgeons, particularly in challenging environments. These systems offer technical solutions and effectively monitor essential data online while operating on delicate anatomical structures in hazardous proximity settings. Robotics is a multidisciplinary field that involves the integration of real-world data with physical systems. It includes various disciplines such as engineering, computer science, biological mechanics, sports science, biomedical science, neurology, cognitive science, and others. The coordination of sensors, motors, and humans in robotics presents challenges in diverse settings [27].

When incorporating robots into the healthcare industry, it is important to take into account many variables such as design, transportation, and cognitive capabilities, among others. The structure defines the various degrees of human involvement in monitoring the operations of robots throughout standard processes, particularly in the event of experiencing system failures. The movement consists of predetermined routes for movement within the given surroundings. Intelligence may be defined as an inherent capacity or expertise that enables individuals to do tasks with a level of competence or knowledge that is lower than what would typically be expected of a person undertaking a comparable task [28]. Robots also help reduce infections by sanitizing healthcare facilities. Automation in drug formulation and dispensing reduces pharmaceutical errors and ensures precise dosing in pharmacies. Overall, healthcare robotics is improving patient care, operational efficiency, and sustainability through optimizing resource use and expanding healthcare access.

2.2.1 Robotics for enhancing health and quality of life

The previous study identified two research topics on technological acceptability, including humanoids. There are two distinct types of research in this field. The first type focuses on conducting experiments with a specific group of individuals, such as children or older adults, to investigate the functionalities of robots. The second type of research is concerned with the broader social acceptance of humanoids and the factors that influence their widespread use in everyday life, particularly among older individuals. The societal acceptability of humanoids in elderly care will play a significant role in determining their widespread deployment [29]. The first investigation area pertains to the operational capabilities of robots and incorporates a specific study. The use of robots has significant promise in facilitating autonomy and enhancing health-related results among the elderly population, while simultaneously alleviating the responsibilities borne by caregivers. The primary objective of robots is to assist elderly individuals in achieving and maintaining a high level of independent living and functioning. Robotic technology has the potential to assist elderly people in many duties, such as lifting, capturing, and moving objects. Additionally, robots may serve as reminders for medicine intake, identify health concerns, evaluate the situation, monitor and encourage physical activity, and fulfill social requirements using contact. Broekens et al. [30] made a distinction between two categories of robots that support the functioning of older persons. These categories include robots employed as rehabilitation workers and robots employed as social workers. Rehabilitation robots are mostly used to provide physical assistive technology functionalities. These devices are not designed to interact with older people, and as a result, they are not seen as social beings. Instances of such robotic systems include intelligent wheelchairs [31], prosthetic limbs [32], mobile and ambulatory robots, automated beds [33], dynamic orthoses [34], and exoskeletons [35]. On the other hand, social robots may be classified as either service workers or associated employees.

Service employees are employed to manage fundamental activities associated with self-sufficiency, including, but not limited to, meal preparation and personal hygiene, assistance with movement and orientation, provision of dietary guidance, and monitoring of overall health. Companion employees prioritize the enhancement of the physical and psychological welfare of elderly individuals. Frequently, social robots possess the capability to execute several tasks concurrently, serving as service robots to assist elderly people in many aspects of their lives while also functioning as auxiliary robotics to provide companionship. As per the assertions made by Kate Darling, a social robot may be defined as a physically embodied and self-governing entity that engages in communication and interaction with a human being on an emotional plane. Furthermore, social robots adhere to the rules of social conduct, possess a range of cognitive states, and adjust their behavior based on the knowledge acquired via interactions [36]. The significance of their character is crucial since social robots are designed to engage with others on an emotional level. This kind of contact relies on visual and tactile perception, in addition to verbal communication [37]. Table 2.1 displays the essential data about the chosen robots, including their fundamental characteristics and primary use. Four distinct categories of advantages associated with the use of robots were identified, including the effects on quality of life (QL), social implications (S), functional considerations (F), and ethical problems (E).

Table 2.1: Robotic applications for fundamental health supportive care [23].

S. no.	Robotics for human healthcare	Feature	Reference
1.	ARMAR III	The robot was created to assist in "human-centered" settings, including households. The upper torso has a humanoid design designed to communicate with people and control environment-based objects.	[38]
2.	Cody	A robotic assistant has been designed to help caregivers in their duties of patient hygiene, particularly in the area of bed baths. It uses a flexible arm and little pressure to produce "wiping movements" like those done during bed baths.	[39]
3.	PR2	The design of a mobile humanoid robot is intended to assist in various activities inside human environments. The individual can manipulate items and engage in many activities, including arranging tables and unloading a dishwasher.	[40]
4.	PaPeRo	Utilized for conversation; identifies voice and provides speech synthesis; understands several users at once; responds to touch; searches the internet for answers to queries asked; is used to improve senior citizens' happiness.	[41]
5.	RIBA	Robot has a high load capacity that can lift and move patients. Designed to look like a massive teddy bear. Transports patients with "human-like" limbs. The device is outfitted with tactile sensors that are capable of detecting the location of touch with the patient. This enables the device to make appropriate adjustments to its motion.	[42] [43]
6.	Care-O-Bot 3	The mobile robot assistant is specifically designed to provide comfort to individuals in common environments. The robotic system comprises an upper limb with a gripper designed to manipulate objects, a tray that facilitates the transportation and transfer of items, and a flexible torso that enables butler-like movements such as bowing and nodding.	[44]

Table 2.1 (continued)

S. no.	Robotics for human healthcare	Feature	Reference
7.	ROSE	Home care robot that can be controlled remotely. The robot exhibits the ability to perform many actions, including object manipulation, door manipulation, and cleaning operations. Robot Rose is a multipurpose robot that intends to assist healthcare providers while also giving clients more control, encouraging them to exercise more, helping them to retain social connections, and increasing their feeling of safety.	[45]
8.	ASIMO	A humanoid robot the size of a person. The individual can walk, run, and climb stairs, carry objects, open doors, and push vehicles. Helps elderly people live more independently by performing household tasks on their behalf; learns to recognize its owners and their preferences; responds to its name and comes when called; understands hundreds of words and phrases;	[46, 47]
9.	Pearl	A robot designed to support the elderly in their everyday lives, be it reminding them to eat, drink, take their medications, or use the restroom, or assisting them in navigating their surroundings.	[48]
10.	Hair washing robotic	The development of a robotic assistant specifically designed to help caretakers in washing a patient's hair.	[49]
11.	Robotic nursing assistant	The primary objective of this design is to assist caregivers in effectively managing physically challenging tasks such as the lifting and transporting of patients. The upper torso has a humanoid shape.	[50]
12.	Pepper	Independently walks and lives, recognizes people of all ages, responds to their environment based on their emotions, and uses expression, touch, and emotion interfaces to gauge its user's state of mind. Used not just to encourage elderly people to go out and exercise, but also to accompany them on their walks and give them a hand while they are walking.	[51]

The field of robotics includes studies that investigate both the humanistic characteristics of robots and their ability to exhibit realistic behavior. The study conducted by Lazzeri et al. focused on investigating the facial movements observed in a robot named Eva. The researchers emphasized the significance of facial expressions in conveying emotional information, which plays a crucial role in facilitating social interactions [52]. The second study area pertains to the degree of societal acceptability and understanding of the usefulness of robots in everyday life. The implementation of technology needs to be conducted sustainably. It is crucial to first comprehend the underlying reasons of older adults in their decision to either embrace or reject a novel technology [51]. Hence, an enhanced comprehension of the variables that predict individuals' views about the qualities of a technology has the potential to promote the extensive adoption of such technology [53].

Baganzi and Lau assert that there is a significant lack of understanding about the elements that influence the long-term adoption and sustainability of a given technology. For a technology to be effectively and sustainably used, it is essential to conduct research that elucidates the societal views of trust and risk [54]. Sustainable acceptance of technology means that it may enhance social and economic growth by giving people access to applications that help them deal with social problems and by making it easier to come up with new ideas [55]. The Global System for Mobile Association claims that newly accepted technologies should foster ecologically friendly communities [56]. The effective adoption of technology in a sustainable manner necessitates the identification of key components that influence the entire process [57].

Broadbent et al. [58] argue that an essential component in enhancing the adoption of healthcare robots is a comprehensive analysis of human requirements to integrate them with the robot's work, appearance, and activity. Two distinct sets of criteria were identified as being essential for the future of healthcare robots. The first set of criteria was associated with the individual characteristics of users, including age, needs, gender, cognitive capacity, education level, experience, and culture. The second set of parameters pertains to the qualities of robots, including their appearance (such as humanness and facial features), size, gender, ergonomics, role, and "personality" [58]. The outcomes of research conducted in the area of social psychology have facilitated the categorization of elements into three distinct groups: organizational, technical, and individual. The capacity to fulfill the requirements of the human or humanoid user is essential to the success of any relationship between robots and human beings [58]. Deligianis et al. [59] found that building and maintaining trust is crucial in human-robot interactions. In this kind of connection, trust depends on the person, the robot, and the surroundings [60].

2.2.2 Robotics application in the healthcare sector

There is no need for therapy in prevention; however, any observed abnormality might be diagnosed by a robot [25]. As an example, Toyota has developed the Balanced Training Assist, a two-wheel robot that presents interactive games on a monitor. These games are generated based on the input data received by the machine, while the patient engages in weight-shifting movements [61]. The utilization of robots in surgical procedures offers several advantages over the performance of physicians, particularly in situations involving small spaces. These advantages include enhanced accuracy, reliability, and consistency. Examples of surgical procedures where robots can be particularly beneficial include microsurgery, minimally invasive surgeries, nanobots, remote surgery facilitated by the Internet of things (IoT) systems, robotics-assisted surgery, and various other medical interventions. Nanobots, referred to as steerable surgeons, are used in the treatment of ophthalmic ailments. These nanobots are constructed using flat nickel components and are controlled using magnetic fields with the use of extrinsic electromagnetic coils [62]. Healthcare robotics is used in the provision of medical treatment for the elderly population. Their functions include monitoring, helping healthcare personnel, facilitating physical exercises, and performing other medical professional tasks [63]. For example, Cody, a humanoid portable operator, uses a direct physical interface (DPI) to assist a healthcare professional in directing robot mobility, which in turn directly touches the human body and reacts to the patient's movements [64].

Rehabilitation services are often administered either in a healthcare institution or within the individual's home environment. These services include various therapeutic interventions aimed at maintaining and enhancing muscular function and motor coordination. The conduct shown by individuals has a significant role in the manifestation and progression of mental diseases. Cerebral and neuromuscular problems may result in impaired brain functioning, potentially leading to extremity impairments. The Hybrid Assistive Limb (HAL), a robotic device with cyborg-like features, has the potential to enhance and augment the physical capabilities of its user by providing assistance to the lower limbs during mobility, according to the specific requirements of the individual [65, 66]. Robotic technology has the potential to facilitate many daily living tasks, including their execution, aiding those with limited mobility, and even serving as a means of organ replacement via the use of intelligent prosthetic devices. For instance, the electronic wheelchair known as Friend has many components such as a robotic arm, a computer, sensors, and joysticks. These elements function together inside the system to facilitate the user's ability to read a book and turn its pages [67].

2.2.2.1 IoT for the health sector

The IoT may be broadly characterized as the interconnectivity of computer devices that are integrated into common things, facilitating the exchange of data between these devices. The concepts behind IoT applications may be simply defined as follows: data collection, data conversion, storing data, and data interpretation. A variety of interrelated gadgets, including sensors, monitoring, detection devices, equators, and cameras, are used to gather data. The input undergoes a conversion process from analog to digital form, facilitating further processing. The accumulation of data is facilitated by the use of cloud-based services for storage purposes. Ultimately, the use of advanced statistical methods in data processing provides healthcare practitioners with the essential information required for effective decision-making [68–70].

While the ideas mentioned may also be applied to the human healthcare staff, IoT plays a crucial role in enabling the seamless transmission of data, hence enabling real-time decision-making. It is widely assumed that healthcare-related IoT applications would center on research, medical practice, and the management of patients. The use of IoT within the healthcare sector facilitates the monitoring of chronic illnesses, provision of care for older individuals, and management of healthcare systems, among several other applications. IoT in the healthcare sector may be categorized into two main approaches: service-centered or application-centered, as shown in Figure 2.1. The potential influence of IoT applications on healthcare may be seen at the intersection of insurance and industry with healthcare [68, 71–73]. As regards patients, the IoT infrastructure mostly comprises wearable gadgets. Wearable devices potentially include several monitoring capabilities such as oxygen saturation, blood pressure, pulse/heart rate, and glucose level. The specific selection of these parameters is contingent upon the patient's medical history and the monitoring requirements [73–75]. In the realm of medical professionals, the IoT provides a means of establishing instantaneous connectivity between doctors and their patients, as well as facilitating communication with other colleagues and the various facilities inside their clinic or laboratory. A cardiologist may get notification on an arrhythmia impacting a patient under their care, whereas a diabetologist may be notified of a patient's hypoglycemic condition. In both scenarios, patients have access to expert medical advice and assistance. Simultaneously, medical practitioners can evaluate the degree to which patients comply with prescribed treatments or regimens. Neglecting therapy may lead to blood pressure rises, although device monitoring can also be a factor. The transfer of this data to healthcare practitioners in distant locations, such as a physician monitoring a nursing home or a specialist visiting with patients in isolated towns, is also very significant [76, 77].

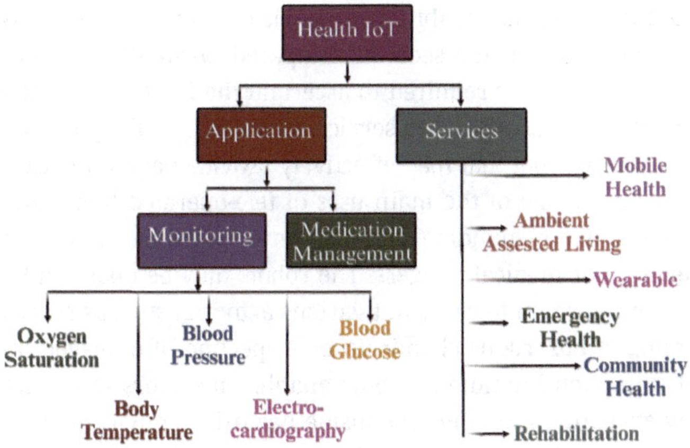

Figure 2.1: Health IoT.

2.2.2.2 Tele-operated humanoid robots

A tele-operated service refers to a system in which human operators remotely control and manage a service or device using telecommunications technology. The robot is an autonomous machine that is operated remotely by a human operator and is designed to carry out various activities, often in unpredictable and unregulated settings [78]. These robots have a bipedal structure that includes limbs, legs, and a cranium to replicate the human form. Compared to completely autonomous robots, tele-operated humanoid robots need human guidance to execute their motions and activities, rendering them a significant asset in scenarios that require human skill, decision-making, and flexibility. Tele-operation allows an operator to perform actions from a distant location, simulating the operator's presence as if they were physically there. This is achieved by replicating the operator's manipulations from a distance. One instance that demonstrates the application of advanced technology in the field of medical surgery includes the Da Vinci Robot [78]. One of the main uses of tele-operated humanoid robots is in dangerous situations, like responding to a disaster or working in an area that could have nuclear, chemical, or medical dangers. The robots may be operated by human individuals from a distant location in order to carry out various activities such as search and rescue operations, handling of hazardous chemicals, or inspection of damaged infrastructure. The use of tele-operated humanoid robots enables operators to ensure their safety while effectively executing intricate jobs inside hazardous environments. The use of tele-operated human-like robots in the healthcare sector gives a promising prospect for mitigating the elevated risk associated with medical professionals, including physicians, nurses, and patients. The transmission of infectious diseases has been extensively studied [79, 80].

This study examines the use of tele-operated robots as a means to enable healthcare professionals to do some tasks while maintaining a safe distance from patients who are contaminated. Studies indicate that patients exhibit lower levels of confidence towards the robot in situations when the surgery is not seen, as compared to situations where the procedure is accessible. Further study is required to ascertain the full degree of the impact. Nevertheless, the design of the tele-operated service robot [78] has the potential to be modified to create a healthcare robot that may effectively provide help in the care of patients in a variety of situations. One of the main uses of tele-operated humanoid robots is in dangerous situations, like responding to a disaster or working in an area that could have nuclear, chemical, or medical dangers. The robots may be operated by human individuals from a distant location to carry out various activities such as search and rescue operations, handling of hazardous chemicals, or inspection of damaged infrastructure. The use of tele-operated humanoid robots enables operators to ensure their safety while effectively executing intricate jobs inside hazardous environments. Tele-operated humanoid robots have shown their usefulness in several healthcare categories, including telemedicine and surgical procedures. Surgeons possess the capability to exert control over robotic systems, enabling them to execute surgical procedures with a notable level of accuracy, even while situated at a considerable distance from the operating site. The availability of medical knowledge to neglected communities or in emergency cases requiring quick surgical intervention might be very advantageous.

2.2.2.3 Robotic device for pain relief

Humanoid robots are used as a means of implementing a techno-psychological diversionary technique for children, to mitigate the experience of pain resulting from tension and anxiety during medical treatment. Multiple studies [81–83] confirm that kids are happier during medical procedures if a robot is involved. The objective is to divert the focus of youngsters from the discomfort caused by the needle towards an engaging pastime [83, 84]. According to the ideas of the attentional capacity hypothesis, the distracting stimulus has to be more intense than the painful one to get the child's attention. Music and cartoons have been shown to be useful in lowering pain and anxiety in children undergoing various medical procedures [82, 85, 86]. It seems that these diversions may not always exert sufficient influence to divert children's attention from the experience of pain. It is presently thought that multisensory methods, which use all three senses (sight, sound, and touch), may be more effective in relieving pain than single-sensory ones [82, 85]. Considering the varied outcomes, it may be important to use more engaging types of diversion to engage children in activities during medical procedures. Another category, implantable neuromodulation devices, represents a more invasive yet highly effective method for targeted pain relief. These devices, such as spinal cord stimulators, are surgically implanted and function by delivering electrical pulses directly to the spinal cord. The process effectively suppresses pain signals before their transmission to the brain. These devices have particular significance for patients who are grappling with persistent back pain or neuropathic diseases.

Robot-assisted rehabilitation technologies, such as exoskeletons or limb rehabilitation systems, are crucial in the treatment of pain throughout the process of recovery. These gadgets serve to enhance the process of physical therapy and rehabilitation, aiding patients in the restoration of strength and movement after accidents or surgical procedures, hence often leading to less discomfort and enhanced general functionality. For pediatric pain therapy, experts advise that technological and psychological distractions for children be examined, in addition to the various therapies that are now available. When a humanoid robot is developed with humanistic qualities and designed to implement psychological methods, there is potential for it to effectively alleviate procedural pain and discomfort in children [83].

2.3 The role of robots in achieving the sustainable development goals (SDGs)

The members of the panel agreed that robotics has the potential to contribute to the attainment of Sustainable Development Goals (SDGs), although with varying degrees of direct impact. The potential of these technologies to assist in different capabilities was emphasized by researchers [87]. The United Nations adopted the 2030 Agenda for Sustainable Development in September 2015, which includes 17 SDGs with specific targets to address a wide range of global challenges. The goals were developed through a collaborative and inclusive process involving governments, civil society, and international organizations. Each goal was formulated to address a critical aspect of sustainable development, to improve the well-being of people and the planet by 2030. The numbers assigned to each SDG simply represent their order in the list of 17 goals, and they do not carry any particular significance beyond that. SDG 1, for example, is "No Poverty," SDG 2 is "Zero Hunger," and so on, up to SDG 17, which is "Partnerships for the Goals" (Figure 2.2). The following arguments exclusively address the possibilities mentioned during the panel discussion. The list of sectors and applications where robots may support the SDGs is not complete, but it does serve that purpose.

Robotics, as described in [88], was found to be very important to SDG 3 ("Good Health") in particular. Healthcare robots now play a significant role in enhancing the standard of medical treatment [89, 90]. Access to medication may be enhanced through the use of drones to deliver medications [91]. Additionally, social robots have the potential to encourage proper hand hygiene practices [92] and support the progress of children diagnosed with autism [93]. Robotics may also make a substantial contribution to SDG 8 ("Labor Productivity and Financial Growth") (Figure 2.2) by replacing people in laborious or dangerous activities [94], particularly in the setting of aging populations [95]. Additionally, they may be used to instruct employees of all skill levels, from low-skilled individuals to highly specialized professionals like cardiovascular surgeons, or to enhance workplace ergonomics [96–98]. One of the panelists also introduced an idea that uses robots to provide technical education to women in countries with low female employment participation rates. The investigators discovered that technology often served as a more effective medium for women compared to a male instructor, surpassing cultural barriers and enhancing their comfort levels. This was particularly evident in the context of traditionally male-dominated skills like plumbing [99, 100]. This is a significant step in the advancement of SDG 5 ("Gender Equality").

The use of robots in education has the potential to enhance the standard of education, as stated in SDG 4, which focuses on "Quality Education." An illustrative example involves the implementation of underwater robots by educational institutions. By employing these robots, educational institutions enable students to explore the oceans, thereby facilitating their understanding of ecosystems, resources, and pollution. This educational approach also contributes to the advancement of SDG 14, which pertains to "Life below Water" [101]. A significant focus has been placed on the use of robots in the prevention and management of emergencies [102], with the aim of minimizing their adverse effects on the SDGs. Robots not only facilitate access to hazardous areas but also accelerate the process of rehabilitation. For example, it is possible to accelerate the process of conducting surveys to ensure the safety of the environment before the restart of economic operations [103]. The potential for surveying and mapping was identified as a means to support the generation of inexpensive and environmentally friendly energy, as mentioned in SDGs 7 and 13, which focus on renewable energy and climate action, respectively. This approach enables the strategic placement of infrastructure, such as offshore wind farms [104].

Similarly, they may monitor pollutants and safeguard ecosystems [105] (SDGs 14 and 15, "Life below Water" and "Life on Land," respectively). The panelists discussed the role that robots may play in the inspection and maintenance of the facilities that support our society (SDG 9, which is titled "Innovation and Infrastructure"), such as dams, bridges, pathways, and power plants. According to previous research [104], the study focused on the examination of sports and plants, vehicles, and transportation systems in cities. They may assist in reducing traffic congestion, improving efficiency, and enhancing the overall transportation experience for individuals. Additionally, robots have the potential to contribute to the development of smart cities by integrating with existing infrastructure and providing innovative solutions for urban mobility challenges. Several studies have examined the significance of robotics in promoting responsible consumption and production, specifically SDG 12, which pertains to sustainable consumption. These studies have primarily focused on the application of robotics in meat production [106]. Additionally, there has been a broader exploration of proactive robot systems that hope to impact human behavior and minimize excessive consumption of food, water, and energy.

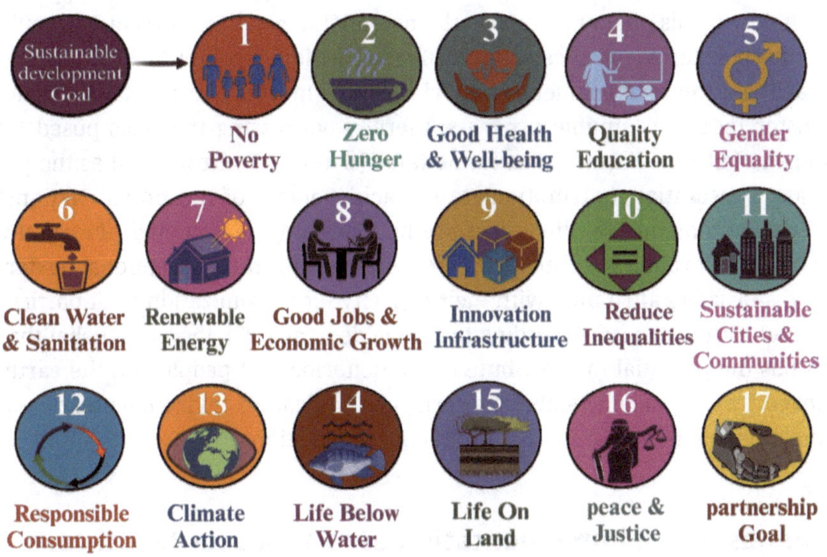

Figure 2.2: A graphical illustration of robotics prospects in the context of the United Nations Sustainable Development Goals (SDGs) [107] (open access in MDPI).

2.4 Sustainable development using soft robotics

Soft robotics has many different advantages compared to conventional robotics, rendering it highly suitable for addressing the concept of CPA (Paris Climate Agreement) and attaining SDGs [108–110]. Through the utilization of the adaptable nature, versatility, biocompatibility/degradability, and the ability to be (re)programmed with soft materials, along with the incorporation of physical as well as embodied intelligence (EI), these robots possess the potential to be employed in the monitoring and restoration of intricate environments, the establishment of warning systems for urban regions, and the gathering of data about alterations in ecological diversity and animals behavior [111–114]. In addition, these devices possess the ability to replicate the motion and conduct behaviors of different animals and plants. Consequently, they may be used to facilitate the processes of pollination, seed dispersion, and soil aeration, thus promoting the regeneration and restoration of ecosystems in remote and inaccessible areas [115–117]. Soft robots are already addressing important difficulties in real-world applications that match the SDGs and CPA [118–122]. Creating low-cost, environmentally friendly, durable, and self-piloting robots that do not depend on hard components is only one of the challenges that the field of soft robotics currently confronts despite these advances [123]. In order to effectively use the capabilities of soft robotics in advancing the SDGs and facilitating beneficial climate actions, it is essential to engage in careful preparation and establish appropriate governance structures, while also garnering the support of diverse stakeholders. This phenomenon has the potential to generate current economic prospects in the short term (specifically, by 2030) and medium term (by 2050). Additionally, it can facilitate the establishment of novel business models and regulatory structures that foster more sustainable progress, thereby addressing the risks posed by climate change to both the Earth and its inhabitants [124–125]. Soft robotics has the potential to make substantial contributions to the achievement of the United Nations' SDGs by offering automation solutions that are both sustainable and environmentally friendly [126–128]. Soft robotics technologies are renowned for their exceptional dexterity, sensitivity, and safety attributes, with their capacity for customization to fit particular activities and circumstances. According to Mengaldo et al. [127], there is a belief that soft robotics has the potential to contribute to the betterment of people and the earth. This is achieved through solving global concerns and mitigating environmental degradation via the implementation of specific productive applications.

2.5 Challenges in the implementation of robotics for healthcare sustainability

Robotic surgery is expanding and gaining popularity in several medical fields nowadays. Numerous healthcare facilities exhibit a strong interest in the emerging iterations of robotic surgical models, which hold promise for addressing the functional and cost constraints inherent in current robotic systems. According to projections provided by Wintergreen Research, the global market for robotic surgery is expected to reach roughly USD 20 billion by the year 2021. Through the use of information technology, a robotic system may possess the ability to autonomously perform physical tasks and engage in interactions across both the physical and digital worlds. A robotic system can integrate inputs from the physical environment, establish a comprehensive understanding, and afterward respond by taking into account its preprogrammed instructions as well as the newly acquired information. In a study [129] authors indicates that there is a deficiency in sensitivity when it comes to perception, control, and interaction insights. Over the last decade, robotic systems like the da Vinci surgical robot have been widely implemented across the globe. There has been an improvement in the technical quality of procedures, resulting in a reduction in both the duration of hospital stays and the postoperative need for rehabilitation. Various robotic systems can sterilize equipment, collect blood, and monitor patient vitals [130, 131]. Furthermore, robotic systems could assist in logistics and supply chain management [73]. Robotic approaches to patient monitoring and eldercare facilities may increase the strength of the healthcare personnel by carrying out routine, and often laborious, duties or assisting the mobility of patients who are disabled [132, 133]. A new study aims to create robotic behavioral architecture for children and mentally disabled persons [134]. Additionally, robots provide advantageous cost-effectiveness characteristics.

Robots, according to hospital executives, may reduce the annual cost of human labor by as much as 65% [135]. Robots possess the capability to transcribe and store vital medical data, thereby reducing the likelihood of errors. Additionally, they aid healthcare professionals in patient diagnosis and provide support to less experienced healthcare workers in administering treatment, thereby diminishing the reliance on higher-skilled professionals [136, 137]. However, health robotics problems must be considered. In summary, these concerns include technological reliance and dehumanizing healthcare [133, 138]. A comprehensive analysis of recent advancements and implementations of robotics in the healthcare sector critically examines the existing concerns and challenges observed in this field.

2.5.1 Learning-based adaptability and change

Older persons have varying degrees of degradation in a variety of functions, including hearing and mobility. As a result of the widespread occurrence of chronic illnesses, individuals may have impairments in several functional areas, leading to restrictions in their capacities. For robots to provide help in walking or other kinds of motor support at an individual level, they must have the capability to adapt to variations among persons resulting from diverse combinations of deterioration and impairment [139–141]. If a robot is supposed to help a patient get around a nursing home, it needs to be able to adapt to the individual's speed of walking, hearing and speaking abilities for speaking orders, health status in case of emergency, and the possible need to stop for a restroom break or a drink of water along the way. Robots need to possess machine learning capabilities [139, 141] to achieve flexibility. There are numerous research topics within this domain, such as the study of human behavior within domestic settings and the identification of abnormal human behavior for emergency response purposes [142]. To ensure reliable and secure functioning, robots must possess the capability to adjust to their operational surroundings, as well as to the individuals utilizing them. This is particularly crucial in residential settings, where the environment is characterized by a lack of structure and a high degree of variability, in contrast to industrial environments. Adaptability and the ability to make changes through learning are the factors to success in both professional and private life. This means being able to accept change, learn new skills and knowledge, and change direction in response to changing conditions. The concept includes a dedication to continuous learning, promoting adaptability in mental procedures and analytical approaches, and cultivating perseverance in response to challenges. The adoption of an adaptable attitude is not only conducive to personal growth and development, but it also plays a crucial role in the field of business. The balance between change and stability allows people and organizations to use learning to traverse a dynamic and ever-changing terrain, leading to greater satisfaction and accomplishment.

2.5.2 Socio-emotional intelligence

The primary objective of service robots, particularly in the healthcare sector, is not to replace nurses or other medical professionals, but instead to augment, support, and collaborate with human users. The significance of effective communication and interaction between humans and robots is emphasized by these criteria, as it is crucial for the achievement of successful and valuable robotic assistance and applications. Social intelligence, a subset of artificial intelligence (AI), includes the social skills that people experience while engaging with robots. It has been identified as a crucial aspect of robotic design, particularly in the environment of healthcare and eldercare settings [143]. Many studies in the field of human-robot interaction (HRI) have shown that enhanced social intelligence, as perceived by users, leads to improved communication effectiveness and thus, greater acceptability among users [144]. The research conducted in this field involves the investigation of a robot's capacity to understand the intentions and feelings of the user and afterward react to them by exhibiting emotions. Additionally, it explores the robot's capability to show its personality by integrating emotions into speech and generating facial expressions [139]. Additional concerns within the field of social intelligence include the ability to recognize and react to both verbal and nonverbal signals, including elements such as speech, gestures, facial expressions, and body language. The integration of social intelligence into robotic systems facilitates enhanced HRI characterized by increased consistency and accessibility. The ability of robots to comprehend and react suitably to these signals generates a feeling of comfort and ease in human users. The addition of emotion perception and response capabilities is crucial for the advancement of healthcare robots. This might include the identification of stress, nervousness, pain, or pleasure in individuals and providing appropriate empathy and support in response. An instance of technological advancement may be seen in the form of a robot capable of identifying troubled patients and providing them with comforting words or actions. This innovation has the potential to greatly enhance the emotional well-being of the patient.

2.5.3 Software architecture

Common functions found in robots often include navigation, localization, organizing, and obstacle avoidance. Furthermore, a healthcare robot often requires certain functions and HRI modules, including voice recognition and generation, as well as user identification, to effectively carry out its intended services. As the field of robotics advances, there is a growing need for robots to possess an increasing array of capabilities to effectively handle services and activities of greater complexity. The role of software architecture in coordinating and executing various function modules is becoming more crucial in ensuring reliable operation. The objective of software architecture development is to enhance the dependable execution of modules and the effective communication between them, as shown in the cases of Minerva and Care-O-bots [145].

The software architecture used in healthcare robots plays a pivotal role in facilitating the harmonious integration and coordination of essential functionalities such as navigation, localization, organization, and obstacle avoidance. Nevertheless, the defining characteristic of healthcare robots is their inherent need for efficient modules that facilitate HRI. The modules contain fundamental functionalities such as speech recognition and generation, user identification, and intuitive interfaces. Within the healthcare domain, software architecture has a crucial role in not only guaranteeing the technical effectiveness of these operations but also in influencing patient care and the user experience. This technology facilitates efficient communication between robots and both patients and medical personnel, allowing for the understanding and appropriate response to user requirements, while also ensuring the safeguarding of confidential medical information. Therefore, the design and robustness of the software architecture play a crucial role in maximizing the capabilities of healthcare robots. This includes improving their functional capacities and their ability to provide compassionate and reliable help in complicated healthcare environments.

2.6 Conclusion

In conclusion, promoting the adoption of robotics for sustainability in healthcare provides a compelling vision for the future of healthcare delivery. The incorporation of robotic technology into healthcare systems holds promise for bolstering sustainability via the optimization of resource allocation, the enhancement of patient outcomes, and the mitigation of difficulties arising from an aging population and shortages in the healthcare workforce. The significance of robots in achieving the SDGs was explored at different levels, yielding valuable information to be discussed in the robotics community. Robotics provide several prospects for facilitating the achievement of the SDGs across different fields, including healthcare, infrastructure, disaster response, ecological monitoring, responsible production and consumption, and even gender equality. They may also have unintended consequences that might exceed any beneficial results, including changes to the employment market, environmental implications, moral issues, data handling challenges, safety concerns, and effects on norms and values. Limited understanding among decision makers about robotics, as well as the insufficient knowledge within the robotics community regarding the SDGs poses a hindrance to fully realizing the potential benefits associated with these prospects. Furthermore, it is essential to implement a change in the problem-solving methodology by including human-centered, multidisciplinary systemic thinking. This shift is crucial to guarantee the sustained applicability of robotic interventions to the SDGs.

Enhancing the societal perception of robotics can be facilitated through the education of both children and the general public. This can be achieved by incorporating sustainability principles into robotics education programs, implementing sustainability assessments, and organizing workshops and initiatives. These efforts are expected to increase awareness about the SDGs within the robotics community. Health robots improve the standard of treatment, increase the adaptability of the healthcare profession, and make a substantial contribution to educational and research efforts. The connectivity of IoT is crucial in ensuring the availability of information collected by robots for healthcare practitioners. In the future, healthcare systems are expected to include more advanced and cost-effective technology, such as IoT-integrated robotic systems, capable of operating efficiently in diverse environments and performing challenging tasks. Robots have the potential to contribute significantly to the optimization of resource allocation, cost reduction, and the provision of accessible, high-quality healthcare. This may be achieved via the automation of repetitive work, the assistance provided to medical staff, and the improvement of patient care. However, successful adoption is dependent on overcoming several challenges, such as legal constraints, ethical issues, cost-effectiveness, and guaranteeing seamless human-robot cooperation. Healthcare providers, engineers, policymakers, and other stakeholders must collaborate across disciplines to actualize the full potential of robotics in healthcare. With ongoing innovation, study, and dedication to ethical and responsible implementation, the use of robots in healthcare can be a key part of building a sustainable and adaptable healthcare environment that meets the needs of both current and future generations.

References

[1] Vänni K. Robot applications in communication. Smart technology solutions support the elderly to continue living in their own homes. Reports of the Ministry of the Environment. 2017;7(2017):44–51.

[2] Sun CB, Wang YY, Liu GH, Liu Z. Role of the eye in transmitting human coronavirus: What we know and what we do not know. Frontiers in Public Health. 2020 Apr 24;8:155.

[3] Wang WT, Cheng SY, Huang LY. Technology-based service encounters using self-service technologies in the healthcare industry. International Journal of Human-Computer Interaction. 2013 Jan 1;29(3):139–55.

[4] Springer Medizin. Roboter in der Pflege – Ein Ausweg aus dem Personalnotstand?. Geriatrie-Report. 2019 Jun;14:6–7.

[5] Čaić M, Mahr D, Oderkerken-Schröder G. Value of social robots in services: Social cognition perspective. Journal of Services Marketing. 2019 Jun 18;33(4):463–78.

[6] Dana LP, Salamzadeh A, Hadizadeh M, Heydari G, Shamsoddin S. Urban entrepreneurship and sustainable businesses in smart cities: Exploring the role of digital technologies. Sustainable Technology and Entrepreneurship. 2022 May 1;1(2):100016.

[7] Chen F, Liu G. Population aging in China. In: Uhlenberg P (eds.), International Handbook of Population Aging. International Handbooks of Population (vol. 1, pp. 157–72). Springer: Dordrecht; 2009 Apr 29. https://doi.org/10.1007/978-1-4020-8356-3_8

[8] Höpfl F, Peisl T, Greiner C. Exploring stakeholder perspectives: Enhancing robot acceptance for sustainable healthcare solutions. Sustainable Technology and Entrepreneurship. 2023 Jul;29:100045.

[9] Arent K, Cholewiński M, Domski W, Drwięga M, Jakubiak J, Janiak M, Kreczmer B, Kurnicki A, Stańczyk B, Szczęśniak-Stańczyk D. Selected topics in design and application of a robot for remote medical examination with the use of ultrasonography and auscultation from the perspective of the remedy project. Journal of Automation Mobile Robotics and Intelligent Systems. 2017 Jun 2;11(2):82–94.

[10] Nawrat Z. Polski robot kardiochirurgiczny. Postępy robotyki/Przemysłowe i medyczne systemy robotyczne/Praca zbiorowa pod redakcją Krzysztofa Tchonia/Warszawa 2005r. str. 2005;117:129.

[11] Joseph A, Christian B, Abiodun AA, Oyawale F. A review on humanoid robotics in healthcare. In: MATEC Web of Conferences 2018 (Vol. 153, p. 02004).

[12] Iftikhar M, Majid MJ, Muralindran M, Thayabaren G, Vigneswaran R, Brendan TT. OTOROB: Robot for orthopedic surgeon-roboscope: A non-interventional medical robot for telerounding. In: 2011 5th International Conference on Bioinformatics and Biomedical Engineering 2011 May 10 (pp. 1–5).

[13] Balasubramanian S, Klein J, Burdet E. Robot-assisted rehabilitation of hand function. Current Opinion in Neurology. 2010 Dec 1;23(6):661–70.

[14] Nef T, Mihelj M, Riener R. ARMin: A robot for patient-cooperative arm therapy. Medical & Biological Engineering & Computing. 2007 Sep;45:887–900.

[15] Kaczmarski M, Granosik G. Rehabilitation robot rrh1. Archive of Mechanical Engineering. 2011;58(1):103–13.

[16] Mariappan M, Ganesan T, Ramu V, Iftikhar M. Safety system and navigation for orthopaedic robot (OTOROB). In: Intelligent Robotics and Applications: 4th International Conference, ICIRA 2011, Aachen, Germany, December 6–8, 2011, Proceedings, Part II 4 2011 (pp. 358–67).

[17] Mukai T, Hirano S, Nakashima H, Kato Y, Sakaida Y, Guo S, Hosoe S. Development of a nursing-care assistant robot RIBA that can lift a human in its arms. In: 2010 IEEE/RSJ International Conference on Intelligent Robots and Systems 2010 Oct 18 (pp. 5996–6001).

[18] Urdiales C, Fernandez-Carmona M, Peula JM, Annicchiarricco R, Sandoval F, Caltagirone C. A collaborative control scheme for haptics-based blind wheelchair driving. Prace Naukowe Politechniki Warszawskiej. Elektronika. 2010(175, t. 1):3–22.

[19] Belter JT, Segil JL, Dollar AM, Weir RF. Mechanical design and performance specifications of anthropomorphic prosthetic hands: A review. Journal of Rehabilitation Research and Development. 2013 May 30;50(5):599.
[20] Panzirsch M, Weber B, Rubio L, Coloma S, Ferre M, Artigas J. Telehealthcare with humanoid robots: A user study on the evaluation of force feedback effects. In: 2017 IEEE World Haptics Conference (WHC) 2017 Jun 6 (pp. 245–50). IEEE.
[21] Chen TL, King CH, Thomaz AL, Kemp CC. An investigation of responses to robot-initiated touch in a nursing context. International Journal of Social Robotics. 2014 Jan;6(1):141–61.
[22] Napper SA, Seaman RL. Applications of robots in rehabilitation. Robotics and Autonomous Systems. 1989 Nov 1;5(3):227–39.
[23] Li Z, Moran P, Dong Q, Shaw RJ, Hauser K. Development of a tele-nursing mobile manipulator for remote care-giving in quarantine areas. In: 2017 IEEE International Conference on Robotics and Automation (ICRA) 2017 May 29 (pp. 3581–86).
[24] Chen AI, Balter ML, Maguire TJ, Yarmush ML. Real-time needle steering in response to rolling vein deformation by a 9-DOF image-guided autonomous venipuncture robot. In: 2015 IEEE/RSJ International Conference on Intelligent Robots and Systems (IROS) 2015 Sep 28 (pp. 2633–38).
[25] Butter M, Rensma A, Boxsel JV, Kalisingh S, Schoone M, Leis M, Gelderblom GJ, Cremers G, Wilt MD, Kortekaas W, Thielmann A. Robotics for healthcare. Final report.
[26] Morel CM, Acharya T, Broun D, Dangi A, Elias C, Ganguly NK, Gardner CA, Gupta RK, Haycock J, Heher AD, Hotez PJ. Health innovation networks to help developing countries address neglected diseases. Science. 2005 Jul 15;309(5733):401–04.
[27] Taylor RH. A perspective on medical robotics. Proceedings of the IEEE. 2006 Sep;94(9):1652–64.
[28] Bakhru V. The future of medical robotics: How technology will improve health outcomes. Becker's Hospital Review. 2017 Dec 7;7:2017.
[29] Vincze M, Zagler W, Lammer L, Weiss A, Huber A, Fischinger D, Koertner T, Schmid A, Gisinger C. Towards a robot for supporting older people to stay longer independent at home. In: ISR/ Robotik 2014; 41st International Symposium on Robotics 2014 Jun 2 (pp. 1–7).
[30] Broekens J, Heerink M, Rosendal H. Assistive social robots in elderly care: A review. Gerontechnology. 2009 Apr 1;8(2):94–103.
[31] Gomi T, Griffith A. Developing intelligent wheelchairs for the handicapped. Assistive Technology and Artificial Intelligence: Applications in Robotics, User Interfaces and Natural Language Processing. 2006 May;31:150–78.
[32] Bhattacharyya S, Konar A, Tibarewala DN. A differential evolution-based energy trajectory planner for artificial limb control using motor imagery EEG signal. Biomedical Signal Processing and Control. 2014 May 1;11:107–13.
[33] Choi H, Park JO, Ko SY, Park S. Deflection analysis of a robotic bed on the applied loads and its postures for a heavy-ion therapeutic system. In: 9th International Conference on Robotic, Vision, Signal Processing and Power Applications: Empowering Research and Innovation 2017 (pp. 343–50).
[34] Yan T, Cempini M, Oddo CM, Vitiello N. Review of assistive strategies in powered lower-limb orthoses and exoskeletons. Robotics and Autonomous Systems. 2015 Feb 1;64:120–36.
[35] Kazerooni H. Exoskeletons for human power augmentation. In: 2005 IEEE/RSJ International conference on intelligent Robots and Systems 2005 Aug 2 (pp. 3459–64).
[36] Darling K. Extending legal protection to social robots: The effects of anthropomorphism, empathy, and violent behavior towards robotic objects (April 23, 2012). In Calo, Froomkin, Kerr (eds.), Robot Law. Edward Elgar; 2016. We Robot Conference 2012, University of Miami, Available at SSRN: https://ssrn.com/abstract=2044797 or http://dx.doi.org/10.2139/ssrn.2044797
[37] Campa R. Roboty społeczne i pRaca socJalna. Zeszyty Pracy Socjalnej. 2016;21(2):65–75.

[38] Asfour T, Regenstein K, Azad P, Schroder J, Bierbaum A, Vahrenkamp N, Dillmann R. ARMAR-III: An integrated humanoid platform for sensory-motor control. In: 2006 6th IEEE-RAS International Conference on Humanoid Robots 2006 Dec 4 (pp. 169–75).

[39] King CH, Chen TL, Jain A, Kemp CC. Towards an assistive robot that autonomously performs bed baths for patient hygiene. In: 2010 IEEE/RSJ International Conference on Intelligent Robots and Systems 2010 Oct 18 (pp. 319–24).

[40] Bergerman M, Van Henten E, Billingsley J, Reid J, Mingcong D. IEEE robotics and automation society technical committee on agricultural robotics and automation. IEEE Robotics & Automation Magazine. 2013 Jun 1;20(2):20–23.

[41] Yinger OS, Gooding LF. A systematic review of music-based interventions for procedural support. The Journal of Music Therapy. 2015 Mar 1;52(1):1–77.

[42] Mukai T, Onishi M, Odashima T, Hirano S, Luo Z. Development of the tactile sensor system of a human-interactive robot "RI-MAN". IEEE Transactions on Robotics. 2008 Apr 4;24(2):505–12.

[43] Mukai T, Hirano S, Yoshida M, Nakashima H, Guo S, Hayakawa Y. Tactile-based motion adjustment for the nursing-care assistant robot RIBA. In: 2011 IEEE International Conference on Robotics and Automation 2011 May 9 (pp. 5435–41).

[44] Reiser U, Connette C, Fischer J, Kubacki J, Bubeck A, Weisshardt F, Jacobs T, Parlitz C, Hägele M, Verl A. Care-O-bot® 3-creating a product vision for service robot applications by integrating design and technology. In: 2009 IEEE/RSJ International Conference on Intelligent Robots and Systems 2009 Oct 10 (pp. 1992–98).

[45] Doelling K, Shin J, Popa DO. Service robotics for the home: A state of the art review. In: Proceedings of the 7th International Conference on PErvasive Technologies Related to Assistive Environments 2014 May 27 (pp. 1–8).

[46] Karar AS, Said S, Beyrouthy T. Pepper humanoid robot as a service robot: A customer approach. In: 2019 3rd International Conference on Bio-engineering for Smart Technologies (BioSMART) 2019 Apr 24 (pp. 1–4).

[47] Ejdys J, Halicka K. Sustainable adaptation of new technology – The case of humanoids used for the care of older adults. Sustainability. 2018 Oct 18;10(10):3770.

[48] Pollack ME, Brown L, Colbry D, Orosz C, Peintner B, Ramakrishnan S, Engberg S, Matthews JT, Dunbar-Jacob J, McCarthy CE, Thrun S. Pearl: A mobile robotic assistant for the elderly. In: AAAI Workshop on Automation as Eldercare 2002 Aug (Vol. 2002, pp. 85–91).

[49] Hirose T, Fujioka S, Mizuno O, Nakamura T. Development of a hair-washing robot equipped with scrubbing fingers. In: 2012 IEEE International Conference on Robotics and Automation 2012 May 14 (pp. 1970–75).

[50] Hu J, Edsinger A, Lim YJ, Donaldson N, Solano M, Solochek A, Marchessault R. An advanced medical robotic system augmenting healthcare capabilities-robotic nursing assistant. In: 2011 IEEE International Conference on Robotics and Automation 2011 May 9 (pp. 6264–69).

[51] Piezzo C, Suzuki K. Feasibility study of a socially assistive humanoid robot for guiding elderly individuals during walking. Future Internet. 2017 Jul 8;9(3):30.

[52] Lazzeri N, Mazzei D, Ben Moussa M, Magnenat-Thalmann N, De Rossi D. The influence of dynamics and speech on understanding humanoid facial expressions. International Journal of Advanced Robotic Systems. 2018 Jul 4;15(4):1729881418783158.

[53] Chen CF, Xu X, Arpan L. Between the technology acceptance model and sustainable energy technology acceptance model: Investigating smart meter acceptance in the United States. Energy Research & Social Science. 2017 Mar 1;25:93–104.

[54] Baganzi R, Lau AK. Examining trust and risk in mobile money acceptance in Uganda. Sustainability. 2017 Dec 3;9(12):2233.

[55] Kikulwe EM, Fischer E, Qaim M. Mobile money, smallholder farmers, and household welfare in Kenya. PloS one. 2014 Oct 6;9(10)::e109804.

[56] Bakare BI, Ekanem IA, Allen IO. Appraisal of global system for mobile communication (GSM) in Nigeria. American Journal of Engineering Research. 2017;6(6):97–102.
[57] Jiang W, Huang Y, Sha A. A review of eco-friendly functional road materials. Construction and Building Materials. 2018 Dec 10;191:1082–92.
[58] Broadbent E, Stafford R, MacDonald B. Acceptance of healthcare robots for the older population: Review and future directions. International Journal of Social Robotics. 2009 Nov;1:319–30.
[59] Deligianis C, Stanton CJ, McGarty C, Stevens CJ. The impact of intergroup bias on trust and approach behavior towards a humanoid robot. Journal of Human-robot Interaction. 2017 Dec 1;6(3):4–20.
[60] Hancock PA, Billings DR, Schaefer KE, Chen JY, De Visser EJ, Parasuraman R. A meta-analysis of factors affecting trust in human-robot interaction. Human Factors. 2011 Oct;53(5):517–27.
[61] Dang VA, Vu Khanh Q, Nguyen VH, Nguyen T, Nguyen DC. Intelligent healthcare: Integration of emerging technologies and Internet of Things for humanity. Sensors. 2023 Apr 22;23(9):4200.
[62] Vince G, Wilson C. The rise of the miniature medical robots. New Scientist. 2009 Nov 18;204 (2735):50–3.
[63] Rayan RA, Tsagkaris C, Zafar I. 1 IoT-Integrated Robotics. Robotic Technologies in Biomedical and Healthcare Engineering. 2021 Jun;29:1.
[64] Grifantini K. Robotic Nurse Washes Human. MIT Technology Review. 2010 Nov 10. https://www.technologyreview.com/2010/11/10/199054/robotic-nurse-washes-human/
[65] Han T, Anderson F, Irani P, Grossman T. Hydroring: Supporting mixed reality haptics using liquid flow. In: Proceedings of the 31st Annual ACM Symposium on User Interface Software and Technology 2018 Oct 11 (pp. 913–25).
[66] Lazarte M. Enter the first cyborg-type robot. ISOfocus. 2014 Sep;106:16–21.
[67] Psiha MM, Vlamos P. IoT applications with 5G connectivity in medical tourism sector management: Third-party service scenarios. In: GeNeDis 2016: Geriatrics 2017 (pp. 141–54).
[68] Latif S, Qadir J, Farooq S, Imran MA. How 5G wireless (and concomitant technologies) will revolutionize healthcare. Future Internet. 2017 Dec 11;9(4):93.
[69] O'brolcháin F, De Colle S, Gordijn B. The ethics of smart stadia: A stakeholder analysis of the Croke Park project. Science and Engineering Ethics. 2019 Jun 15;25(3):737–69.
[70] Mittelstadt B. Ethics of the health-related internet of things: A narrative review. Ethics and Information Technology. 2017 Sep;19(3):157–75.
[71] Bandyopadhyay D, Sen J. Internet of things: Applications and challenges in technology and standardization. Wireless Personal Communications. 2011 May;58:49–69.
[72] Gopal G, Suter-Crazzolara C, Toldo L, Eberhardt W. Digital transformation in healthcare–architectures of present and future information technologies. Clinical Chemistry and Laboratory Medicine (CCLM). 2019 Mar 1;57(3):328–35.
[73] Özdemir V, Hekim N. Birth of industry 5.0: Making sense of big data with artificial intelligence, "the internet of things" and next-generation technology policy. Omics: A Journal of Integrative Biology. 2018 Jan 1;22(1):65–76.
[74] Gope P, Hwang T. BSN-Care: A secure IoT-based modern healthcare system using body sensor network. IEEE Sensors Journal. 2015 Nov 20;16(5):1368–76.
[75] Chai PR, Zhang H, Jambaulikar GD, Boyer EW, Shrestha L, Kitmitto L, Wickner PG, Salmasian H, Landman AB. Internet of things buttons to measure and respond to restroom cleanliness in a hospital setting: A descriptive study. Journal of Medical Internet Research. 2019 Jun 19;21(6):e13588.
[76] Stefano GB, Kream RM. The micro-hospital: 5G telemedicine-based care. Medical Science Monitors Basic Research. 2018;24:103.
[77] van Osch M, Bera D, van Hee K, Koks Y, Zeegers H. Tele-operated service robots: ROSE. Automation in Construction. 2014 Apr 1;39:152–60.
[78] Riek LD. Healthcare robotics. Communications of the ACM. 2017 Oct 24;60(11):68–78.

[79] Kraft K, Smart WD. Seeing is comforting: Effects of teleoperator visibility in robot-mediated health care. In: 2016 11th ACM/IEEE International Conference on Human-Robot Interaction (HRI) 2016 Mar 7 (pp. 11–18).
[80] Yasemin M, Kasımoğlu Y, Kocaaydın S, Karslı E, İnce EB, İnce G. Management of dental anxiety in children using robots. In: 2016 24th Signal Processing and Communication Application Conference (SIU) 2016 May 16 (pp. 237–40).
[81] Beran TN, Ramirez-Serrano A, Vanderkooi OG, Kuhn S. Humanoid robotics in health care: An exploration of children's and parents' emotional reactions. Journal of Health Psychology. 2015 Jul;20(7):984–89.
[82] Beran TN, Ramirez-Serrano A, Vanderkooi OG, Kuhn S. Reducing children's pain and distress towards flu vaccinations: A novel and effective application of humanoid robotics. Vaccine. 2013 Jun 7;31(25):2772–77.
[83] McCaul KD, Malott JM. Distraction and coping with pain. Psychological Bulletin. 95(1984):516–33. Pain. 1985 Nov 1;23(3):315.
[84] DeMore M, Cohen LL. Distraction for pediatric immunization pain: A critical review. Journal of Clinical Psychology in Medical Settings. 2005 Dec;12:281–91.
[85] Klassen JA, Liang Y, Tjosvold L, Klassen TP, Hartling L. Music for pain and anxiety in children undergoing medical procedures: A systematic review of randomized controlled trials. Ambulatory Pediatrics. 2008 Mar 17;8(2):117–28.
[86] Hitka M, Schmidtová J, Lorincová S, Štarchoň P, Weberová D, Kampf R. Sustainability of human resource management processes through employee motivation and job satisfaction. Acta Polytechnica Hungarica. 2021 Jan 1;18(2):7–26.
[87] Haidegger T. Taxonomy and standards in robotics. In: Encyclopedia of Robotics (pp. 1–10). 2021.
[88] Kyrarini M, Lygerakis F, Rajavenkatanarayanan A, Sevastopoulos C, Nambiappan HR, Chaitanya KK, Babu AR, Mathew J, Makedon F. A survey of robots in healthcare. Technologies. 2021 Jan 18;9(1):8.
[89] Haidegger T. Autonomy for surgical robots: Concepts and paradigms. IEEE Transactions on Medical Robotics and Bionics. 2019 Apr 25;1(2):65–76.
[90] Courtney P, Royall PG. The role of medical logistics drones in support of SDG 3–Lessons for the robotics community. In: Proc. IEEE IROS 2021 Workshop on the Role Robot. Achieving UN's Sustainable Development Goals 2021 Oct.
[91] Sasidharan S, Pasupuleti D, Das AM, Kapoor C, Manikutty G, Rao RB. Haksh-e: An autonomous social robot for promoting good hand hygiene among children. In: IROS 2021 Workshop on The Roles of Robotics in Achieving the UN's Sustainable Development Goals 2021 Oct.
[92] Lemaignan S, Newbutt N, Rice L, Daly J, Charisi V. UNICEF guidance on AI for children: Application to the design of a social robot for and with autistic children. arXiv preprint arXiv:2108.12166. 2021 Aug 27.
[93] Khamis A. Minefield mapping using distributed mobile sensors. In: 2016 International Conference on Robotics and Automation for Humanitarian Applications (RAHA) 2016 Dec 18 (pp. 1–6).
[94] Boesl DB, Bode M, Liepert B, Greisel S. Structured megatrends research as foundation for future-oriented research-planning and R&D roadmapping in robotics. In: 2017 International Conference on Current Trends in Computer, Electrical, Electronics and Communication (CTCEEC) 2017 Sep 8 (pp. 24–30).
[95] Akshay N, Deepu S, Rahul ES, Ranjith R, Jose J, Unnikrishnan R, Bhavani RR. Design and evaluation of a Haptic simulator for vocational skill Training and Assessment. In: IECON 2013-39th Annual Conference of the IEEE Industrial Electronics Society 2013 Nov 10 (pp. 6108–13).
[96] Sasi D, Bhavani RR. Characterization of expertise to build an augmented skill training system for construction industry. In: 2018 IEEE 18th International Conference on Advanced Learning Technologies (ICALT) 2018 Jul 9 (pp. 116–18).

[97] Mai V, Vanderborght B, Haidegger T, Khamis A, Bhargava N, Boesl DB, Gabriels K, Jacobs A, Moon A, Murphy R, Nakauchi Y. The role of robotics in achieving the United Nations Sustainable Development Goals – The experts' meeting at the 2021 IEEE/RSJ IROS workshop [industry activities]. IEEE Robotics & Automation Magazine. 2022 Mar 22;29(1):92–107.
[98] Jose J, Unnikrishnan R, Marshall D, Bhavani RR. Haptic simulations for training plumbing skills. In: 2014 IEEE International Symposium on Haptic, Audio and Visual Environments and Games (HAVE) Proceedings 2014 Oct 10 (pp. 65–70). IEEE.
[99] Bhavani RR, Sheshadri S, Maciuika LA. Addressing the first teachers: Education and sustainable development for children, families, and communities through vocational education, technology, and life skills training for women. Children and Sustainable Development (pp. 319–334). Ecological Education in a Globalized World. 2017. DOI:10.1007/978-3-319-47130-3_25.
[100] Hemsley B, Darcy S, Given F, Murray BR, Balandin S. Going thirsty for the turtles: Plastic straw bans, people with swallowing disability, and Sustainable Development Goal 14, Life Below Water. International Journal of Speech-Language Pathology. 2023 Jan 2;25(1):15–19.
[101] Jorge VA, Granada R, Maidana RG, Jurak DA, Heck G, Negreiros AP, Dos Santos DH, Gonçalves LM, Amory AM. A survey on unmanned surface vehicles for disaster robotics: Main challenges and directions. Sensors. 2019 Feb 8;19(3):702.
[102] Jean GV. Robots primed for disaster operations. National Defense. 2008 Nov 1;93(660):34–35.
[103] Denoncourt J. Companies and UN 2030 sustainable development goal 9 industry, innovation and infrastructure. Journal of Corporate Law Studies. 2020 Jan 2;20(1):199–235.
[104] Ferreira F, Miškovic N. Addressing the sustainability of the Adriatic Sea using marine robotics. In: Proc. IEEE IROS 2021 Workshop on Role Robot. Achieving UN's Sustainable Development Goals 2021 Oct (p. 2).
[105] Mason A, Korostynska O, Cordova-Lopez LE, Esper I, Romanov D, Ross S, Takacs K, Heidegger T. Meat Factory Cell: Assisting meat processors address sustainability in meat production. In: 2021 IEEE 21st International Symposium on Computational Intelligence and Informatics (CINTI) 2021 Nov 18 (pp. 000103–08). IEEE.
[106] Grosinger J. Fostering sustainable consumption through the development of proactive human-centered robot systems. In: Proc. IEEE IROS 2021 Workshop on the Role Robot. Achieving UN's Sustainable Development Goals 2021 Oct (p. 1).
[107] Mabkhot M, Ferreira P, Maffei A, Podržaj P, Mądziel M, Antonelli D, Lanzetta M, Barata J, Boffa E, Finžgar M, Paśko Ł. Mapping industry 4.0 enabling technologies into United Nations sustainability development goals. Sustainability. 2021 Feb 27;13(5):2560.
[108] Matthews HD, Wynes S. Current global efforts are insufficient to limit warming to 1.5 C. Science. 2022 Jun 24;376(6600):1404–09.
[109] Meinshausen M, Lewis J, McGlade C, Gütschow J, Nicholls Z, Burdon R, Cozzi L, Hackmann B. Realization of Paris Agreement pledges may limit warming just below 2 C. Nature. 2022 Apr 14;604(7905):304–09.
[110] Laschi C, Mazzolai B, Cianchetti M. Soft robotics: Technologies and systems pushing the boundaries of robot abilities. Science Robotics. 2016 Dec 6;1(1):eaah3690.
[111] Rossiter J, Winfield J, Ieropoulos I. Eating, drinking, living, dying and decaying soft robots. In: Soft Robotics: Trends, Applications and Challenges: Proceedings of the Soft Robotics Week, April 25–30, 2016, Livorno, Italy 2017 (pp. 95–101). Springer International Publishing.
[112] Mazzolai B, Laschi C. A vision for future bioinspired and biohybrid robots. Science Robotics. 2020 Jan 22;5(38):eaba6893.
[113] Sitti M. Physical intelligence as a new paradigm. Extreme Mechanics Letters. 2021 Jul 1;46:101340.
[114] Valdes S, Urza I, Pounds P, Singh S. Samara: Low-cost deployment for environmental sensing using passive autorotation. In: Robotics: Science and Systems Workshop on Robotics for Environmental Monitoring 2012 Jul.

[115] Hartmann F, Baumgartner M, Kaltenbrunner M. Becoming sustainable, the new frontier in soft robotics. Advanced Materials. 2021 May;33(19):2004413.

[116] Kim BH, Li K, Kim JT, Park Y, Jang H, Wang X, Xie Z, Won SM, Yoon HJ, Lee G, Jang WJ. Three-dimensional electronic microfliers inspired by wind-dispersed seeds. Nature. 2021 Sep 23;597(7877):503–10.

[117] Naclerio ND, Hawkes EW. Simple, low-hysteresis, foldable, fabric pneumatic artificial muscle. IEEE Robotics and Automation Letters. 2020 Feb 26;5(2):3406–13.

[118] Amend J, Cheng N, Fakhouri S, Culley B. Soft robotics commercialization: Jamming grippers from research to product. Soft Robotics. 2016 Dec 1;3(4):213–22.

[119] Ng CS, Tan MW, Xu C, Yang Z, Lee PS, Lum GZ. Locomotion of miniature soft robots. Advanced Materials. 2021 May;33(19):2003558.

[120] Elfferich JF, Dodou D, Della Santina C. Soft robotic grippers for crop handling or harvesting: A review. IEEE Access. 2022 Jul 14;10:75428–43.

[121] Li M, Pal A, Aghakhani A, Pena-Francesch A, Sitti M. Soft actuators for real-world applications. Nature Reviews Materials. 2022 Mar;7(3):235–49.

[122] Giordano G, Murali Babu SP, Mazzolai B. Soft robotics towards sustainable development goals and climate actions. Frontiers in Robotics and AI. 2023 Mar 17;10:1116005.

[123] He C, Zhang Y, Schneider A, Chen R, Zhang Y, Ma W, Kinney PL, Kan H. The inequality labor loss risk from future urban warming and adaptation strategies. Nature Communications. 2022 Jul 6;13(1):3847.

[124] Shukla PR, Skea J, Slade R, Al Khourdajie A, Van Diemen R, McCollum D, Pathak M, Some S, Vyas P, Fradera R, Belkacemi M. Climate change 2022: Mitigation of climate change. Contribution of working group III to the sixth assessment report of the Intergovernmental Panel on Climate Change. 2022;10:9781009157926.

[125] Grau Ruiz MA, O'Brolchain F. Environmental robotics for a sustainable future in circular economies. Nature Machine Intelligence. 2022 Jan;4(1):3–4.

[126] Guenat S, Purnell P, Davies ZG, Nawrath M, Stringer LC, Babu GR, Balasubramanian M, Ballantyne EE, Bylappa BK, Chen B, De Jager P. Meeting sustainable development goals via robotics and autonomous systems. Nature Communications. 2022 Jun 21;13(1):3559.

[127] Mengaldo G, Renda F, Brunton SL, Bächer M, Calisti M, Duriez C, Chirikjian GS, Laschi C. A concise guide to modeling the physics of embodied intelligence in soft robotics. Nature Reviews Physics. 2022 Sep;4(9):595–610.

[128] Patel AR, Patel RS, Singh NM, Kazi FS. The vitality of robotics in the healthcare industry: An Internet of Things (IoT) perspective. In: Internet of Things and Big Data Technologies for Next-Generation Healthcare (pp. 91–109). Springer: Germany; 2017. DOI:10.1007/978-3-319-49736-5_5

[129] Dervaderics J. The dawn of robotic surgery–from the roots up to the da Vinci telemanipulator system. Orvosi Hetilap. 2007 Dec 1;148(49):2307–13.

[130] Koh DH, Jang WS, Park JW, Ham WS, Han WK, Rha KH, Choi YD. Efficacy and safety of robotic procedures performed using the da Vinci robotic surgical system at a single institute in Korea: Experience with 10000 cases. Yonsei Medical Journal. 2018 Oct 1;59(8):975–81.

[131] Bedaf S, Gelderblom GJ, De Witte L. Overview and categorization of robots supporting independent living of elderly people: What activities do they support and how far have they developed?. Assistive Technology. 2015 Apr 3;27(2):88–100.

[132] Alaiad A, Zhou L. The determinants of home healthcare robots adoption: An empirical investigation. International Journal of Medical Informatics. 2014 Nov 1;83(11):825–40.

[133] Cao HL, Gómez Esteban P, Albert DB, Simut R, Van de Perre G, Lefeber D, Vanderborght B. A collaborative homeostatic-based behavior controller for social robots in human–robot interaction experiments. International Journal of Social Robotics. 2017 Nov;9:675–90.

[134] Broadbent E, Kerse N, Peri K, Robinson H, Jayawardena C, Kuo T, Datta C, Stafford R, Butler H, Jawalkar P, Amor M. Benefits and problems of health-care robots in aged care settings: A comparison trial. Australasian Journal on Aging. 2016 Mar;35(1):23–29.
[135] Challacombe B. Ben Challacombe in defense of robots. The Health Service Journal. 2013 Dec 1;123(6378):16–17.
[136] Yoon SN, Lee D. Artificial intelligence and robots in healthcare: What are the success factors for technology-based service encounters?. International Journal of Healthcare Management. 2019 Jul 3;12(3):218–25.
[137] Fracastoro G, Borzellino G, Castelli A, Fiorini P. Robotics in general surgery: Personal experience, critical analysis and perspectives. Chirurgia Italiana. 2005 Nov 1;57(6):687–93.
[138] López ME, Bergasa LM, Barea R, Escudero MS. A navigation system for assistant robots using visually augmented POMDPs. Autonomous Robots. 2005 Jul;19:67–87.
[139] Barea R, Bergasa LM, López E, Escudero MS, Leon C. Face recognition for social interaction with a personal robotic assistant. In: EUROCON 2005-The International Conference on" Computer as a Tool" 2005 Nov 21 (Vol. 1, pp. 382–85). IEEE.
[140] Tamura T, Yonemitsu S, Itoh A, Oikawa D, Kawakami A, Higashi Y, Fujimooto T, Nakajima K. Is an entertainment robot useful in the care of elderly people with severe dementia? The Journals of Gerontology Series A: Biological Sciences and Medical Sciences. 2004 Jan 1;59(1):M83–85.
[141] Schulz D, Burgard W, Fox D, Cremers AB. Tracking multiple moving targets with a mobile robot using particle filters and statistical data association. In: Proceedings 2001 ICRA. IEEE International Conference on Robotics and Automation (Cat. No. 01CH37164) 2001 May 21 (Vol. 2, pp. 1665–70).
[142] Heerink M, Krose B, Evers V, Wielinga B. The influence of a robot's social abilities on acceptance by elderly users. In: ROMAN 2006-The 15th IEEE International Symposium on Robot and Human Interactive Communication 2006 (pp. 521–26).
[143] Nejat G, Ficocelli M Can I be of assistance? The intelligence behind an assistive robot. In: 2008 IEEE International Conference on Robotics and Automation 2008 May 19 (pp. 3564–69).
[144] Heerink M, Kröse B, Wielinga B, Evers V. Human-Robot user studies in eldercare: Lessons learned. In: Proceedings ICOST, Belfast 2006 (pp. 31–38).
[145] Thrun S, Bennewitz M, Burgard W, Cremers AB, Dellaert F, Fox D, Hahnel D, Rosenberg C, Roy N, Schulte J, Schulz D. MINERVA: A second-generation museum tour-guide robot. In: Proceedings 1999 IEEE International Conference on Robotics and Automation (Cat. No. 99CH36288C) 1999 May 10 (Vol. 3).

Chapter 3
Design of healthcare sustainability: innovation and challenges

Abstract: The sustainability of healthcare innovations has received less attention despite substantial empirical, theoretical, and practical contributions to their development and implementation. However, several healthcare advances fail to maintain their effectiveness. There is a need to enhance the level of clarity regarding the concept of innovation sustainability to guide the development of knowledge in the field. The relationship between sustainability and healthcare is evident since the condition of the environment significantly impacts the well-being of the general population. A comprehensive theoretical model for sustainability in healthcare has been constructed, with consideration of the recognized sustainability practices in the healthcare sector. The field of implementation science was developed to address the difficulties of implementing concepts supported by evidence into reality. The main goal is to establish the sustainability of innovations for long-term continuation and integration after overcoming the difficulty. Tools and metrics have been developed to evaluate the sustainability of innovations in professional and educational environments. The changing environment of sustainability in the healthcare sector, with a focus on innovation and the many problems that healthcare systems around the world face affect each other in important ways. The chapter explores the growing importance of innovation in healthcare as a means of addressing the growing environmental, economic, and social concerns people face. Sustainability in healthcare innovation is a complex concept that may have various meanings in various situations and is inconsistently employed. Furthermore, it is proposed that sustained innovations are influenced by a variety of innovation-, context-, leadership-, and process-related preconditions or factors. The chapter includes innovative healthcare solutions like artificial intelligence, telemedicine, and Internet of things devices, as well as sustainable practices including green healthcare facilities and supply chain optimization. These developments may help enhance healthcare while decreasing expenses and limiting damage to the environment. The chapter emphasizes the significant challenges that healthcare sustainability faces, such as the increasing expenses of healthcare, differences in healthcare accessibility, and ethical issues related to developing technology. It provides a detailed review of the present status of sustainability in the healthcare sector by conducting a thorough analysis of both innovation and difficulties. The healthcare industry is complicated and must adapt to new trends without enough resources and structures. The need for a new approach to new challenges in the healthcare industry offers a stimulating application field for design disciplines. The chapter provides an overview of the strategy and technique used by the systemic innovation design.

https://doi.org/10.1515/9783111436432-003

3.1 Introduction

The healthcare industry is undergoing significant transformations that provide novel challenges to the existing framework having an impact on several aspects, particularly in industrialized nations. Implementing evidence-based therapies or technology advances that provide high-quality, affordable care is needed to meet the healthcare needs of elderly people [1]. Studies indicate that implementing evidence-based practices is often unsustainable, leading to unreliable outcomes [2–4]. The sustainability model developed by the National Health Service Institution for Innovation and Improvement was developed specifically to assist teams in the implementation of novel practices within their respective work environments, to effectively address that problem. The purpose of this initiative is to facilitate teams in identifying and evaluating essential factors within their specific environment that influence the potential for long-term adoption of a new practice [5]. Additionally, it seeks to encourage teams to take appropriate measures promptly to enhance the probability of successful implementation. In this sense, "sustainability" refers to the maintenance or adoption of an innovative method by an enterprise to a level that it becomes embedded in everyday operations and continues to provide the intended results [6–8]. As businesses, such as in the healthcare sector, experience growth, numerous challenges have the potential to restrict their growth. These problems include rising operating costs, more compact environmental regulations and laws, and more competition. These problems have compelled service firms to reduce waste, enhance efficiency, and strive for competitiveness and profitability, all while adhering to environmental regulations and guidelines. The concept of sustainability has become a significant strategic focus for businesses [9, 10]. Consequently, firms have devised novel approaches to incorporate sustainability principles into different functional areas [11].

The application of the idea of sustainability has been effectively implemented in manufacturing organizations. However, the examination of this topic in service companies has only been conducted in recent times [12, 13]. Within the framework of healthcare organizations, the concept of sustainability has significant importance and is a crucial factor for healthcare professionals to consider [14, 15]. Therefore, it is essential that the novel technology meets the requirements established by healthcare organizations and is also appropriate for patients who engage with both the technology and the healthcare professional. Furthermore, the occurrence of pandemics such as COVID-19 requires an enhanced need and accelerated advancement of digital technology within the healthcare industry, enabling the remote monitoring of health-related data. There is a particular need for services targeted toward the elderly population and those with chronic illnesses to effectively achieve the objective of implementing social distancing measures within society. Nevertheless, healthcare institutions are inherently complicated and frequently lack a history of collaborative innovation with small firms [16]. Small firms suffer many intricate obstacles while aiming to build digital technologies and various healthcare services, such as eHealth technology. The healthcare resources in Western nations are under substantial pressure as a result of the ongoing and projected growth of the elderly population. Therefore, a need exists for the development of healthcare services that are characterized by increased innovation, efficiency, and quality [17]. Healthcare delivery should focus on accessibility to patients and their residences, aiming to minimize the need for hospitalization. Therefore, local primary healthcare facilities need to assume greater responsibility in delivering healthcare services to the general population, while also ensuring that patients' various care requirements are adequately addressed inside the comfort of their own homes. It requires the use of novel digital technologies and innovative work methodologies.

Primary healthcare facilities have the potential to contribute to the continuing transformation process by actively participating in the development of new and suitable digital technologies and services. Such collaboration often takes place with small firms. Innovation in the field of healthcare is the act or method of creating novel concepts and instruments that suggest fresh approaches to healthcare delivery. Innovation involves more than just the reorganization of work practices at a single primary healthcare center. It involves the development of digital technology and novel work techniques that may be distributed and used across several healthcare centers [18–20]. Innovations must increase productivity, and be simple to utilize for patients and the healthcare sector, as well as from a commercial and technical perspective [21]. The innovations must possess long-term sustainability, thus making a positive contribution to the environment, while also being financially sustainable and beneficial to society [22]. The field of implementation studies has arisen in response to the difficulties associated with integrating evidence-based innovations into practical applications [23]. Implementing healthcare innovations is challenging, hence several models, frameworks, and theories have been established to guide funders, [24] stakeholders, practitioners, doctors, and researchers [25–27] Sustaining innovations within the environments of practice is crucial to promoting their long-term inclusion and continuation after the challenge of converting innovations into practice has been overcome [28].

Sustainability is an aim for long-term success in the company. Businesses first prioritized economic factors above environmental and social concerns in their sustainability bottom-line strategy [29, 30]. The concept of sustainable development in the healthcare sector is now an issue of controversy. This is mostly due to the absence of clarity about language and the identification of elements that contribute to the sustainability of innovations [31, 32]. For small firms to ensure their survival and sustainability, they must develop a flexible strategy that effectively responds to the requirements of the healthcare sector [30]. The potential success of innovation initiatives is subject to variability, which poses a risk for small firms. A study has explored innovations that are sustainable in small firms [29, 33]. The study of innovation in small firms has been conducted using the resource-based view framework. This framework suggests that innovation is facilitated by the use of suitable resources and that the strategy and resources of small enterprises have a significant role in determining their level of innovation [34]. The establishment of collaborative partnerships between the healthcare industry and its patients is crucial in the development of new technologies for primary healthcare, with a focus on long-term sustainability. To find sustainable innovations and capitalize on economic prospects, small firms must use both internal and external knowledge and skills [31, 32] for the sustained well-being and economic success of human communities [35]. The healthcare industry's environmental impact has grown more evident in past years [36].

The interconnection between sustainability and healthcare arises from the recognition that the condition of the environment significantly impacts the well-being of the general population. The global business sector has acknowledged the physical and financial implications of the sustainability issue [37]. The incorporation of sustainability measurements and indicators is essential for adopting an integrated systems approach to effectively tackle sustainability concerns within the healthcare sector. However, for the transition from conventional to sustainable healthcare, an analysis of interconnections among sustainability metrics may be used to establish the order of activities to be performed. Simultaneously, to facilitate effective policy development in the healthcare sector, it is important to ascertain the primary indicators of sustainability.

3.2 Definition and characteristics of sustainability

The concept of sustainability includes the ability of many systems, such as those related to the environment, economy, and society, to survive and continue over an extended period while maintaining a condition of equilibrium and cohesion. The concept consists of three core dimensions: environmental sustainability, which emphasizes responsible resource management and the reduction of ecological impacts; economic sustainability, which strives for long-term prosperity and equitable distribution of wealth; and social sustainability, aimed to establish fair and inclusive societies that meet current needs while safeguarding the needs of future generations. The concept of "sustainability" includes a wide range of implications and interpretations. It suggests both the predictability of change that is established and the dynamism of change that is continuous. The word "sustainability" has been suggested by several writers as the best way to describe a long-lasting invention since it "incorporates the basic principles of permanence and duration without confining its manifestations to any single form" [38, 39].

Analysis showed that more than half [40] and up to 65% [41] of studies did not give a clear meaning of "sustainability." Table 3.1 has a few examples that demonstrate the wide range of meanings. Various definitions of sustainability highlight different aspects of innovations, including their associated benefits, their ability to face challenges, and their impact on their environment over time [38, 40, 42]. Although there is no universal definition of sustainability, authors from health promotion, management, and healthcare literature suggest integrating all three elements in comprehensive definitions [43].

The characteristics of sustainability in innovation are derived from the three primary parts, namely benefits, routinization/institutionalization, and development, as

Table 3.1: Different sustainability definitions.

Definition	Source	Authors	References
– The degree to which early attempts to achieve acceptance of an idea have been accomplished. – The extent to which a change program is continued after the original resources given by a changing agency are ended.	Sociology; diffusion of innovation	Rogers	[44]
– The continuity of program components that were created and executed during earlier phases is assessed in terms of whether they are sustained or discontinued after the completion of original financing or other driving factors. – Maintaining program components and activities beyond original financing to achieve desired results for the program and population.	Public health; healthcare	Scheirer Dearing and Scheirer	[45]
– Innovation adoption requires adaptability from people, companies, and healthcare delivery systems throughout time.	Healthcare	Davies and Edwards	[46]
– The ability of programs to consistently address community concerns.	Promotion of health	Mancini and Marek	[47]
– Integration of an adaptive preventative system and sustainable innovation into continuing operations for the benefit of several stakeholder groups is the goal of this procedure.	Health promotion	Johnson et al.	[39]
– Complex systems include programs, health issues, drivers, and stakeholders, interacting dynamically in every given situation. – Able to stay at a certain level or rate.	Public and community health; healthcare	Gruen et al.	[42]
– The process of maintaining new working techniques, performance targets, and progress trajectories over a certain period.	Healthcare; medical treatment	Buchanan et al. (2005, 2007)	[48, 49]

defined in various definitions. The health promotion study emphasizes the significant importance placed on involving health benefits as a primary aspect of sustainability, with a particular focus on enhancing health outcomes. The focus on organizational design and operation has attracted the most attention to innovation continuance in management. Sustainability involves recognizing the intricate relationships between these dimensions, taking a long-term view, promoting resource efficiency, ensuring equity and social justice, promoting adaptability and resilience, striving for balanced decision-making, accepting innovation and technology, emphasizing transparency and accountability, engaging communities, and constantly improving sustainability efficacy. The importance of focusing on both personal and communal capacity has been recognized within the larger framework of community development since capacity building is an essential concept of this field of study. For example, in the framework of program continuation, the American population's health discipline exhibited a stronger emphasis on the concept of "institutionalization," while the literature concerning sustained advancements in developing countries presented a more dynamic conceptualization [38, 39]. Developed nations have more advanced health systems, allowing for the integration of innovative programs. The continuous benefits should be the primary focus of sustainability "independent of specific activities . . . or the style in which they are given" [47] since a few experts have proposed that only "successful" programs are worth preserving [50]. There was a general agreement among experts that an essential component of sustaining innovation involves consistently attaining goals, achieving long-lasting improvements in positive outcomes, and maintaining a continuous perception of achievement [39, 49, 51, 52]. Sustainability has also been described more briefly as "heightened problem salience" [45]. Maintaining issue resolution or goal achievement, independent of the initial invention, is a possible aspect of sustainability [40, 53] While benefits were widely recognized as essential for the concept of sustainability, there was often a lack of clarity around the particular scope of benefits. According to Gruen et al. [42], one specific perspective is that innovation sustainability is determined by the presence of positive and measurable outcomes that demonstrate the achievement of the innovation's aims. Studies on the benefits of sustainability have mostly focused on objective results for clients or customers [45]; providers; or systems [41]. One alternative perspective argues that the attainment of advantages, as well as the perception of these benefits (which may not always be supported by formal assessment), is crucial for maintaining long-term change [50]. The significance of recognizing the advantages of changes is evident in several fields, ranging from psychology to implementation science [51, 54–56]. As an example, the concept of "observability" refers to the extent to which the beneficial outcomes of implementing an invention are apparent to the individuals or groups involved in it [44]. These concepts demonstrate the varying viewpoints on the role of benefits in the sustainability of innovations.

3.3 Healthcare sustainability: environmental, social, and economic

The topic of sustainability in healthcare encompasses several disciplines, including medical science, management of operations, and sustainability studies. The concept of sustainable development, as described in the study, received acceptance in the business field, leading to the development of corporate sustainability [57]. The International Institution of Sustainability and Development defines corporate sustainability as the implementation of business strategies and practices that address the present needs of the operation and its stakeholders, while also safeguarding, maintaining, and improving the natural and human resources that will be required in the future [58]. The provided definition underscores the significance of addressing the requirements of both social and nonsocial stakeholders, while also achieving a harmonious equilibrium between the economic, environmental, and social aspects of company performance. The implementation of ecologically friendly healthcare procedures has acquired significant support within the sector. However, it is important to highlight that the healthcare sector is committed to environmental sustainability, as shown by the study conducted by Marimuthu et al. [14]. This dedication to eco-friendliness positions the healthcare business as more environmentally conscious compared to various other sectors. The stated emphasis has resulted in the implementation of circular practices, the incorporation of integrated design principles in healthcare facilities, the reduction and effective management of medical waste, and the adoption of sustainable procurement strategies. In the field of healthcare, circular practices encompass various activities such as recycling, repair, reuse, remanufacturing, and refurbishment of medical products [59, 60]. Additionally, these practices involve the efficient utilization of resources and the adoption of electric vehicles for transportation, which contributes to the reduction of carbon emissions associated with this sector [61].

Furthermore, the conservation of natural resources, including water conservation, is also emphasized [62]. Consequently, this phenomenon has resulted in the prolongation of the lifespan of goods [63], the conceptualization of products as services [64], the practice of product sharing [65], and the mitigation of waste via the use of waste reduction strategies in product design. The impact of the physical surroundings on well-being and health in medical care conditions is a significant factor. Integrated facilities design, also known as green design, incorporates lean principles and involves the collaboration of all stakeholders to achieve substantial improvements compared with conventional design methods [66–68]. The concept of green design promotes the use of environmentally friendly sources of energy, such as solar, wind, biomass, tidal, and the effective management of energy resources [69]. Furthermore, there are resources available to provide advice on the reduction and elimination of adverse environmental effects in healthcare facilities. These resources include evidence-based design principles, Leadership Certification in Energy and Environmental Design Healthcare certification, and the Green Guide for Healthcare, among others [70, 71].

The focus of this study is the reduction and efficient management of medical waste. Several studies [72] have examined this issue. These efforts aim to control pollution by minimizing or preventing the release of healthcare-related emissions into the air, land, and water. Previous investigation [73] has also explored the environmental impact of medical waste management. Efficient waste management plays a crucial role in safeguarding the cleanliness and safety of healthcare staff and communities. The use of waste reduction strategies in operational processes may contribute to the reduction of operating expenses. The correlation between environmental efficiency and economic efficiency is often seen, resulting in the generation of value for stakeholders [74].

The substantial potential benefit of sustainable procurement to promoting environmental sustainability may be seen when considering the significant size and impact of the medical industry based on its percentage share of a country's gross domestic product (GDP) [75]. Therefore, the implementation of green procurement efforts that prioritize procurement criteria plays a significant role in mitigating particular environmental hazards. This is achieved via the use of established methods such as product labeling (e.g., ISO 14001 series), energy efficiency assessments (e.g., energy star rating), chemical safety evaluations, and procurement scorecards. By implementing these measures, it is possible to reduce the environmental impact of the healthcare sector [74]. In the field of healthcare, social sustainability refers to the capacity of a healthcare system to augment the quality of life and promote the overall well-being of a community. The concept of occupational sustainability in healthcare services relates to the capacity to effectively adapt to evolving healthcare operations while ensuring the preservation of the health and overall well-being of healthcare employees [76]. Numerous researchers have made significant contributions to the investigation of indicators of social sustainability. These indicators include patient satisfaction, encompassing factors such as safety, security, and comfort of patients, as well as the quality of services and trust in the services provided. Additionally, employee satisfaction, which encompasses aspects such as workforce education, training, and employment, has also been examined. Other indicators explored include accessibility, availability (24×7), and affordability of services. Subsequently, the concept of sustainable health has also been studied for social sustainability [77–79].

Several social indicators may be used in multiple methodologies, approaches, and practices within collaborative healthcare information exchange initiatives to enhance the cost-effectiveness and quality of healthcare services [80]. The utilization of home-based telemedicine services and home-based surgery, along with developed healthcare innovations such as limited to noninvasive procedures with online monitoring and alarm systems [81], and by utilizing Internet of things technologies, Lo Presti et al. [82] address the challenges of medical care accessibility, cost-effectiveness, effectiveness, and individual satisfaction. These advancements signify progress toward healthcare care 4.0. Additionally, the implementation of personalized medications through 3D printing technology contributes to a more intelligent and sustainable healthcare system [83, 84]. To ensure universal access, funding for premium-priced medicines is necessary. Vaccination plays a crucial role in promoting sustainable health, while sustainable medical education serves as an effective strategy for implementing sustainability measures in the healthcare sector [85, 86]. Lastly, appropriate workforce development and the adoption of "green" human resource management practices contribute to the overall sustainability of the healthcare industry [87–91]. The concept of medical tourism involves individuals seeking healthcare services beyond the national border. Additionally, they propose that developing a culture of sustainability inside healthcare organizations may serve as an effective implementation technique to achieve social sustainability [92].

3.4 Conceptual framework of healthcare sustainability

The integration of principles of sustainability into the strategic business objectives of the healthcare sector is important. Consequently, across an extensive examination of current studies and theories regarding sustainability in organizations, specifically corporate social responsibility (CSR), theory of stakeholders, commercial sustainability, and green economics, this study evaluates a comprehensive theoretical framework, as represented in Figure 3.1, which includes sustainability practices within the healthcare sector. These practices implemented under the team-based learning (TBL) approach prompt the healthcare system to observe and effectively handle the value-added aspects associated with them. The model incorporates a variety of environmental, social, and economic indicators and sub-indicators derived from these behaviors and suggests the following.

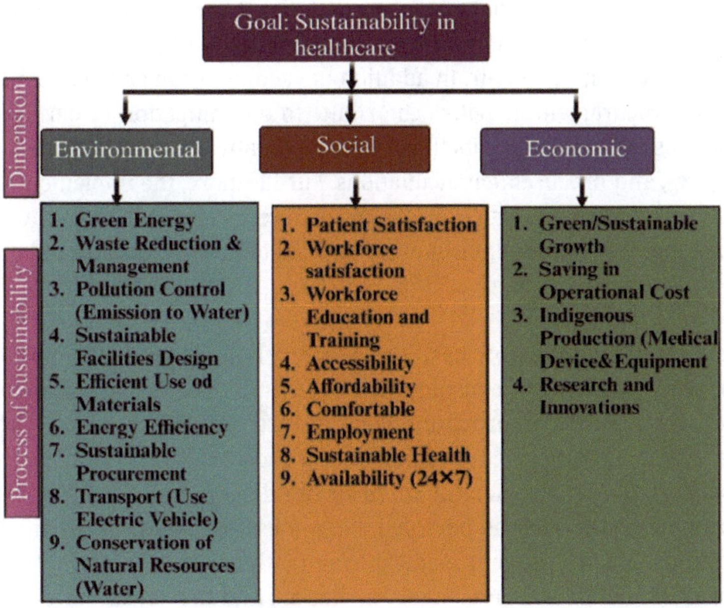

Figure 3.1: Conceptual model of healthcare sustainability.

To enhance the ecological sustainability of the healthcare sector, it is essential for developing nations to adopt sustainable and green economies, similar to those implemented by developed countries such as the United States and China, from a policy approach. Circular practices include a range of activities that involve the recycling, repair, reuse, remanufacturing, and restoration of medical goods. Additionally, these practices promote the effective exploitation of resources, including the preservation of natural resources. These activities promote the prolongation of product life span, the conceptualization of goods as services, the adoption of product-sharing models, and the implementation of waste prevention strategies, including the integration of waste reduction principles into product design. Using green or sustainable design principles in healthcare facilities aims to enhance the overall environmental, social, and economic aspects of healthcare performance [93].

The reduction of the environmental impact produced throughout the entire lifespan of healthcare services can be achieved by implementing various strategies. These include selecting suitable locations for healthcare facilities, adopting green energy sources, implementing water conservation measures, improving energy efficiency, utilizing recycled and low-emission materials, minimizing waste generation, maximizing the use of natural daylight, promoting the consumption of locally sourced food, using environmentally friendly cleaning products, and incorporating electric vehicles for transportation purposes. The reduction and effective handling of medical waste play a crucial role in maintaining hospital cleanliness and ensuring the safety of both personnel and communities. Additionally, these practices help regulate pollution by minimizing or preventing the release of dangerous pollutants, such as chemicals and gases, into the environment, including land, water, and air. In addition to yielding social and economic advantages, sustainable procurement activities contribute to the mitigation of particular environmental hazards using product labeling, environmentally friendly machinery, chemical safety measures, and procurement evaluations. Furthermore, the implementation of environmentally sustainable practices in the healthcare sector leads to a reduction in emissions, resulting in a cleaner environment that has a direct impact on public health:

- The primary indications of a sustainable healthcare sector should include the development of health consequences, health consumption, and cost of treatment. These indicators should be based on a focus on providing “health” care rather than only “sick” care. Therefore, attention must be placed on maintaining good health. The basic importance of universalizing preventive healthcare has been demonstrated by the implementation of many measures such as personal cleanliness practices, heightened awareness, vaccination protocols, frequent medical examinations, and the promotion of healthcare via activities like as exercise, yoga, meditation, and the consumption of nutritious food. In addition to the core principle of “not harm,” enhancing the accessibility of high-quality healthcare and minimizing the cost of healthcare delivery have the potential to enhance patient happiness. Moreover, in addition to the aspects of productivity and job quality, there exists a prevailing notion that the level of contentment among healthcare employees has considerable influence over the satisfaction received by patients.

- The advancement of the healthcare sector toward green growth is essential for the establishment of a green economy. This includes the adoption of healthcare practices that are less harmful to the environment and more efficient in their use of resources.

The implementation of green policies may play an essential part in facilitating this transformation and creating a more sustainable healthcare environment. Furthermore, it is crucial for emerging countries to engage in substantial domestic production of medical instruments, technology, and diagnostic equipment that correspond to global quality standards, hence leading to cost-effective healthcare services. Moreover, the healthcare sector must accelerate its reinvention process and focus on the advancement, innovation, and adaptation of care delivery models that promote integrated and value-based care. This can be achieved through the implementation of emerging concepts such as the "bouquet hospital," "budget hospital," and public-private partnership models. These models have garnered significant attention in developing economies and are expected to shape the future of healthcare [93].

3.4.1 Indicators of sustainability in the healthcare sector

The theoretical representation of sustainability in healthcare that has been suggested serves as a fundamental framework for further investigation using the analytic hierarchy process and interpretive structural modeling approaches. The TBL, or Triple Bottom Line, encompasses an organizational framework that consists of a range of practices, activities, and approaches. The model is subsequently subjected to evaluation by healthcare professionals, who suggested the categorization of 27 sustainability activities into 12 metrics for analysis. Some of the specialists had involvement in private healthcare and other sectors within the healthcare services business, therefore demonstrating significant familiarity with established sustainability techniques. Sustainability metrics were developed by categorizing a set of activities and establishing a comprehensive taxonomy for them if feasible, while maintaining the overarching sustainability aspect associated with each activity. Similarly, a total of 27 practices have been restructured and categorized into 12 measures based on the guidance provided by the experts. It is important to note that there were no additions or removals of any specific practices.

Table 3.2 provides a comprehensive summary of metrics that have been categorized into three aspects of sustainability: environmental, social, and economic. These measurements have been classified based on their respective qualities and attributes. The measurement of sustainability within the healthcare industry involves a thorough examination of several aspects, including environmental, economic, and social factors. Healthcare institutions engage in environmental monitoring to assess their energy consumption, water use, waste generation, and emissions of greenhouse gases. Their objective is to reduce their ecological impact by using energy-efficient technology and adopting sustainable waste management practices. From an economic perspective, indicators primarily center on healthcare expenses, expenditures as a proportion of GDP, and financial stability. These indicators place significant emphasis on the effective deployment of resources and the containment of costs within the healthcare sector. In a

Table 3.2: Indicators for sustainability in healthcare [93].

Sustainability	Indicators	Description	References
Environmental	Circular methods	It is the idea of maximizing the amount of time that resources are put to productive use by recovering and reusing such resources by the "cradle-to-cradle" methodology.	[60, 78]
	Facility planning	The term relates to the structural configuration of a hospital establishment, including its technological infrastructure, equipment, and their impact on ensuring patient safety. There is a need to establish an atmosphere that fosters genuine healing, including elements such as natural surroundings, enough sunshine, and a continuous supply of fresh air.	[66, 68, 70, 94]
	Procurement for sustainability	The process involves integrating social and environmental aspects with financial considerations to inform decisions regarding purchases. The process includes making choices by considering life-cycle costs and the corresponding environmental and social risks and benefits. The measures include the discontinuation of mercury and PVC plastic, along with the replacement of hazardous compounds.	[74, 75, 95, 96]
	Reduction and management of waste	The primary objective of waste reduction is to significantly decrease the amount of trash that is ultimately disposed of by various methods such as reusing, recycling, and implementing other initiatives. Approximately 80% of all healthcare waste may be effectively managed and disposed of using standard municipal trash processes. Improper disposal of the remaining 20% might potentially give rise to health hazards. The methods of waste disposal exhibit variability based on the nature of the waste, the characteristics of the local environment, the technological resources at hand, the associated costs and financial considerations, as well as the level of societal acceptability.	[97–102]

Table 3.2 (continued)

Sustainability	Indicators	Description	References
Economic	Green development	The integration of environmentally sustainable techniques into healthcare services is represented as the establishment of a green hospital. The concept of a green economy may be defined as one that is characterized by a minimal carbon footprint, efficient use of resources, and equality in terms of social aspects.	[69, 103]
	Reduced operational expenses and increased profits	Healthcare providers globally emphasize the reduction of overall care delivery costs as a critical objective since it is the only means to enhance financial performance. The economic dimension encompasses strategies aimed at enhancing cost efficiency, generating revenues, and fostering research and development activities. Cost savings may be obtained via green growth, local medical instrument and device manufacture, noninvasive methods, and telemedicine.	[41, 77, 104–106]
	Research and development	Research and innovations play an integral part in facilitating transformative change as they have the capacity to significantly contribute to fundamental shifts toward sustainability. Within the field of healthcare, initiatives aimed at enhancing various aspects such as clinical practice, nursing practice, and cost reduction are classified as improvement programs.	[107–110]
	Indigenous manufacturing	The medical technology operations in India rely heavily on imports, with just 30% of the country's need for medical equipment being fulfilled by domestic manufacture, and even then, predominantly in the low-end category. India is often referred to as the global leader in pharmaceutical production and is rapidly becoming a significant center for both innovative research and manufacturing activities.	[93]

Table 3.2 (continued)

Sustainability	Indicators	Description	References
Social	Satisfaction of patients	Patient satisfaction refers to the degree to which individuals indicate contentment with the healthcare services they receive. The measurement of care quality provides healthcare personnel with valuable information about multiple aspects of healthcare, including the efficacy of their care interventions as well as their capacity for empathy. The level of the major success measures for hospitals is closely associated with it.	[14, 77, 111]
	Affordability	Affordability is mostly determined by a person's income, discretionary expenditure, and perceived worth of products and services. A healthcare facility must be made accessible and inexpensive to all individuals, operating continuously on a 24-hour basis.	[78, 112–114]
	Staff satisfaction	The significance of job satisfaction in the retention of healthcare personnel has been recognized. Studies have shown that it influences the quality of healthcare provision, patient satisfaction, and staff turnover.	[76, 90, 114]

clinical context, the efficacy of healthcare services is evaluated by quality indicators such as patient outcomes and satisfaction ratings. The indicators of social sustainability in the context of healthcare include fairness in healthcare access, decreased healthcare inequities, and community participation. These indicators are indicative of a dedication to inclusive and equitable healthcare practices. In aggregate, these indicators serve as a collective framework that directs healthcare systems toward a more sustainable and comprehensive approach to the delivery of healthcare services [93].

3.5 Sustainable healthcare design

Applying the design thinking method to difficult challenges within the healthcare industry may have several advantages that are becoming more widely accepted. According to [115] design thinking has the potential to provide a clearly defined and identifiable framework for incorporating more creative, transdisciplinary, and human-centered approaches into healthcare administration, innovation, and practice. Three primary components of design thinking can contribute to the improvement of healthcare: (i) empathy, which involves the examination of both explicit and implicit user needs to comprehend them and develop suitable solutions; (ii) multidisciplinary teamwork with various stakeholders; and (iii) fast prototyping to evaluate multiple preliminary concepts and identify the most effective ones [115]. Empathy is an important feature that sets design thinking apart from other conventional techniques in the healthcare field. Design thinking prioritizes the needs and experiences of users, particularly patients and caregivers, as the foundation upon which the system is subsequently developed [116]. The primary source of issues within the healthcare industry, particularly in the delivery of treatments, may be attributed to a flawed system design. This design was based on the assumption that individuals would conform to it, rather than the system adapting to meet their needs. There are numerous opportunities for the effective implementation of the design thinking approach in the healthcare sector. These opportunities encompass a wide range of areas, including the redesign of tools and environments, the improvement of information and service flow between healthcare professionals and patients [117], and the reimagining of the treatment experience itself [118]. A detailed study by the State of the Art of Design in Health [119] shows that the main effects (in terms of number) of design involvement in healthcare in the UK are the creation of digital items and physical products. These subjects mostly refer to health informatics, telehealth, telecare, telemedicine, mobile apps, and websites that showcase the many possibilities that digital technology provides to the healthcare industry. According to Chamberlain et al. [119] assistive technology, such as wearable gadgets, leads the latter category, followed by medical products.

Sustainable healthcare has experienced evolutionary development throughout time. Initially, the concept of healthcare was primarily associated with economic factors. However, subsequent efforts to conceptualize healthcare have recognized the significance of addressing the needs of both patients and workers, as well as considering the economic and environmental implications, and the social and cultural dimensions. The integrated strategy aims to establish a healthcare system that operates in collaboration with individuals and the environment, while also ensuring the preservation of resources for future generations [120]. Despite the previous requirements, the current discourse around sustainability in the healthcare sector has been focused on financial considerations, with less regard for other important dimensions such as social and environmental aspects [14].

As healthcare is a complicated system, the idea of sustainable healthcare (SH) should take into consideration the many factors that affect the business, the environment, and society. The findings derived from the examination of existing literature underscore the significance of adopting a comprehensive perspective toward sustainability within the healthcare industry. However, this aspect is infrequently addressed, with the focus primarily centered on management and environmental factors [14]. When considering therapy, it is essential to have a comprehensive perspective of SH that encompasses not just the social, economic, and environmental consequences. The analysis should also address the needs of the individuals receiving care and the individuals providing care, taking into account their respective contexts. This analysis should examine the interactions established between these individuals and their contexts, as well as the impacts and consequences that arise from these interactions. However, in industrialized nations with high healthcare standards, the problem of environmental sustainability is receiving more attention in tandem with the economic one. However, despite the growing interest in healthcare that sustainability is generating, SH lacks a consensus definition. The heart of the systemic design method will be described in the next paragraph, where the concepts of this approach will be examined. The aim of SH is impacted by a multitude of factors that include several levels, ranging from institutional and financial management to the engagement and understanding of healthcare professionals. The key elements that have emerged from the analysis of the literature encompass long-term strategy and the ability of the medical field to adapt, the necessity of developing policies to promote SH, the organizational dedication of healthcare professionals and their training in improving sustainability in the workplace, the importance of involving all stakeholders interested in the issue, and the empowerment of patients to take responsibility for their health [120–124]. The design has many potential applications in healthcare, but its integration and purpose are still unclear [125]. With a focus on the implications for the role of designers, Freire and Sangiorgi [126] provide an intriguing study of the UK healthcare sector's development in connection with the shift in economic paradigm. Initially, the healthcare service was organized based on a progressive paradigm, whereby information was possessed by a limited number of specialists, patients were seen as consumers, and the primary emphasis was placed on the effectiveness of the service. The transition from mass production to mass customization seen in the industrial sector was also evident in the healthcare industry, as there was an increased focus on addressing patients' requirements and delivering individualized treatments. Recent years have seen an acceleration of the shift from mass customization to mass collaboration and open invention, ushering in a patient-centered, co-creative healthcare paradigm. The role of designers has transformed in the transition from the linear paradigm, where they were primarily responsible for creating products and services, to a more facilitative role. In the current environment, designers serve as intermediaries, connecting various stakeholders and offering tools and methodologies to effectively evaluate the intricate nature of contemporary healthcare challenges [126]. The complexity arises, in part, due to the unique and distinct nature of each case (each patient) within the healthcare sector, hence posing challenges in devising a standardized solution. The position of the designer has undergone significant changes in recent decades due to the growing recognition of the need to promote sustainable development. Achieving this objective requires systemic modifications to the economic framework, exerting influence on several stakeholders operating at various levels. Designers play a pivotal role in facilitating the shift toward sustainability by actively participating in both the consumption and production processes. Their contribution is essential in promoting the environmental integrity of goods and services, as well as in enabling the adoption of new sustainable lifestyles [127].

3.6 Factor affecting sustainability in healthcare

The sustainability of healthcare depends on a multidimensional interaction of several elements, including environmental, economic, social, and healthcare-specific characteristics. The achievement of environmental sustainability is dependent upon reducing the ecological impact resulting from healthcare activities, which entails the reduction of energy consumption, waste creation, and the use of deleterious substances. Economic sustainability includes the strategic allocation of resources, effective cost management, and the establishment of long-term financial viability to preserve the accessibility and affordability of healthcare services. Social sustainability includes the promotion of fair and just distribution of healthcare services, the enhancement of the physical and mental well-being of healthcare professionals, and the active involvement of communities in healthcare initiatives. In addition, healthcare-related variables such as improvements in medical technology, medications, and treatment procedures have the potential to influence sustainability by exerting an effect on prices and the usage of resources. The accomplishment of sustainability in the healthcare sector involves an essential collaboration of stakeholders, the establishment of constitutional frameworks, and the easy incorporation of sustainable practices into healthcare systems. Several aspects serve as indicators of sustainability within the healthcare industry. The elements under consideration include lean management [128], patient satisfaction [129], employee satisfaction [130], continuous improvement [131], CSR [132], brand [133], and accreditation [134].

3.6.1 Healthcare lean management

The lean approach was developed by the Toyota Production System (TPS) as a means of addressing operational inefficiencies in the company [135, 136]. The approach of waste minimization, derived from lean principles, has been increasingly used in several service industries, including healthcare. Lean management has been embraced as a process development strategy in medical facilities, including the principles of waste reduction. The use of lean management principles has expanded beyond the automotive and industrial sectors. The implementation of lean management is undertaken to attain strategic objectives, facilitate operational elements, and ascertain the long-term sustainability of changes [137]. One viable strategy for addressing the growing need for medical treatments is the use of lean principles to expand healthcare capacity. This method focuses on eliminating several forms of waste, such as errors, overproduction, excess inventory, superfluous movement, overprocessing, and waiting [138]. This study examines the significance of waste reduction strategies within the healthcare industry. Lean management in the healthcare sector is a methodical strategy that aims to optimize operational processes, minimize inefficiencies, and enhance overall effectiveness, all while concurrently improving the standard of patient care. The process includes the removal of unnecessary procedures, reduction of inactive intervals, and optimization of resource use, resulting in financial benefits and enhanced patient results. Lean concepts, including continuous improvement, standardized work, and value stream mapping, are used in healthcare environments to optimize operations, minimize mistakes, and improve patient satisfaction. Lean management plays a crucial role in improving the efficiency, responsiveness, and sustainability of healthcare organizations amidst growing demands and limited resources. This is achieved through a focus on value enhancement, elimination of nonvalue-added activities, and the cultivation of a collaborative and continuous learning culture among healthcare professionals.

3.6.2 Continuous development

In modern times, the sustained well-being and prosperity of an organization are dependent upon the organization's dedication to ongoing improvement [139–141]. To maintain competitiveness, it is essential to implement this particular vision since expectations are subject to ongoing change and evolution [142, 143].

The process of implementing solutions, as well as producing and testing ideas, is essential for continuous improvement [144]. Previous studies have shown different results on the impact of continuous improvement strategies on performance [144, 145]. The selection of sub-factors was aided by an investigation of relevant studies and expert opinions. This included considering the execution of solutions, such as the integration of new technologies, improvement in workflow, as well as areas of innovation [144, 145] and business continuity [146]. Continuous improvement plays a vital role in promoting sustainability within the healthcare sector. It involves employing a comprehensive strategy to enhance operational processes, optimize the allocation of resources, and improve the quality of patient care. The process includes a constant commitment to the identification and resolution of inefficiencies, the reduction of mistakes, and the continual pursuit of possibilities to improve the quality of healthcare services. The use of a process-driven technique not only yields enhanced patient outcomes but also has a direct influence on cost reduction, thus enhancing the long-term financial sustainability of healthcare. In addition, the promotion of a culture centered on ongoing enhancement within healthcare institutions promotes flexibility, creativity, and the capacity to promptly address emerging healthcare issues. This guarantees that healthcare systems can maintain their resilience, effectiveness, and long-term viability despite constantly changing demands and limitations.

3.6.3 Patient satisfaction

The problem of patient satisfaction has significance for both healthcare professionals and patients since the latter serve as the clients within the healthcare sector [147]. Healthcare professionals consistently make constant attempts to ensure the satisfaction of their patients. The significance of assessing and sustaining customer satisfaction is underscored by the competitive nature of the healthcare sector [148]. The pleasure of healthcare clients is of the greatest significance for healthcare providers, as it directly impacts the profitability and future sustainability of the healthcare field. To attain client satisfaction, medical professionals must effectively determine the desires and issues of patients, choose appropriate methods to meet those requirements, and develop strategies to address any unexpected challenges [149].

There are other variables that contribute to patient satisfaction, such as dependability, compassion, responsiveness, and empathy [149], efficiency, consistency, accessibility of services, confidence, physician behavior, and results [150]. Additional elements that should be taken into consideration include expertise, customization, fundamental amenities, academic credibility, and communications [151], personnel performance, and facility quality [152]. According to the opinions of experts, the factors employed to measure patient satisfaction in the present research were the professionalism of physicians, the availability of appointments, the quality of the facility, and the core services provided. Patient satisfaction is an essential parameter used by healthcare practitioners to evaluate the quality of treatment and services they deliver. Positive patient experiences play a crucial role in establishing a favorable reputation for healthcare institutions and cultivating patient loyalty, which may result in heightened usage of healthcare services and enhanced financial viability. On the other hand, unpleasant interactions with patients may lead to harming the reputation of the healthcare organization and a decreasing patient loyalty, hence causing significant financial and operational consequences. Consequently, healthcare providers possess a strong interest in guaranteeing elevated levels of patient contentment to maintain their competitiveness and general sustainability in a fiercely competitive healthcare environment. Furthermore, it is important to include the perspective of patients, since their satisfaction plays a significant role in influencing both their general well-being and treatment results. A happy patient is more likely to follow medical advice, treatment plans, and follow-up visits, resulting in improved health results. In addition, a positive patient experience can substantially reduce the tension and anxiety associated with healthcare interactions, which is especially essential for individuals contending with illness or chronic conditions. Most patients trust the healthcare system more when they feel noticed, respected, and involved in their care. This makes them happier with the services they receive generally. Consequently, patient satisfaction is not solely a measure of consumer service, but rather a fundamental aspect of patient-centered care that directly influences health outcomes and patient participation in their health management. For healthcare practitioners to offer high-quality treatment and patients to have pleasant healthcare experiences that improve health and well-being, patient satisfaction is important.

3.6.4 Employee satisfaction

Employee satisfaction may be defined as the amount of contentment an employee has with his or her employment, or the extent to which an employee enjoys his or her work [153]. The personnel within the healthcare business are considered to be its primary assets since they play a crucial role in achieving the industry's aims and objectives. There is a wide range of elements that contribute to employee happiness, which are both diverse and numerous [154]. The factors that have been categorized by the well-known two-factor theory [155] are an example of satisfaction. According to the predictions made by equity theory [156, 157], relative terms are an additional component of work satisfaction.

The study incorporates many significant factors that impact employee happiness, including reward [158], professional growth [159], conduct package [160], encouragement and assistance [161], and empowerment [162].

The satisfaction of employees is a complex and crucial factor in the achievement of organizational success, as it impacts not only the welfare and motivation of the workforce but also overall efficiency and the accomplishment of organizational objectives. It comprises several elements, which include job satisfaction, the work environment, remuneration, and prospects for professional growth. The satisfaction of workers with their occupations contributes to the development of a good workplace culture, which may provide substantial advantages. High levels of employee satisfaction are associated with increased productivity and performance. Satisfaction among workers has been shown to positively correlate with increased levels of motivation, engagement, and commitment to their respective positions. Individuals tend to allocate more effort, exceed expectations, and actively participate in the process of resolving issues and fostering creativity. The enhanced level of dedication shown by individuals may have a positive impact on their work performance, leading to enhanced results for the business. Additionally, satisfied workers have a lower likelihood of leaving their jobs, which saves on the expenses of recruiting and training while keeping the knowledge and skills of the organization. Additionally, the level of employee happiness significantly influences workplace morale and the broader company culture. A workforce that is both satisfied and engaged has been shown to create an environment characterized by positivity and collaboration. When workers experience a sense of worth, respect, and support in their respective positions, they tend to exhibit enhanced communication skills, collaborate well with other employees, and maintain a favorable attitude. This not only improves collaboration and cooperation but also helps create a more welcoming and peaceful workplace, all of which are critical in today's competitive labor market.

Additionally, satisfied workers are more likely to connect themselves with the firm's goals and values, and this contributes to the success of the organization over the long run. They promote the organization's reputation and attract prospective customers, clients, and even future employees. Employee satisfaction is not limited to personal happiness; it significantly influences the entire well-being and success of the firm. Promoting employee satisfaction is a strategic requirement for firms trying to fulfill their goals, establish a great work environment, and maintain long-term success.

3.6.5 Corporate social responsibility (CSR)

CSR may be described as the dedication of a company to enhance the overall welfare of the community by engaging in voluntary business activities and allocating corporate resources for this purpose [163]. Furthermore, the enhanced competitive position and increased value derived from the enhancement of economic, social, and environmental variables via proactive and voluntary actions undertaken by corporations may be identified as CSR [164]. CSR is defined by the World Business Council for Sustainable Development as an organization's continuing commitment to act ethically, advance economic growth, and better the lives of its employees, their families, and the communities in which they operate. Sustainability and CSR have emerged as pervasive concerns that are extensively incorporated into the operational and strategic aspects of several firms. CSR is widely regarded as a responsible business approach and an effective marketing tool that enables companies to establish and maintain a competitive advantage [165, 166]. Competitiveness, as described by investigators, refers to the capacity of a business or corporation to adequately remunerate its personnel and generate exceptional returns for its shareholders based on its long-term profitability [167]. It utilizes the three primary components of CSR, namely social, economic, and environmental aspects, as sub-criteria [164].

CSR is a basic commitment by corporations to contribute to the benefit of society and the environment beyond profit maximization. This commitment involves the use of company resources and discretionary procedures to support programs that improve community well-being, tackle social and environmental issues, and preserve ethical standards. CSR comprises a wide variety of activities and initiatives aimed at positively impacting the communities in which a company operates. Donating to charities, sponsoring community activities, and supporting education and healthcare are examples. In addition, CSR consists of ethical business practices that prioritize environmental sustainability, equitable labor standards, and responsible sourcing. Companies can reduce their ecological footprint, guarantee ethical treatment of employees and suppliers, and contribute to the creation of a more equitable and sustainable world by incorporating these principles into their operations. CSR is a strategic need for modern organizations, not just a moral responsibility. CSR activities can improve a company's reputation, increase stakeholder trust, and generate a competitive advantage. An effective CSR program may draw in socially aware customers, investors, and workers, which can boost market share, open up new investment possibilities, and facilitate the hiring of top talent. Also, CSR projects often save money and make operations more efficient because responsible and sustainable practices cut down on waste, protect resources, and make business processes better overall. CSR is a corporate commitment to increase community well-being and contribute to a more sustainable and ethical society via management practices and resource allocation. Beyond its ethical significance, CSR is increasingly recognized as a strategic approach that can generate numerous benefits, such as a strengthened reputation, a competitive advantage, and increased operational efficiency. As social and environmental problems change, CSR is likely to stay an important strategy for businesses that want to make money while also being socially and environmentally responsible.

3.7 Innovation in healthcare sustainability as a long-term perspective

The standard of local demand is significant [168, 169] because the same local interactions apply to innovation concerns as they do to sustainability challenges. A small business must examine the local environment to grow into a creative and innovative competitor [170, 171]. Enterprises and associations with demanding local consumers have the greatest chance of prospering in a global market. Considering local factors and understanding business community dynamics is essential [172]. According to Muzzi and Albertini [173], it has become clear over the last few decades that healthcare businesses depend on networking to find new ideas and access to greater knowledge bases [174, 175]. Sustainability work in corporations and other organizations has historically emerged at the intersections of environmental requirements and competitiveness. Porter and van der Linde [176, 177] say that the connection between environmental rules and competition should not be seen as a fixed one, but as one that changes over time. Severe environmental regulations can steer businesses and organizations in several positive directions. It can lead to new ideas to improve production, which makes an enterprise more competitive.

This often results in better and more effective utilization of resources [178–180]. Environmental restrictions are essential for driving ecological advances, according to Porter and van der Linde's [176] concepts. An innovation may be a product or a service, but the true innovation is frequently how the product or service is utilized, and how it affects the organization and the patient's healthcare. While the underlying technology may not be novel, it offers care providers novel opportunities and perspectives in patient care. Digital solutions can enhance the potential for receiving healthcare services in the comfort of one's residence or geographically distant areas. The term "company" refers to an organization or business entity that engages in commercial activities. Innovation may be defined as the development of novel procedures and methodologies within the realm of primary healthcare, particularly about the introduction of new goods. The delivery of a solution or platform is only a means to an end, since the true essence of innovation lies in its service-oriented nature. A consensus was reached among a number of the firms that were questioned. Since the healthcare industry is governed by laws and regulations, the same ideas may be applied to it. For example, innovations that use eHealth based on cloud technology are subject to certain rules [181].

The genesis of innovation arises from the dynamic interplay of individuals and is driven by a comprehensive, enduring outlook aimed at aiding patients via the provision of novel approaches to monitor, organize, and enhance their rehabilitation from ailments. Companies encourage the need to develop prospective innovations based on the effect of new functions in current or new products and the possible business case. Organizations emphasize the need to actively engage in consumer feedback to avoid developing products or services that are not aligned with the needs and preferences of the healthcare sector. It is important to establish a harmonious equilibrium after considering the level of effort necessary for the development of the novel function, its prospective ramifications on the healthcare sector, and its marketability feasibility. The process of selling a product and successfully implementing it requires a significant amount of time. Many firms often discuss extended turnover periods and intricate procurement processes and regulations. Long-term, sustainable innovation is a purposeful and dynamic way to handle the changing world's complicated concerns. The process includes a sequence of interrelated phases that go beyond the conventional understanding of innovation.

The initial stage involves the identification of possibilities and problems that are associated with environmental, social, or economic sustainability. It involves comprehensive study, active involvement of stakeholders, and an extensive understanding of worldwide patterns and growing requirements. Sustainable inventors evaluate both short-term and long-term commercial prospects. Sustainable innovation incorporates sustainability into each stage of the process of making a product or service. It is important to present and investigate various theoretical frameworks related to sustainability, innovation, and adaptation locally to effectively assess and understand diverse strategies for attaining sustainability, innovation, and local adaptation. The prevalent approach in the study of sustainable innovation is quantitative, with surveys being the prevailing technique used for data collecting. The use of a case study design allows for the development of an extensive understanding of innovation inside small firms [182]. The design process encompasses several factors of greatest significance, such as the incorporation of eco-friendly materials, the optimization of energy efficiency, and the minimization of waste generation. During the process of development, there is continuous monitoring and refinement of sustainability indicators and effect evaluations to ensure compatibility with long-term sustainability objectives. A long-term view on sustainable innovation also includes market uptake and scalability. Innovations must demonstrate both short-term value and the potential for long-term impact and widespread adoption. The achievement of sustainability-focused solutions often necessitates the engagement of diverse stakeholders, such as governmental bodies, nongovernmental organizations, and local communities, to surmount obstacles and provide a conducive environment. Long-term sustainable innovation is dynamic and diverse, going beyond immediate benefits. The strategy encompasses a comprehensive perspective that incorporates concepts of sustainability across the whole process, including the conceptualization, development, implementation, and scaling stages. Sustainable innovation addresses current issues while providing a more sustainable and resilient future for future generations by using an integrated approach. Healthcare innovations often have an impact on both the medical personnel and the patients. Therefore, it is essential to consider both of these stakeholders when designing products. Different perspectives and information may provide a sense of belonging and inclusion while dividing groups. The objective is to enhance communication among various levels of care providers, enhance accessibility to professionals, and improve care quality. The improvements examined in this research primarily focus on enhancing the efficiency of patient examination, facilitating early detection of illnesses, and improving prognostic capabilities for patients. The primary objective of these novel goods and services is to enhance operational effectiveness within the healthcare sector, thereby generating economic advantages for the whole healthcare system.

3.8 Conclusion

The healthcare industry is being challenged by new trends for which it is not yet equipped to find solutions. The medical sector is seeing a growing complexity as numerous parties engage in new responsibilities and the parameters of the system expand due to the introduction of new technology, tools, information flows, and changes in the delivery of treatment. These novel difficulties call for an entirely novel approach to resolving current challenges. The discussion has revolved around the advantages that arise from the use of design thinking in addressing intricate problems and devising efficient solutions, as well as the pivotal role that designers play in facilitating the shift toward sustainable development, and the conceptual frameworks for sustainability and evaluation methods found in the healthcare sector. Regarding the moment at which an invention achieves sustainability, there is still some disagreement. The inclusion of sustainability considerations in the plans of healthcare professionals is crucial for the long-term development and preservation of favorable results in the field of innovation. Every sustainability initiative should take into account the fundamental criteria identified as facilitators of sustainability. The use of various conceptual approaches and evaluation tools within diverse healthcare environments has facilitated the customization and development of a distinct conceptual method for the field of pharmacy. The framework under consideration was developed using the characteristics that have been recognized as moderators of sustainability in the conceptual approaches within the healthcare field. The issue of sustainability has become an important issue in contemporary society. On a global scale, the phenomenon of unsustainable development has been a significant factor in the deterioration of ecological systems and the effects of human challenges, thus reducing the capacity of both ecosystems and social structures to maintain human existence. The healthcare sector has been implicated in recent environmental degradation and harm, yet leaders of change are already emerging from this sector. Healthcare professionals have the potential to act as examples of sustainability and play a crucial role in promoting the required shift toward more sustainable practices. The study evaluates a comprehensive theoretical model for sustainability in the healthcare sector, which is based on the concept of the triple bottom line. The model provides a representation of the degree to which sustainable measures are implemented within the healthcare sector. These practices prompt the healthcare sector to track and manage value-added activities. The model is designed to include a variety of environmental, social, and economic indicators and sub-indicators. Improving healthcare sustainability requires prevalent indigenous manufacture of medical technology equipment. Enhancing healthcare research may lead to commercial innovations that improve value generation. Environmental, social, and technical innovations, together with green policies, may transform healthcare ecosystems for sustainability and resilience. Sustained and integrated expenditures can address healthcare challenges such as cost, quality, time, access, growth, and environmental quality.

References

[1] Fitzgerald L. The diffusion of innovations: The translation and implementation of evidence-based innovations. In: Challenging Perspectives on Organizational Change in Health Care 2017 Apr 21 (pp. 31–48).

[2] Scheirer MA. Is sustainability possible? A review and commentary on empirical studies of program sustainability. American Journal of Evaluation. 2005;26(3):320–47.

[3] Stirman SW, Kimberly J, Cook N, Calloway A, Castro F, Charns M. The sustainability of new programs and innovations: A review of the empirical literature and recommendations for future research. Implementation Science. 2012;7(17):2–4.

[4] Ham C, Kipping R, McLeod H. Redesigning work processes in health care: Lessons from the National Health Service. Milbank Quarterly. 2003;81(3):415–39.

[5] Maher L, Gustafson D, Evans A. NHS sustainability model. NHS Institute for Innovation and Improvement; 2010. Available from: http://www.institute.nhs.uk/sustainability.

[6] Pluye P, Potvin L, Denis J-L. Making public health programs last: Conceptualizing sustainability. Evaluation and Program Planning. 2004;27(2):121–33.

[7] Levitt B, March JG. Organizational learning. Annual Review of Sociology. 1988;14(1):319–38.

[8] Dearing JW, Kee KF, Peng TQ. Historical roots of dissemination and implementation science. Dissemination and Implementation Research in Health: Translating Science to Practice. 2012 May 24;55:71.

[9] Silvius G, Schipper R. Developing a maturity model for assessing sustainable project management. The Journal of Modern Project Management. 2015;3(1):112.

[10] Visser WAM. Sustainability reporting in South Africa. Corporate Environmental Strategy. 2002;9 (1):79–85.

[11] Holliday C. Sustainable growth, the DuPont way. Harvard Business Review. 2001;79(8):129–32.

[12] Hussain M, Khan M, Al-Aomar R. A framework for supply chain sustainability in the service industry with confirmatory factor analysis. Renewable and Sustainable Energy Reviews. 2016;55:1301–12.

[13] Stylos N, Vassiliadis C. Differences in sustainable management between four-and five-star hotels regarding the perceptions of three-pillar sustainability. Journal of Hospitality Marketing and Management. 2015;24(8):791–825.
[14] Marimuthu M, Paulose H, Paulose H. Emergence of sustainability-based approaches in healthcare: Expanding research and practice. Procedia - Social and Behavioral Sciences. 2016;224:554–61.
[15] Rabin BA, Brownson RC. Terminology for dissemination and implementation research. Dissemination and Implementation Research in Health: Translating Science to Practice. 2017 Nov 10;2:19–45.
[16] Horgan D, Van Kranen HJ, Morré SA. Optimizing SME potential in modern healthcare systems: Challenges, opportunities and policy recommendations. Public Health Genomics. 2018;21(1–2):1–17.
[17] Arocena R, Sutz J. Innovation systems and developing countries. DRUID (Danish Research Unit for Industrial Dynamics Working Paper. 2002;2(05).
[18] Lundvall BA. Product innovation and user-producer interaction. The Learning Economy and the Economics of Hope. 1985;19:19–60.
[19] Wassrin S, Lindgren I, Melin U. Open innovation contests for improving healthcare – An explorative case study focusing on challenges in a testbed initiative. In: Proceedings of the International Conference on Electronic Government, Thessaloniki, Greece, August 30; 2015 Sept 2 (pp. 91–104).
[20] Côté-Boileau É, Denis JL, Callery B, Sabean M. The unpredictable journeys of spreading, sustaining and scaling healthcare innovations: A scoping review. Health Research Policy and Systems. 2019;17(1):84.
[21] Ciani O, Armeni P, Boscolo PR, Cavazza M, Jommi C, Tarricone R. De innovation: The concept of innovation for medical technologies and its implications for healthcare policy-making. Health Policy and Technology. 2016;5(1):47–64.
[22] Emilsson M, Ernstson U, Gustavsson L, Svensson A. Sustainable innovations in small enterprises for the transformation of the primary healthcare sector. Sustainability. 2020 Aug 7;12(16):6391.
[23] Eccles MP, Mittman BS. Welcome to implementation science. Implementation Science. 2006;1(1):1.
[24] Wandersman A, Duffy J, Flaspohler P, Noonan R, Lubell K, Stillman L, et al. Bridging the gap between prevention research and practice: The interactive systems framework for dissemination and implementation. American Journal of Community Psychology. 2008;41(3–4):171–81.
[25] Greenhalgh T, Robert G, Macfarlane F, Bate P, Kyriakidou O. Diffusion of innovations in service organizations: Systematic review and recommendations. Milbank Quarterly. 2004;82(4):581–629.
[26] Simpson DD, Flynn PM. Moving innovations into treatment: A stage-based approach to program change. Journal of Substance Abuse Treatment. 2007;33(2):111–20.
[27] Moullin JC, Sabater-Hernández D, Fernandez-Llimos F, Benrimoj SI. A systematic review of implementation frameworks of innovations in healthcare and resulting generic implementation framework. Health Research Policy and Systems. 2015;13(1):16.
[28] Crespo-Gonzalez C, Garcia-Cardenas V, Benrimoj SI. The next phase in professional services research: From implementation to sustainability. Research in Social and Administrative Pharmacy. 2017;13(5):896–901.
[29] Carter CR, Rogers DS. A framework of sustainable supply chain management: Moving toward new theory. International Journal of Physical Distribution & Logistics Management. 2008;38(5):360–87.
[30] Elkington J. Partnerships from cannibals with forks: The triple bottom line of 21st-century business. Environmental Quality Management. 1998;8(1):37–51.
[31] Eichler GM, Schwarz EJ. What sustainable development goals do social innovations address? A systematic review and content analysis of social innovation literature. Sustainability. 2019 Jan 19;11(2):522.
[32] Akerlund KM. Prevention program sustainability: The state's perspective. Journal of Community Psychology. 2000 May;28(3):353–62.
[33] Henriques A, Richardson J. The Triple Bottom Line: Does It All Add Up. Routledge: London; 2013.

[34] Slaper TF, Hall TJ. The triple bottom line: What is it and how does it work. Indiana Business Review. 2011;86(1):4–8.
[35] Fiksel J, Eason T, Frederickson H. A framework for sustainability indicators at EPA. A report of national risk management research laboratory, sustainable technology division. United States Environmental Protection Agency; 2012.
[36] Hensher M, McGain F. Health care sustainability metrics: Building a safer, low carbon health system. Health Affairs (Millwood). 2020;39(12):2080–87.
[37] Stigson B, Stigson P. A future resource and pollution constrained world – An agenda for a new partnership between business, governments and academia. Technological Forecasting and Social Change. 2015 Sep 1;98:255–59.
[38] Shediac-Rizkallah MC, Bone LR. Planning for the sustainability of community-based health programs: Conceptual frameworks and future directions for research, practice and policy. Health Education Research. 1998;13(1):87–108.
[39] Johnson K, Hays C, Center H, Daley C. Building capacity and sustainable prevention innovations: A sustainability planning model. Evaluation and Program Planning. 2004;27(2):135–49.
[40] Wunder S, Engel S, Pagiola S. Taking stock: A comparative analysis of payments for environmental services programs in developed and developing countries. Ecological Economy. 2008 May 1;65 (4):834–52.
[41] De Vries H, Bekkers V, Tummers L. Innovation in the public sector: A systematic review and future research agenda. Public Admin. 2016 March;94(1):146–66.
[42] Gruen RL, Elliott JH, Nolan ML, Lawton PD, Parkhill A, McLaren CJ, et al. Sustainability science: An integrated approach for health-programme planning. Lancet. 2008;372(9649):1579–89.
[43] Savaya R, Spiro SE. Predictors of sustainability of social programs. American Journal of Evaluation. 2012;33(1):26–43.
[44] García-Avilés JA. Diffusion of innovation. The International Encyclopedia of Media Psychology. 2020 Sep;8:1–8.
[45] Scheirer MA, Dearing JW. An agenda for research on the sustainability of public health programs. American Journal of Public Health. 2011;101(11):2059–67.
[46] Straus S, Tetroe J, Graham I, Zwarenstein M, Bhattacharyya O, Shepperd S. Knowledge Translation in Health Care: Moving from Evidence to Practice. CMAJ : Canadian Medical Association journal = journal de l'Association medicale canadienne. 2010; 182:E94–8. DOI:10.1503/cmaj.081335
[47] Mancini JA, Marek LI. Sustaining community-based programs for families: Conceptualization and measurement. Family Relations. 2004;53(4):339–47.
[48] Buchanan D, Fitzgerald L, Ketley D, Gollop R, Jones JL, Lamont SS, et al. No going back: A review of the literature on sustaining organizational change. International Journal of Management Reviews. 2005;7(3):189–205.
[49] Buchanan FL, Ketley D. The Sustainability and Spread of Organizational D Change: Modernizing Healthcare. London, UK: Routledge; 2007.
[50] Racine DP. Reliable effectiveness: A theory on sustaining and replicating worthwhile innovations. Administration and Policy in Mental Health. 2006;33(3):356–87.
[51] Fleiszer AR, Semenic SE, Ritchie JA, Richer MC, Denis JL. The sustainability of healthcare innovations: A concept analysis. Journal of Advanced Nursing. 2015 Jul;71(7):1484–98.
[52] Bowman CC, Sobo EJ, Asch SM, Gifford AL. Measuring persistence of implementation: QUERI Series. Implementation Science. 2008;3:13.
[53] Scheirer MA, Hartling G, Hagerman D. Defining sustainability outcomes of health programs: Illustrations from an online survey. Evaluation and Program Planning. 2008;31(4):335–46.
[54] Grol R, Wensing M, Eccles M. Improving Patient Care: The Implementation of Change in Clinical Practice. Philadelphia: Elsevier, 2005.

[55] Michie S, Johnston M, Abraham C, Lawton R, Parker D, Walker A, et al. Making psychological theory useful for implementing evidence-based practice: A consensus approach. Quality and Safety in Health Care. 2005;14(1):26–33.
[56] Greenhalgh T, Macfarlane F, Barton-Sweeney C, Woodard F. "If we build it, will it stay?" A case study of the sustainability of whole-system change in London. Milbank Quarterly. 2012;90(3):516–47.
[57] WCED SW. World commission on environment and development. Our Common Future. 1987 Jan 29;17(1):1–91.
[58] Kaasa SM. The UN Commission on Sustainable Development: Which mechanisms explain its accomplishments? Global Environmental Politics. 2007 Aug 1;7(3):107–29.
[59] Jameton A, McGuire C. Toward sustainable health-care services: Principles, challenges, and a process. International Journal of Sustainability in Higher Education. 2002;3(2):113–27.
[60] Vogt J, Nunes KRA. Recycling behaviour in healthcare: Waste handling at work. Ergonomics. 2014;57(4):525–35.
[61] Mackenzie J. The old care paradigm is dead, long live the new sustainable care paradigm: How can GP commissioning consortia meet the demand challenges of 21st Century healthcare?. London Journal of Primary Care (Abingdon). 2011;4(1):65–69.
[62] McGain F, Naylor C. Environmental sustainability in hospitals: A systematic review and research agenda. Journal of Health Services Research & Policy. 2014;19(4):245–52.
[63] Ertz M, Patrick K. The future of sustainable healthcare: Extending product lifecycles. Resources, Conservation and Recycling. 2020 Feb 1;153:104589.
[64] Giusti M, Persiani N, Silvia F, Ricci R. Planning and Managing of Diagnostic Imaging Technologies in Healthcare: The Introduction of Managed Equipment Service for the System's Sustainability (pp. 325–38). KnE Social Sciences: Dubai; 2020.
[65] Elabed S, Belal A, Shamayieh A. Sustainability of Medical Equipment in the Healthcare Industry: An Overview (pp. 1–4). IEEE Publications: Tripoli, Lebanon; 2019.
[66] Bilec MM, Geary M, Ries RJ, Needy KL, Cashion MK. A method for quantifying the benefits of greening a healthcare facility. Engineering Management Journal. 2010;22(3):3–11.
[67] Shepley MM. Developing evidence for sustainable healthcare. Building Research & Information. 2010;38(3):359–61.
[68] Pinzone M, Lettieri E, Masella C. Sustainability in healthcare: Combining organizational and architectural levers. International Journal of Engineering, Business and Management. 2012;4:1–9.
[69] Franco A, Shaker M, Kalubi D, Hostettler S. A review of sustainable energy access and technologies for healthcare facilities in the Global South. Sustainable Energy Technologies and Assessments. 2017;22:92–105.
[70] Buffoli M, Capolongo S, Noia MD, Gherardi G, Gola M, et al. Chapter 3: Healthcare sustainability evaluation systems. In: S Capolongo, et al. (eds.), Improving Sustainability during Hospital Design and Operation, Green Energy and Technology (pp. 23–29). Switzerland: Springer International Publishing; 2015. 978-3-319-14036-0-1.
[71] Stevanovic M, Allacker K, Vermeulen S. Hospital building sustainability: The experience in using qualitative tools and steps towards the life cycle approach. Procedia Environment Sciences. 2017;38:445–51.
[72] Campion N, Thiel CL, Woods NC, Swanzy L, Landis AE, Bilec MM. Sustainable healthcare and environmental life-cycle impacts of disposable supplies: A focus on disposable custom packs. Journal of Cleaner Production. 2015;94:46–55.
[73] Saad SG. Integrated environmental management for hospitals. Indoor and Built Environment. 2003;12(1–2):93–98.
[74] Messelbeck J, Whaley M. Greening the health care supply chain: Triggers of change, models for success. Corporate Environmental Strategy. 1999;6(1):39–45.

[75] Gelderman CJ, Semeijn J, Vluggen R. Development of sustainability in public sector procurement. Public Money Management. 2017;37(6):435–42.

[76] Balka E, Messing K, Armstrong P. Indicators for all: Including occupational health in indicators for a sustainable healthcare system. Policy and Practice in Health and Safety. 2006;4(1):46–61.

[77] Faezipour M, Ferreira S. Applying systems thinking to assess sustainability in healthcare system of systems. International Journal of System of Systems Engineering. 2011;2(4):290–308.

[78] Gupta N, Sharma V. Exploring employee engagement – A way to better business performance. Global Business Review. 2016 June;17(3):45S–63S.

[79] Lifvergren S, Huzzard T, Docherty P. A development coalition for sustainability in healthcare. ResearchGate, 292710138. 2009.

[80] Maghsoudi T, Cascón-Pereira R, Beatriz Hernández Lara A. The role of collaborative healthcare in improving social sustainability: A conceptual framework. Sustainability. 2020;12(8):3195.

[81] Ferrando C, Belda FJ. A new generation of sustainable healthcare system. Trends in Anaesthesia and Critical Care. 2017;15:1–2.

[82] Lo Presti LL, Testa M, Marino V, Singer P. Engagement in healthcare systems: Adopting digital tools for a sustainable approach. Sustainability. 2019;11(1):220.

[83] Godman B, Malmstrom RE, Gustafsson LL. Are new models needed to optimize the utilization of new medicines to sustain healthcare systems? Expert Review of Clinical Pharmacology. 2014;8 (1):77–94.

[84] Godman B, Novakovic T, Tesic D, Oortwijn W, Martin AP, Parker M, et al. Addressing challenges for sustainable healthcare in Central and Eastern Europe. Expert Review of Pharmacoeconomics and Outcomes Research. 2016;16(6):685–87.

[85] Teherani A, Nishimura H, Apatira L, Newman T, Ryan S. Identification of core objectives for teaching sustainable healthcare education. Medical Education Online. 2017;22(1):1386042.

[86] Walpole SC, Barna S, Richardson J, Rother HA. Sustainable healthcare education: Integrating planetary health into clinical education. Lancet Planet Health. 2019;3(1):e6–7.

[87] Ramadevi D, Gunasekaran A, Roy M, Rai BK, Senthilkumar SA. Human resource management in a healthcare environment: Framework and case study. Industrial and Commercial Training. 2016;48(8):387–93.

[88] Ahmed T, Shirahada K. Toward a sustainable healthcare service system in a limited resource context: Case study of Bangladesh Rural Advancement Committee's healthcare system. International Journal of Healthcare Management. 2017;12:1389512.

[89] Srinivasan V, Chandwani R. HRM innovations in rapid growth contexts: The healthcare sector in India. International Journal of Human Resource Management. 2014;25(10):1505–25.

[90] Pinzone M, Guerci M, Lettieri E, Huisingh D. Effects of green training on pro-environmental behaviors and job satisfaction: Evidence from the Italian healthcare sector. Journal of Cleaner Production. 2019;226:221–32.

[91] Pinzone M, Guerci M, Lettieri E, Redman T. Progressing in the change journey towards sustainability in healthcare: The role of 'Green' HRM. Journal of Cleaner Production. 2016;122:201–11.

[92] Ramirez B, West DJ, Costell MM. Development of a culture of sustainability in health care organizations. Journal of Health Organization and Management. 2013;27(5):665–72.

[93] Guandalini I. Sustainability through digital transformation: A systematic literature review for research guidance. Journal of Business Research. 2022;148:456–471. DOI:10.1016/j.jbusres.2022.05.003.

[94] Fonseca A, Abreu I, Guerreiro MJ, Abreu C, Silva R, Barros N. Indoor air quality and sustainability management – Case study in three Portuguese healthcare units. Sustainability. 2018;11(1):101.

[95] Chiarini A, Opoku A, Vagnoni E. Public healthcare practices and criteria for a sustainable procurement: A comparative study between UK and Italy. Journal of Cleaner Production. 2017;162:391–99.

[96] Ossebaard HC, Lachman P. Climate change, environmental sustainability and health care quality. International Journal for Quality in Health Care. 2021;33(1):24–26.

[97] Sherman J, Le C, Lamers V, Eckelman M. Life cycle greenhouse gas emissions of anesthetic drugs. ResearchGate, 223973615. 2012.

[98] Chauhan A, Singh A. Modeling the drivers of healthcare waste management in India: A policy perspective. Management of Environmental Quality: An International Journal. 2017;25(3):3–5.

[99] Eckelman MJ, Sherman J. Environmental impacts of the U.S. health care system and Effects on Public Health. PLOS ONE. 2016;11(6):e0157014.

[100] Eckelman MJ, Sherman JD. Estimated global disease burden from US health care sector greenhouse gas emissions. American Journal of Public Health. 2018;108(S2):S120–22.

[101] Leite H, Bateman N, Radnor Z. Beyond the ostensible: An exploration of barriers to lean implementation and sustainability in healthcare. Production Planning and Control. 2019;31(1):1–18.

[102] Lindsay CF, Kumar M, Juleff L. Operationalising lean in healthcare: The impact of professionalism. Production Planning Control. 2020;31(8):629–43.

[103] Zhu Q, Johnson S, Sarkis J. Lean Six Sigma and environmental sustainability: A hospital perspective. International Journal - Supply Chain Forum. 2018;19(3):2–6.

[104] Parmar P. Outsourcing non-clinical processes in the healthcare industry: Contributions to long term sustainability. International Journal of Healthcare Management. 2015;8(2):65–67.

[105] Jadhav S, Yeravdekar R, Kulkarni M. Cross-border healthcare access in South Asian Countries: Learnings for sustainable healthcare tourism in India. Procedia - Social and Behavioral Sciences. 2014;157:109–17.

[106] Zadeh RS, Xuan X, Shepley MM. Sustainable healthcare: Existing challenges and future directions for an environmental, economic, and social approach to sustainability. Facilities. 2016;34(5/6):264–288.

[107] Shaw KM. Leveraging affordable innovation to tackle India's healthcare challenge. IIMB Management Review. 2017;30(1):4–8.

[108] Aquino RP, Barile S, Grasso A, Saviano M. Envisioning smart and sustainable healthcare: 3D-printing technologies for personalized medication. Futures. 2018;103:35–50.

[109] Gupta P, Armbruster R. Innovations in healthcare: A reality. International Journal of Innovation Science. 2011;3(2):101–04.

[110] Bukve O, Haug K, Aslaksen AB, Plessen CV, Hovlid E. Sustainability of healthcare improvement: what can we learn from learning theory? BMC Health Services Research. 2012;12(235):150–175.

[111] AlJaberi OA, Hussain M, Drake PR. A framework for measuring sustainability in healthcare systems. International Journal of Healthcare Management. 2020;13(4):276–85.

[112] Palozzi G, Schettini I, Chirico A. Enhancing the sustainable goal of access to healthcare: Findings from a literature review on telemedicine employment in rural areas. Sustainability. 2020;12(8):3318.

[113] Celdran AH, Perez MG, Clemente FJG, Perez GM. Sustainable securing of medical cyber-physical systems for the healthcare of the future. Sustainable Computing: Informatics and Systems. 2018;234:1–28.

[114] Turcu CE, Turcu CO. Internet of things as key enabler for sustainable healthcare delivery. Procedia – Social and Behavioral Sciences. 2013;73:251–56.

[115] Roberts JP, Fisher TR, Trowbridge MJ, Bent C. A design thinking framework for healthcare management and innovation. Healthcare. 2016;4(1):11–14.

[116] Faste RA. Ambidextrous Thinking. Innovations in Mechanical Engineering Curricula for the 1990s. New York: American Society of Mechanical Engineers, November 1994; 1994.

[117] D'avolio L. Why we need design thinking in healthcare. Information Week. 2015:1–6.

[118] Barbero S, Pallaro A. Systemic design for sustainable healthcare. The Design Journal. 2017 Jul 28;20(sup1):2473–85.

[119] Chamberlain P, Wolstenholme D, Dexter M, Seals E. The State of the Art of Design in Health: An Expert-led Review of the Extent of the Art of Design Theory and Practice in Health and Social Care. Sheffield, UK: Sheffield Hallam University; 2015.

[120] Fischer M. Fit for the future? A new approach in the debate about what makes healthcare systems really sustainable. Sustainability. 2015;7(1):294–312.

[121] Ryan-Fogarty Y, O'Regan B, Moles R. Greening healthcare: Systematic implementation of environmental programs in a university teaching hospital. Journal of Cleaner Production. 2016;126:248–59.

[122] Goh CY, Marimuthu M. The path towards healthcare sustainability: The role of organisational commitment. Procedia – Social and Behavioral Sciences. 2016;224:587–92.

[123] Grose J, Bennallick M, Nichols A, Pahl S, Richardson J. Facilitating sustainable waste management behaviors within the health sector: A case study of the National Health Service (NHS) in Southwest England, UK. Sustainability. 2012;4(4):630–42.

[124] De Francisco Shapovalova N, Meguid T, Campbell J. Health-care workers as agents of sustainable development. The Lancet Global Health. 2015 May 1;3(5):e249–50.

[125] Patrício L, Grenha Teixeira J, Vink J. A service design approach to healthcare innovation: From decision-making to sense-making and institutional change. AMS Review. 2019 Jun 15;9:115–20.

[126] Freire K, Sangiorgi D Service design & healthcare innovation: From consumption to coproduction and co-creation. Paper presented at ServDes. Vol. 2010 – Second Nordic Conference on Service Design and Service Innovation, Linköping, Sweden; 2010.

[127] Vezzoli C. A new generation of designers: Perspectives for education and training in the field of sustainable design. Experiences and projects at the Politecnico di Milano University. Journal of Cleaner Production. 2003;11(1):1–9.

[128] Vinodh S, Arvind KR, Somanaathan M. Tools and techniques for enabling sustainability through lean initiatives. Clean Technologies and Environmental Policy. 2011;13(3):469–79.

[129] Sheth JN, Sethia NK, Srinivas S. Mindful consumption: A customer-centric approach to sustainability. Journal of the Academy of Marketing Science. 2011;39(1):21–39.

[130] Figge F, Hahn T, Schaltegger S, Wagner M. The sustainability balanced scorecard–linking sustainability management to business strategy. Business Strategy and the Environment. 2002;11(5):269–84.

[131] Garcia-Sabater JJ, Marin-Garcia JA. Enablers and inhibitors for sustainability of continuous improvement: A study in the automotive industry suppliers in the Valencia region. Intangible Capital. 2009;5(2):183–209.

[132] Baumgartner RJ. Managing corporate sustainability and CSR: A conceptual framework combining values, strategies and instruments contributing to sustainable development. Corporate Social Responsibility and Environmental Management. 2014;21(5):258–71.

[133] Dauvergne P, Lister J. Big brand sustainability: Governance prospects and environmental limits. Global Environmental Change. 2012;22(1):36–45.

[134] Lejeune C. Is continuous improvement through accreditation sustainable? A capability-based view. Management Decision. 2011;49(9):1535–48.

[135] Lee B-H, H-j J. The mutation of the Toyota production system: Adapting the TPS at Hyundai Motor Company. The International Journal of Production Research. 2007 Aug 15;45(16):3665–79.

[136] Hussain M, Malik M, Al Neyadi HS. AHP framework to assist lean deployment in Abu Dhabi public healthcare delivery system. Business Process Management Journal. 2016;22(3):546–65.

[137] Teich ST, Faddoul FF. Lean management – The journey from Toyota to healthcare. Rambam Maimonides Medical Journal. 2013;4(2):e0007.

[138] Young T, Brailsford S, Connell C, Davies R, Harper P, Klein JH. Using industrial processes to improve patient care. BMJ. 2004;328(7432):162–64.

[139] Deming WE. Out of the Crisis. Cambridge: Massachusetts Institute of Technology Center for Advanced Engineering Study; 1986.
[140] Reed R, Lemak DJ, Mero NP. Total quality management and sustainable competitive advantage. Journal of Quality Management. 2000 Mar 1;5(1):5–26.
[141] Juran JM. Managerial Breakthrough: A New Concept of the Manager's Job. New York: McGraw-Hill; 1964.
[142] Kehoe DF. The Fundamentals of Quality Management. Springer Science & Business Media: London; 2012 Dec 6.
[143] Sitkin SB, Sutcliffe KM, Schroeder RG. Distinguishing control from learning in total quality management: A contingency perspective. Academy Management Review. 1994;19(3):537–64.
[144] Curkovic S, Vickery S, Dröge C. Quality-related action programs: Their impact on quality performance and firm performance. Decision Science. 2000;31(4):885–902.
[145] Douglas TJ, Fredendall LD. Evaluating the Deming management model of total quality in services. Decision Science. 2004;35(3):393–422.
[146] Devlen A. How to build a comprehensive business continuity programme for a healthcare organization. Journal of Business Continuity and Emergency Planning. 2009;4(1):47–61.
[147] Ofili OU. Patient satisfaction in healthcare delivery – A review of current approaches and methods. European Scientific Journal. 2014;10(25):23–31.
[148] Murti A, Deshpande A, Srivastava N, Deshpande A, Srivastava N. Service quality, customer (patient) satisfaction and behavioral intention in health care services: Exploring the Indian perspective. Journal of Health Management. 2013;15(1):29–44.
[149] Tucker JL, Adams SR, Adams SR. Incorporating patients' assessments of satisfaction and quality: An integrative model of patients' evaluations of their care. Managing Service Quality: An International Journal. 2001;11(4):272–87.
[150] Davies AR, Ware JE Jr. Measuring patient satisfaction with dental care. Social Science & Medicine. Part A: Medical Psychology & Medical Sociology. 1981 Dec 1;15(6):751–60.
[151] Ramsaran-Fowdar RR. Identifying health care quality attributes. Journal of Health and Human Services Administration. 2005;27(4):428–43.
[152] Butler D, Oswald SL, Turner DE, Turner DE. The effects of demographics on determinants of perceived health-care service quality: The case of users and observers. Journal of Management in Medicine. 1996;10(5):8–20.
[153] Spector PE. Job Satisfaction [Application]. Assessment, Causes, and Consequences. Sage Publications: United States; 1997.
[154] Parvin MM, Kabir MN. Factors affecting employee job satisfaction of pharmaceutical sector. Australian Journal of Business and Management Research. 2011;1(9):113.
[155] Soliman HM. Motivation-hygiene theory of job attitudes: An empirical investigation and an attempt to reconcile both the one-and the two-factor theories of job attitudes. Journal of Applied Psychology. 1970 Oct;54(5):452.
[156] Adams JS. Towards an understanding of inequity. Journal of Abnormal and Social Psychology. 1963;67(5):422–36.
[157] Vecchio RP. Predicting worker performance in inequitable settings. Academy Management Review. 1982;7(1):103–10.
[158] Stumpf SA, Tymon WGJ Jr, Favorito N, et al. Employees and change initiatives: Intrinsic rewards and feeling valued. Journal of Business Strategy. 2013;34(2):21–29.
[159] Robinson S, Radnor ZJ, Burgess N, Burgess N, Worthington C. Simlean: Utilising simulation in the implementation of lean in healthcare. European Journal of Operation Research. 2012;219(1):188–97.
[160] Chaudhry MS, Sabir HM, Rafi N, et al. Exploring the relationship between salary satisfaction and job satisfaction: A comparison of public and private sector organizations. Journal on Communications. 2011;3(4):1.

[161] Hee Yoon M, Hyun Seo J, Seog Yoon T. Effects of contact employee supports on critical employee responses and customer service evaluation. Journal of Services Marketing. 2004;18(5):395–412.
[162] Arce RM, Diaz PI, Barros SP, Galloway P. Gastrointestinal infections. Act Clinica Belgica. 2009;64:346–48.
[163] Kotler P, Lee N. Corporate Social Responsibility: Doing the Most Good for Company and Your Cause. Wiley; 2006.
[164] Leela V. Corporate social responsibility in hospitals: Need for transparent CSR initiatives for internal and external stakeholders. International Journal of Business Ethics in Developing Economies. 2014;3(1):1–7.
[165] Drumwright ME. Socially responsible organizational buying: Environmental concern as a noneconomic buying criterion. Journal of Marketing. 1994;58(3):1–19.
[166] Maignan I, Ferrell O, Ferrell OC. Antecedents and benefits of corporate citizenship: An investigation of French businesses. Journal of Business Research. 2001;51(1):37–51.
[167] Siudek T, Zawojska A. Competitiveness in the economic concepts, theories and empirical research. Acta Scientiarum Polonorum. Oeconomia. 2014 Mar 30;13(1):91–108.
[168] Martin R, Aslesen HW, Grillitsch M, Herstad SJ. Regional innovation systems and global flows of knowledge. In: Isaksen A, Martin R, Trippl M (eds.), New Avenues for Regional Innovation Systems-theoretical Advances, Empirical Cases and Policy Lessons (pp. 127–47). Cham, Switzerland: Springer; 2018.
[169] Sandström S, Berger S. Näringslivsutveckling I En Föränderlig Värld Globalisering Och Omställning I Värmland. Karlstad, Sweden: Karlstad University Studies; 2010.
[170] Porter ME. Location, competition, and economic development: Local clusters in a global economy. Economic Development Quarterly. 2000;14(1):15–34.
[171] Porter M. The economic performance of regions. Regional Studies. 2003;37(6–7):545–46.
[172] Dubois A, Roto J Making the best of Europe's sparsely populated areas. On making geographic specificity a driver for territorial development in Europe. Nordregio working:Paper 2012:15. Stockholm, Sweden: Nordregio; 2012.
[173] Muzzi C, Albertini S. Communities and managerial competencies supporting SMEs innovation networking: A longitudinal case study. R&D Management. 2015;45(2):196–211.
[174] Gausdal AH, Nilsen ER. Orchestrating innovative SME networks. The case of "Health Innovation". Journal of the Knowledge Economy. 2011;2(4):586–600.
[175] Veronica S, Alexeis GP, Valentina C, Elisa G. Do stakeholder capabilities promote sustainable business innovation in small and medium-sized enterprises? Evidence from Italy. Journal of Business Research. 2020 Oct 1;119:131–41.
[176] Porter ME, Van Der Linde C. Toward a new conception of the environment-competitiveness relationship. Journal of Economic Perspectives. 1995;9(4):97–118.
[177] Porter M, Van der Linde C. Green and competitive: Ending the stalemate. The Dynamics of the Eco-efficient Economy: Environmental Regulation and Competitive Advantage. 1995 Sep;33:120–34.
[178] Blättel-Mink B. Innovation towards sustainable economy – The integration of economy and ecology in companies. Sustainable Development. 1998;6(2):49–58.
[179] Murphy J, Gouldson A, Gouldson A. Environmental policy and industrial innovation: Integrating environment and economy through ecological modernization. Geoforum. 2000;31(1):33–44.
[180] Drake F, Purvis M, Hunt J, Millard D. European businesses, national priorities: Pioneers and laggards of ecological modernization. European Environment. 2003;13(3):164–82.
[181] Gozman D, Willcocks L. The emerging cloud dilemma: Balancing innovation with cross-border privacy and outsourcing regulations. Journal of Business Research. 2019;97:235–56.
[182] Zahoor N, Al-Tabbaa O. Inter-organizational collaboration and SMEs' innovation: A systematic review and future research directions. Scandinavian Journal of Management. 2020;36(2):101109.

Chapter 4
Healthcare sustainability with artificial intelligence: innovations, benefits, and challenges

Abstract: The use of artificial intelligence (AI) is becoming common in several fields such as medicine and has been seen as a potential method to enhance or replace human practitioners in primary healthcare. In healthcare, AI has shown efficiency in academic research but has limited real-world applicability in preventative, diagnostic, and therapeutic areas. By analyzing success criteria for AI adoption, we hope to bridge the gap between current academic achievements and limited practical use in healthcare. The implementation of AI in healthcare has the potential to enhance the accuracy of diagnoses as well as the effectiveness of preventive measures and treatment strategies for patients. Consequently, it is expected to contribute to cost efficiency and equity within healthcare services. However, many challenges need to be addressed to fully realize the benefits of AI in the public sector. These challenges include the lack of AI adoption in public healthcare systems, concerns about patient privacy, and the potential contravention of patient autonomy rights regarding AI applications. To advance sustainability in the healthcare industry, the chapter examines the innovative potential of AI. AI technologies are having an evolutionary effect on the healthcare sector through their ability to enhance the management of resources, improve disease diagnosis and prevention, personalize therapies to individual patients, and improve administrative procedures within healthcare systems. AI applications in the healthcare sector contribute to the efficient use of resources through the prediction of patient admission and the optimization of staff levels. Thus, it leads to a reduction in wastage and energy consumption. In addition, AI plays an essential part in enhancing the identification and prevention of diseases at an early stage by using data analysis techniques. This effectively reduces the overall burden on the healthcare system and minimizes the associated environmental costs. The use of AI in healthcare has resulted in the development of personalized treatment plans, expedited drug discovery processes, and the implementation of telemedicine. These advancements have significantly improved the efficiency of healthcare delivery by reducing the need for unnecessary interventions and eliminating travel requirements. The use of AI in the healthcare sector not only serves to increase the quality of patient care but also plays a significant role in promoting environmental sustainability. This is achieved through the reduction of waste, greater energy efficiency, and more efficient management of the healthcare supply chain. However, the use of AI in the healthcare sector presents ethical and data privacy concerns that need appropriate handling. This chapter discusses the innovative applications of AI in the sustainability of healthcare as well as the critical considerations and challenges to its adoption.

https://doi.org/10.1515/9783111436432-004

4.1 Introduction

Artificial intelligence (AI) was first conceptualized and developed by John McCarthy in the year 1955. It is often characterized as an interdisciplinary field that includes the scientific and technical principles involved in developing intelligent robots [1]. The term "artificial intelligence" denotes the use of computational systems to simulate intelligent behavior, while minimizing the need for human intervention. Rapid advancements in AI are transforming the healthcare industry. It transforms the manual healthcare system into an automated one, in which human beings are responsible for performing the regular activities and duties in medical practice, such as the administration of patients and healthcare resources. When developers construct AI systems to perform duties, the technical challenges of digitizing health services create new challenges [2].

AI has significant potential in enhancing the quality of patient care and reducing the financial burden associated with healthcare expenditures [3]. The anticipated growth in population is projected to stimulate the need for healthcare services. The healthcare industry needs creative approaches to enhance its effectiveness and efficiency while avoiding unnecessary costs. The healthcare business may benefit from rapid advancements in technology, particularly in the areas of AI and robots. AI and robotics in the area of healthcare are rapidly advancing, particularly in the field of early detection and diagnostic applications. AI has experienced a concurrent increase in its capabilities. AI allows individuals to accomplish tasks that are often carried out generally with more efficiency, ease, and cost-effectiveness [4]. The use of AI in health care and modern medicine is a prevalent practice, including a range of applications that vary in terms of accessibility, reliability, and ethical considerations. For example, AI possesses the potential to revolutionize the field of healthcare by providing medical services to regions that lack adequate availability, consequently enhancing health outcomes for disadvantaged individuals [5]. Additionally, AI can address the absence of healthcare providers, thereby effectively catering to the medical requirements of low and middle-income nations and aging and rural communities [6]. Robotic surgical errors that prove deadly, incorrect diagnoses, or harmful suggestions are just a few ways in which AI might negatively impact patients [7–9]. The use of AI in the field of medicine faces many challenges, including social, legal, and economic limitations [10].

Nevertheless, these applications offer a considerable amount of potential. Several challenges associated with the reliability and accountability of predictive AI systems have arisen. Several questions have been raised regarding the changing impact these systems will have on clinical decision-making and the doctor-patient relationship, and the reliability of AI-powered equipment. Researchers and AI developers have proposed that the efficiency of managing specific scenarios might be enhanced if the user (either the physician or the patient) places faith in the AI [11–14]. The healthcare sector is characterized by its complexity since it involves several stakeholders and encompasses functional, environmental, experiential, clinical, and organizational objectives that are subject to transformation [15]. The healthcare industry is a valuable and important sector. Making optimal judgments within this area, such as resource allocation and prescription of treatments and drugs, poses challenges in terms of manual decision-making processes. Healthcare concerns must be comprehensive, transparent, and incorporate social and healthcare components [16]. AI-based computational decision-providing applications offer an intelligent alternative for efficiently managing the healthcare decision-making system promptly. Additionally, these applications assist in the precise prediction and diagnosis of critical medical conditions, the allocation of medical professionals, effective resource management, provision of home care advice, and personalized medication prescriptions, among other functionalities [17]. The use of AI and big data analytics has significantly enhanced advancements in robotics and automation. These developments have profound implications across several domains in real-life scenarios [18, 19].

While AI-based decision-making systems can improve healthcare delivery and diagnosis, ethical, regulatory, and financial factors must be considered for sustainability [20, 21]. Proper consideration of sustainable development objectives is crucial for effective AI adoption in healthcare decision-making. The integration of sustainable development objectives with successful commercial innovation has the potential to address a wide range of social challenges and enhance productivity and economic growth across several sectors, including the medical sector [22]. The emission of carbon by the healthcare sector is a notable concern in several countries. However, the environmental effects of the healthcare industry have received less attention, partly due to the assumption that all healthcare technology now in use is clinically essential, hence rendering carbon emissions ethically irrelevant. Therefore, when assessing the carbon impact of healthcare, the focus is mostly on the institutional dimension, namely the carbon emissions associated with hospital infrastructure [23]. Although hospital care and physician and clinical services are the two greatest carbon contributors to healthcare, surpassing healthcare buildings, this structure avoids responsibility for the environmental effect of healthcare delivery [24].

As AI continues to be more integrated and indispensable in the provision of healthcare, ethical considerations must include the possible adverse effects on the environment. The use of AI results in the generation of carbon emissions, which may be seen at several stages of its lifespan, including programming, development, and utilization. This is mostly attributed to the substantial energy and resource requirements associated with AI technology. For example, training Google's AlphaGo Zero game for 40 days has a carbon footprint of 1,000 h of air travel [25]. Every AI system requires development, operation, and training. Further, the process of extracting minerals, metals, and polymers that are essential for the development of AI-capable systems has significant environmental consequences [26] Adopting AI would considerably aid in clinical decision-making, as shown by other studies [27–30]. AI in the healthcare industry has been a subject of significant debate due to technological progress [31]. According to [32], the phrase "AI in healthcare" encompasses the use of software, machine learning algorithms, or AI to replicate human cognitive processes in the interpretation, analysis, and comprehension of healthcare data. An example of the use of AI-based diagnostic algorithms may be seen in the utilization of such algorithms in the analysis of mammograms. This application improves the identification of breast cancer and functions as an additional assessment for radiologists, providing a supplementary perspective [33]. The use of AI in the healthcare sector has resulted in the development of very effective medical applications [34, 35]. Insilico Medicine has created AI algorithms that can inhibit viral replication [36]. According to Rong et al. [37] pregnant women may get dietary counseling based on their health state and algorithm estimates.

4.2 Implementing AI in healthcare

There are several technical categories of AI applications, including a wide range of functionalities, starting with the processing of natural language and extending to expert systems [38]. Certain medical industries often use specialized types of AI applications, such as image analysis in radiology or dermatology [39]. The identification of impediments to the application of AI has been extensively explored via the examination of real-world case studies. AI adoption must be adjusted to current healthcare processes due to the need to access vast volumes of data under tight privacy standards and management acceptability. Figure 4.1 provides a demonstrative illustration of the integration of AI into regular healthcare procedures [40]. The main challenges identified for the application of AI are the following core concerns: the first issue is technical implementation that does not prioritize privacy; the second issue concerns

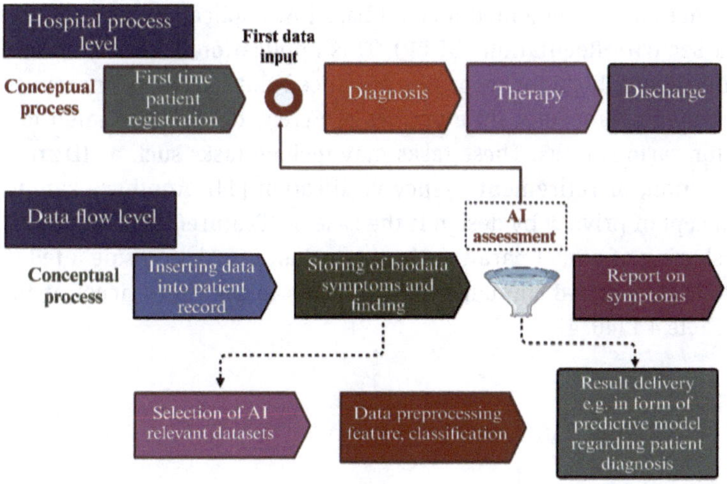

Figure 4.1: The integration of artificial intelligence (AI) into routine healthcare processes [40] (adapted by open access).

the deficiencies in existing regulatory settings; and lastly, there is a lack of comprehensive measurements of the medical and economic impacts. As a point of comparison, the developer of a framework for AI applications in China's telecommunications industry found three success elements. Success determinants include external variables like government participation or vendor alliances, organizational capabilities like management or technical skills, and innovative characteristics like compatibility or relative advantage.

The majority of AI applications in real-life situations face the challenge of balancing practicality and privacy due to their dependence on unrestricted data access. The effective mitigation of this difficulty might be achieved by the implementation of several practical examples that concentrate on indications and are derived from real-world applications of AI. An AI technology known as "persuasive communication tailoring" has been used to provide motivating messages aimed at smoking cessation to adult individuals. The performance of the machine learning-based anti-smoking program was shown to be superior to the previous rule-based system. The algorithm used data from many sources including messages, feedback databases, and user profiles for training purposes [41]. Another example may be seen in the case of MSD, a pharmaceutical firm that developed an AI-powered communication platform using Facebook Messenger to facilitate a chatbot focused on addressing pressing concerns related to immune-oncology. The fundamental conversational dynamic between the physician and the chatbot is not limited to the information included in electronic health records (EHRs). Instead, it functions as an independent tool that prioritizes the retrieval of specific problem-solving data [42]. In addition, there is a growing use of "privacy-by-design" technologies, which seek to include privacy principles in the first stages of developing an AI application [43]. At the institutional level, a health insurance system in Romania has implemented a cloud-based AI application that adheres to the General Data Security Regulation (GDPR). This application utilizes a "SwarmESB-based" architecture, which incorporates enhanced data security measures. Within the cloud architecture, several little entities are created, each dedicated to a particular purpose for various tasks. These tasks may include tasks such as ID copying, job status verification, or retirement agency verification [44]. Another example that supports the concept of privacy by design is the case of "FeatureCloud." This platform facilitates the sharing of model parameters rather than raw data inside a federated AI model [45]. The technical implementation needs to take into account the guidelines given in Table 4.1 [46, 47].

Table 4.1: The guideline should be taken into account throughout the technical implementation.

Discussion paper and request for feedback on the proposed regulatory framework for modifications to artificial intelligence/machine learning (AI/ML)-based software as medical device (SaMD) [46]	**Principles of ethics for AI that can be trusted [47]**
The implementation of quality systems and adherence to good machine learning practices (GMLP) are crucial in ensuring the effectiveness and reliability of machine learning processes. These practices include several aspects such as the use of relevant data, the segregation of training, tuning, and test datasets, as well as the transparency of the output generated.	On April 8, 2019, the European Commission established an autonomous and esteemed expert panel dedicated to the field of artificial intelligence.
The implementation of preliminary pre-market evaluations to ensure safety and efficacy.	The ethical principles that serve as the fundamental basis for the establishment of reliable artificial intelligence (AI) systems include respect for human autonomy, prevention of damage, justice, and explicability.
The monitoring of AI devices is conducted by the implementation of a comprehensive framework that encompasses the creation, validation, and execution of algorithm modifications. This framework includes the use of an "algorithm change protocol" to ensure effective oversight and control.	There are seven essential criteria for achieving dependable AI systems. These include: (1) the presence of human agency and supervision; (2) technological robustness and safety; (3) privacy and data governance; (4) openness; (5) diversity, nondiscrimination, and justice; (6) environmental and social well-being; and (7) accountability.
Reporting on the safety and efficacy of a product in real-world settings after it has been released to the market, to maximize these attributes.	Evaluating the reliability of AI: comprehensive assessment criteria for the development, deployment, and use of AI systems.

4.3 Enhanced AI implementation in success factor categories

To begin the process, several actions have been specified about the establishment of the technical infrastructure. Firstly, it is necessary to customize it according to the specific application sector, distinguishing between indication-focused and institutional-focused applications. Furthermore, the data processing framework must prioritize the establishment of efficient data access and exchange channels, while also including reliable systems to ensure confidentiality. Also, it is essential to include a "privacy-by-design" methodology, ensuring that privacy considerations are integrated from the beginning. The technical infrastructure should possess a considerable level of flexibility to effectively meet developing and upcoming legislative responsibilities. Furthermore, it is essential to establish a well-defined and comprehensive framework for AI policy. This analysis aims to differentiate between the permissionless principle and the precautionary principle, particularly emphasizing the difference between a proactive "quick reaction" strategy that allows for risks and a more cautious "safety first" approach. The system must include guidelines relating to the duration of decision-making, the extent of scenario analysis, and the precision of AI-generated results to evaluate AI judgments prospectively. Thus, it is essential to establish the feasibility of assigning responsibility, liability, and culpability among the many stakeholders, including both human agents and AI entities, within the established framework.

Also, there is a need to enhance the refinement of procedures and metrics used to evaluate the medical and economic implications of AI applications. There is a need to greatly increase the evaluations of both medical and economic impacts. The evaluations should be based on cost-utility estimates, with a specific emphasis on quality-adjusted life years. The use of standardized quality standards, such as the CHEERS and PRISMA criteria, is vital. For example, using a EuroQol-5D questionnaire, these criteria enable the evaluation of outcomes not only by medical professionals but also by institutional stakeholders. Figure 4.2 illustrates success elements for policy, technology, and impact measurement – systematically established success factors, based on current academic research and real-world AI applications that considerably promote the use of AI in healthcare [48]. First, an examination of more real-world AI application examples would have offered additional useful insights for the study. However, it is important to recognize that there is a lack of publicly available information and the presence of confidentiality agreements about both the technical aspects and commercial consequences, dependent on the specific databases used. Hence, investigations re-

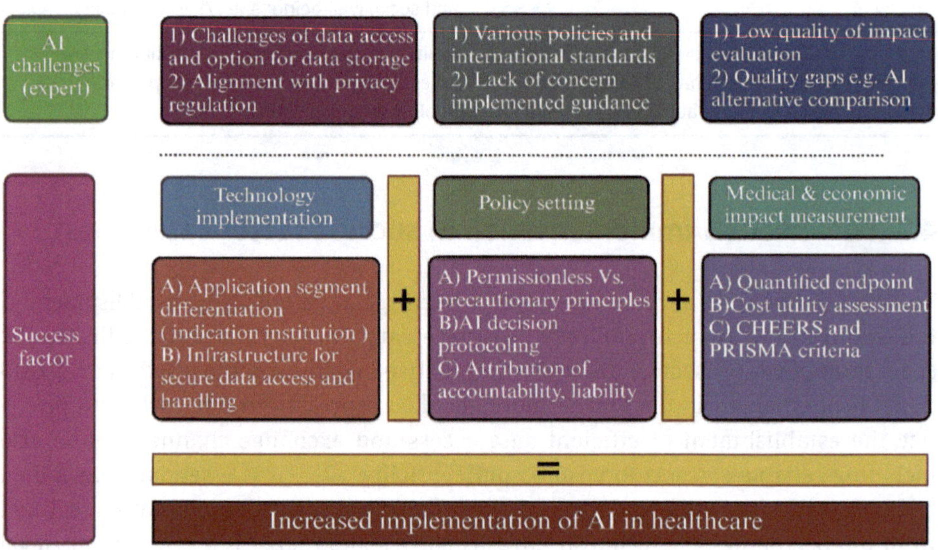

Figure 4.2: An overview of factors facilitating the implementation of artificial intelligence in healthcare [50].

lated to current advancements such as AI policy frameworks often remain unexplored or unpublished. Thirdly, there are significant differences among the various types. For example, an AI-facilitated medication adherence system and an AI-powered robotic surgical program are subject to distinct regulatory, technical, medical, and economic impact evaluation criteria. Consequently, it will be necessary to assign weights to success indicators based on the specific circumstances of digital health and AI within each healthcare system [49].

4.4 The ethical considerations of AI in the healthcare sector

The potential transformation of the healthcare system via expanded deployment of AI and investment in data security may be analyzed from several perspectives. The use of AI is projected to improve healthcare performance and efficiency [27,51]. For example, the use of AI has the potential to improve the accuracy and effectiveness of oncological diagnosis. Additionally, the use of technology will contribute to enhancing the overall quality of the services provided and decreasing their associated expenses. The recent COVID-19 epidemic has shown an increasing dependence on telemedicine, which has resulted in expanded accessibility to healthcare services [52–54]. The Organization for Economic Co-operation and Development (OECD) released a recent study in 2020 acknowledging that the healthcare industry is characterized by inefficiency and wastefulness, leading to its disqualification as an efficient system. As a result, more organizations have switched to AI to help with healthcare system management and service redesign [28].

Cook et al. [55] have introduced an innovative AI development known as epileptic seizure prediction, which is very effective in reducing the effect of epileptic episodes. It was also feasible to accurately predict early stroke using AI and the creation of a unique movement-detecting device [56]. According to Somashekhar et al. [57], IBM Watson is another trustworthy AI system that has the potential to help successfully in the detection of cancer. Technological advancements have contributed to the development of systems that replicate the cognitive capabilities of the human brain [58]. Furthermore, these advancements have facilitated the use of radiosurgery [59]. The implementation of AI has undergone numerous changes, resulting in the emergence of numerous ethical concerns. Several difficulties arise regarding patient safety regarding data [60, 61]. The doctor-patient interaction is also affected since clinicians must depend on the black box AI algorithm to determine diagnoses [62].

AI in healthcare has great potential to enhance diagnostics and efficiency, but it also raises ethical problems. The issue of safety is a significant challenge to the use of AI in the healthcare sector. In using AI algorithms to analyze data from patient medical records and assist physicians in exploring cancer treatment options for their patients, IBM Watson for oncology [59] is a very good example. The issue is that the program was only trained with a small number of "synthetic cancer cases," which were created only by physicians, rather than real patient data [63]. AI algorithms play a crucial role in assisting medical professionals with the diagnosis and treatment of patients. By evaluating patient information, such as with IBM Watson oncology, these algorithms can identify and recommend the most effective treatment choices available. Nevertheless, there have been instances in which the system has shown limitations in delivering accurate diagnoses and recommendations [64]. Another problem occurs when AI systems are instructed on data from a small subset of the population, leading to biased outcomes [65]. In this situation, AI-based systems will not be able to provide relevant and accurate findings for every patient and risk, amplifying preexisting social prejudices [66]. Incorrect diagnosis and treatment will result. AI gives biased findings because that is how technology is different from people. This problem can be fixed by getting a lot of correct and useful information. This case highlights crucial problems about data dependability, validity, openness, and bias in AI development and implementation. Undoubtedly, the trustworthiness, effectiveness, validity, and fairness of any AI system or algorithm taught by humans depend only on the quality and reliability of the data used for training, as well as the integrity and transparency of the developers and practitioners involved. Healthcare organizations face the potential of data breaches, which pose a significant risk since they may quickly weaken patient confidence. Thus, this breakdown of trust may result in patients concealing critical health information from their doctors. According to Landi [67], the number of cyberattacks recorded in the USA reached an all-time high in 2021 [67]. In reality, 45 million people were impacted by healthcare cyberattacks, according to statistics gathered from the US Department of Health and Human Services.

Five key variables might cause or inspire health information theft [68]. In reality, human error is responsible for the loss or lack of data security in 33.5% of situations. In the healthcare industry, the occurrence of data misuse or unauthorized use in non-preauthorized activities accounts for 29.5% of reported occurrences. Physical theft is the third most prevalent concern, accounting for 16.3% of incidents, while hacking is responsible for 14.8% of cases. Malware is considered to be the least common cause, accounting for just 10.8% of occurrences. Goggin's [69] paper provided an actual example of a data security compromise. In 2019, Facebook used AI to gather data from users' posts and then predict their mental well-being and susceptibility toward self-harm. So, Facebook got and kept information about the mental health of its users without their knowledge or permission. The AI firm known as Cense AI has publicly released the stage data, personal information, and health information of over 2.5 million patients in the United States. The data were readily available for public access through the internet without the need for any authentication or credentials. The data was first saved in a storage repository before being placed into the AI system.

AI-based systems solutions have the potential to violate the privacy rights of patients in a majority of situations. The use of AI in various applications gives rise to concerns about user agreements. A contract signed without a face-to-face conversation violates the informed consent procedure [70]. Frequently, individuals fail to allocate sufficient time and consistently violate user agreements [71]. Indeed, having data protection rules in place that appropriately safeguard privacy of people and patients is crucial in the era of big data [72]. Regarding patient privacy, Voigt et al. [73] emphasized that we are living in a knowledge society that is defined by a consistently growing rate of technological advancement. Therefore, it is essential to safeguard patients' data in light of advancements in technology. Voigt and Von dem Bussche [73] emphasize the significance of referencing privacy by design and privacy by default. Privacy protection by design involves collecting just the essential data. Researchers and AI professionals need to possess a comprehensive understanding of these principles to enhance the safeguarding of patient privacy.

4.4.1 The ethical governance of artificial intelligence (AI)

Guidelines for regulating robotics is a thorough study that was produced by the European RoboLaw project. It examines both ethical and legal considerations; the ethical analysis includes rights, liability, and insurance, while the legal considerations include privacy and legal capacity [74]. The British Standards Institution Technical Subcommittee spearheaded the publication of Guide to the Ethical Design and Application of Robots and Robotic Systems in April 2016 [75]. The document covers ethical dangers, mitigation strategies, identifying possible damages, safe design, protective measures, and robot design and application information. There is a significant need for well-defined regulations and specific recommendations about the use of AI in the field of medicine. Furthermore, the guideline establishes seven essential criteria to guarantee the legality, ethics, and resilience of AI systems from both technological and societal perspectives [76]. Safety, privacy, data governance, transparency, nondiscrimination and fairness, environmental and social well-being, and ethical AI development, deployment, and usage are all significant components of medical ethics. They are essential guidelines and standards for the use of AI in medicine, and they include the ethical principles for reliable AI that were discussed. In general, the principles and guidelines have the potential to provide significant concepts and concrete strategies about ethical considerations for the governance guidelines for the next generation of AI. However, there is currently no set of rules or regulations designed with AI in healthcare and medicine. NG-AI governance requires a unique global ethical framework, governance mechanism, and updated rules and principles. It is essential to establish a separate administrative position inside the government to facilitate AI governance and understand the advantages and possible hazards for all parties involved in AI, such as governments, institutions, and people. It is of utmost importance to provide clear elucidation on the consequences that institutes and people may face in the event of contravening the relevant rules, regulations, and standards. Ethical considerations regarding the use of AI in the field of health should be examined from two different perspectives: the collective perspective of mankind as a whole, and the individual level. In the practical implementation of that idea, it is essential to establish a harmonious equilibrium among three interrelated dynamics: the interaction between science and the greater human society, the relationship between people and the broader human society, and the relationship among different persons. The objective of ethical governance is to evaluate and harmonize the relationship among these three associations. The first level of ethical governance often covers philosophical questions about the specific perspective. The subsequent two stages of governance enter into the examination of responsible conduct to validate ethical principles and make reliable decisions by the determinations made at the first level of ethical governance. NG-AI governance requires a unique global ethical framework, governance mechanism, and updated rules and principles. It is essential to establish a separate administrative position inside the government to facilitate AI governance and understand the advantages and possible hazards for all parties involved in AI, such as governments, institutions, and people. It is of the highest priority to provide clear elucidation on the consequences that institutes and people may face in the event of contravening the relevant rules, regulations, and standards. Certain technologies have significantly enhanced the capacity of people to manipulate their natural environment. However, accurately predicting the associated dangers, rewards, and long-term consequences of these technologies is a considerable challenge. Therefore, whereas investigations about certain biological interventions, such as gene editing and stem cell treatment, are now authorized in somatic cells, they are forbidden in human embryos or germ cells [77]. It is often considered to be the fundamental aspect of both biological and AI [78].

Due to present knowledge and experience, experts cannot predict the influence of deep learning in AI on the environment and humanity's future. Furthermore, the integration of AI with biological systems, often referred to as biohybrid systems or biohybrids, has garnered significant attention from several scientific groups [79]. At this level, the evaluation of ethics includes the examination of advantages and disadvantages associated with specialized procedures and products, such as AI, brain-computer interfaces, and robots. The primary emphasis in this particular situation is on their capacity to influence humans, both subjects and patients. The primary objective of ethical governance is to establish and promote ethical behavior while mitigating adverse consequences for all stakeholders involved. The concerns at this level are mostly represented as the possible harm that may arise from AI systems or products, particularly in the area of diagnosis and treatment. However, a portion of their data was dedicated to examining the ethical implications of AI in the healthcare sector. The study sample consisted of healthcare professionals, care providers, select patient groups, and experts in health informatics. The specific findings about the ethical principles of AI are presented in Table 4.2.

Table 4.2: Data related to ethical principles.

S. no.	Year	AI-based technology	Authors	Explicability	Fairness	Human autonomy	Patient privacy
1.	2019	Algorithms that predict DNN	Anderson and Anderson [80].	Black box problem	–	–	–
2.	2019	Integrated AI	Fiske et al. [81]	–	Just providing care	Respect and safeguard patient autonomy	Protection of sensitive health information
3.	2020	Medical diagnosis using machine-learning algorithms	Grote et al. [82]	Machine learning algorithm opaqueness	–	Patient independence	–
4.	2020	Model of AI prediction for readmission Model of AI for predicting mortality	Chen [83].	–	Bias dependent on the form of insurance Significant differences based on race, gender, and insurance coverage	–	–
5.	2020	AI for social benefit	Floridi et al. [84]	Explanation contextualized for the recipient and transparent goals	Justice and environmental equity	Intervention based on the recipient's autonomy	Privacy protection and consent from data subjects
6.	2020	AI decision-making system	Braun et al. [85]	Explanatory ability	Bias may result in discrimination	Informed consent	GDPR, patients as data co-managers
7.	2020	AI-based engineering	Buruk [86]	Explanatory ability	Fairness, diversity and inclusion	Autonomy founded on personal data	Privacy and data predominate

Table 4.2 (continued)

S. no.	Year	AI-based technology	Authors	Explicability	Fairness	Human autonomy	Patient privacy
8.	2020	AI and CDSS (clinical decision support system)	Petkus [87]	_	A CDSS with embedded implicit bias may result in unequal treatment	CDSS can ignore patient preferences	_
9.	2019	AI in healthcare	H. Liyanage et al. [88]	Black box problem.	_	Absence of patient participation in decision-making	_

4.4.2 Ethical governance of artificial intelligence in healthcare

Ethical governance serves as an important addition to legal systems by effectively managing and overseeing the potential risks and rewards associated with scientific research. The exponential growth of AI applications emphasizes the pressing need to establish broad regulations that take into account the ethical and social impacts of AI implementation. The ethical governance of AI in medicine and healthcare is a critical imperative as these technologies become increasingly integrated into patient care. Many significant ethical issues arise within this specific situation. Firstly, transparency and accountability are essential to ensure that decision-making processes of AI algorithms are understandable and traceable by healthcare professionals and patients alike, preventing the "black box" problem. Secondly, data privacy and security must be upheld rigorously to protect sensitive patient information, and ethical data collection practices must be established to prevent bias and discrimination in AI algorithms. Additionally, issues surrounding informed consent, especially when AI is involved in diagnostics and treatment recommendations, demand careful attention. Healthcare professionals must be adequately trained in AI ethics to make informed decisions that prioritize patient welfare. Lastly, legislation and standards should be regularly updated and harmonized across countries to guarantee that AI applications satisfy the highest ethical standards and support patient well-being. Striking a balance between harnessing the potential benefits of AI in healthcare and safeguarding against ethical pitfalls is essential for realizing the full potential of these technologies while maintaining trust and ethical integrity in the medical field.

Research on patient privacy is prevalent in the literature [81, 84–86, 89–93]. The effective development of AI tools relies on the agreement of users to contribute health data since AI-based technology requires training data to generate outputs. Several researches have examined the concept of fairness [91, 94]. The conducted investigations included an examination of several AI-based technologies, such as AI-driven clinical decision support systems (CDSS), AI prediction models, embodied AI, and ML algorithms in medical diagnosis. According to these analyses, discrimination, lack of equality, lack of diversity inclusion, and lack of fair provision of care may result from presumptions during the algorithm creation process. The concept of maintaining patient privacy emerged as an often-explored subject in the literature. The presence of unconscious bias within CDSS has the potential to result in inequitable treatment of patients [87]. This bias can appear in unfair predictions that are influenced by variables such as race, gender, and the specific kind of health insurance coverage held by the patient.

In a retrospective cohort research conducted by the authors [83], two AI prediction models were retrospectively examined in two separate cohort data sets, one from a psychiatric unit and the other from an intensive care unit (ICU). The objective of the investigation was to examine the potential emergence of ethical difficulties in the use of these models. The findings of their research demonstrated that the use of an AI model for forecasting the readmission rate inside a psychiatric unit resulted in notable bias. This bias was mostly attributed to the incorporation of the patient's insurance policy type as a feature within the predictive model. In the same way, the use of an alternative AI model in forecasting ICU mortality rates revealed a notable disparity when including variables such as race, gender, and patient insurance policy type into the predictive model. The study provides evidence of how some unanticipated variables that arise during the developmental stage of an AI system might result in discriminatory outcomes. The proposition was made that the use of algorithms designed to detect and address discrimination might have positive effects in mitigating biases within the healthcare sector [94]. The concept of preventing harm was the subject in the study that received the least amount of attention [81, 82, 84]. The representation of these entities was conducted via the use of a quantitative approach, whereby the number of scholarly investigations that referred to them within the scoping review was measured. Upon examining the use of AI in the healthcare sector, it is evident from the literature that the ethical considerations of justice, preservation of human autonomy, explicability, and patient privacy were consistently investigated. The study included a sample of healthcare professionals, care providers, and select groups of patients. Upon examination of the available data about stakeholder perspectives on the ethical implications of AI implementation, it is significant that it specifically addresses the viewpoints of patients about the ethical concerns surrounding the use of AI in the healthcare sector [90,95]. The paper in question pertained to a modified scoping review that specifically aimed to uncover the available evidence concerning the ethical considerations of AI about individuals with disabilities. There is a significant lack of representation of vulnerable groups in discussions around AI and ethics [95].

4.5 Advantages of AI in the healthcare field

The field of public health is actively working toward achieving optimal health outcomes through the implementation of standards that can modify the diagnosis and treatment of diseases. The integration of AI has had significant and major effects on the practice of medicine, leading to transformative changes in both patient experiences and physicians' daily routines. The application of AI in healthcare has yielded numerous positive outcomes, as demonstrated by various studies [96–100]. Significantly, the use of AI tools and systems has resulted in faster and more accurate diagnoses [96, 97]. For example, intelligent computer programs have been developed to diagnose skin cancer [98], while recently developed algorithms have shown promise in assisting physicians in making improved diagnostics, particularly in the interpretation of mammograms [99, 100]. as well as other medical issues. Additionally, analysis indicates that the use of AI applications is enhancing the capacity of this technology to improve the accuracy of diagnosis and treatment. AI plays a crucial role in enhancing the advantages of health-related activities, particularly in the analysis of unstructured or semi-structured texts derived from e-health reports, online media, and publications. This application of AI enables the evaluation of the impact of suggestions and activities on public health outcomes [101]. Several AI-based models could help in the interpretation of data. The radiographs were obtained by Majkowska et al. [102]. Digital pathology is a relatively new and significant innovation, while radiology imaging was the pioneering field in the use of digital data [103]. According to Bukowski et al. [104], the majority of hospitals have implemented digitalization of their health records as a means of data storage. In the field of healthcare, AI applications have provided valuable assistance to medical professionals, particularly in the area of health information systems. These applications proved useful for tasks such as monitoring and tracking diseases and conditions, offering predictive support for modeling, helping in decision-making processes, and enhancing medical imaging capabilities. Therefore, it can be suggested that the integration of AI in healthcare would provide advantageous outcomes. This is mostly due to the automation of diagnostic processes, which will improve procedures and enhance convenience for patients. AI has the potential to significantly advance the healthcare industry. it has the potential to revolutionize the healthcare industry by establishing a highly effective ecosystem. The application of AI would enable the early detection of illness risks, hence facilitating timely intervention and perhaps saving lives at an earlier stage. However, it is useful to determine if the ideals of developers and practitioners in the field of AI for healthcare applications consistently connect with the values held by physicians.

The application of AI into the public health sector presents a wide range of advantages, including early diagnosis of diseases, improved allocation of resources, personalized healthcare services, and advancements in medical research. These benefits have the potential to improve public health outcomes, decrease healthcare costs, and promote a healthcare system that is both sustainable and efficient. However, it is of greatest significance to acknowledge and address ethical, privacy, and equity concerns to ensure the responsible and adequate use of AI technology for the benefit of all persons and communities.

4.5.1 Enhanced disease diagnosis and early prevention

The use of AI enables healthcare organizations to effectively monitor and analyze a wide range of data sources in real-time, hence enhancing the detection of diseases and facilitating prevention. It includes data provided by various healthcare facilities such as hospitals and clinics, as well as data derived from social media platforms and wearable health devices. AI can recognize irregular patterns and trends, enabling the early identification of possible disease outbreaks, and surpassing the effectiveness of conventional methodologies. During the COVID-19 pandemic, AI algorithms were used to evaluate data related to symptoms that were reported on social media platforms. It enabled the identification of early indicators that pointed toward the emergence of hotspots. The prompt discovery of diseases allows for immediate action by public health authorities, including the implementation of containment strategies, contact tracing efforts, and the allocation of resources. Ultimately, these measures help in effectively limiting the transmission of diseases. The ability of AI to analyze a wide range of data sources is crucial in the timely identification of infectious illnesses. Through the identification of correlations and abnormalities within the collection of data, AI can provide healthcare authorities with timely warning signals. For example, in the case of an atypical increase in the frequency of respiratory symptoms being mentioned on social media within a certain geographical area, AI can identify this as a possible outbreak of an illness. The use of a proactive strategy has the potential to effectively mitigate the transmission of illnesses and ultimately contribute to the preservation of human life [105].

4.5.2 Data-driven decision-making and resource optimization

The healthcare sector has seen a significant increase in the use of AI applications to create new possibilities and resolve existing problems through improved data processing and analysis [106]. The data analysis capabilities of AI play a crucial role in enhancing the efficiency of resource allocation in the public health sector. Predictive analytics models can anticipate patient admissions, thus enabling hospitals to effectively distribute personnel and resources to fulfill the anticipated demand. In addition, algorithms driven by AI can enhance supply chain logistics related to medical equipment, pharmaceuticals, and vaccinations. This optimization process minimizes unnecessary loss and ensures rapid accessibility of these resources, particularly in critical situations. This leads to financial benefits and enhanced efficiency in the provision of healthcare services. Furthermore, AI plays a crucial role in facilitating the effective distribution of field healthcare personnel in regions characterized by a high frequency of diseases, thus guaranteeing the optimal allocation of resources to places that need them the most. Artificial intelligence's ability to evaluate large and complicated datasets is extremely useful for management in the public health sector. Electronic health records (EHR) and other clinical decision-support technologies are now being used to reduce the number of adverse events and mistakes in the healthcare system [107]. In addition, the adoption of scalable digital health interventions can greatly improve the efficiency, efficacy, accessibility, security, and individualization of healthcare [108]. AI can help public health professionals better understand disease patterns, population health, and resource allocation. For example, AI can discover regions with high incidence rates of certain diseases or health conditions by collecting and analyzing EHRs from various healthcare facilities. This data-driven strategy permits policymakers to make informed decisions regarding the allocation of resources, such as medical personnel, apparatus, and supplies.

AI can also identify high-risk populations like chronic disease patients for targeted therapies and healthcare promotion. These data-driven inferences improve the accuracy and efficiency of public health interventions, leading to better health outcomes and more effective use of available resources. Allocating limited resources effectively is a major problem in the healthcare system. AI is essential for this type of optimization. With the use of AI-powered predictive analytics models, medical centers may better prepare for the arrival of new patients by allocating personnel, beds, and other resources. A study of real-world AI applications in healthcare showed that hospitals in developed countries have adopted AI to improve patient diagnosis and treatment, as well as nursing and management effectiveness [109]. AI-based predictive analysis is gaining popularity in healthcare, particularly in decision-making, due to its ability to manage large amounts of heterogeneous data and predict the future [110, 111]. AI-based computational decision-making can be very useful in revealing hidden interactions or abnormalities that are not noticeable to humans when analyzing complex healthcare data in a short time with great precision [112].

4.5.3 Personalized healthcare and research advancements

AI possesses the capacity to significantly transform personalized healthcare through the customization of treatment strategies for specific patients. Machine learning algorithms provide the capability of examining extensive quantities of patient data to determine the most effective treatments and interventions for certain healthcare diseases. This intervention not only enhances patient outcomes but also reduces the use of unnecessary medications and procedures, resulting in decreased healthcare costs and reduced environmental impacts. In addition, AI promotes the process of drug discovery and development through the analysis of molecular data, resulting in a substantial reduction in the duration needed to introduce novel pharmaceuticals to the market. The aforementioned progress holds significant implications in the field of public health, as it facilitates the availability of more efficacious treatments and therapies for a wide range of disorders [113].

4.6 Sustainable development of AI in healthcare

The sustained development of AI in the field of healthcare includes various aspects, one of which involves the implementation of triage algorithms within emergency department settings [114]. The advancement of assessment algorithms has the potential to establish a standardized system for delivering suitable medical interventions, hence facilitating the objectives of healthcare. The system has the potential to enhance fairness in wait times, aligning with the biological principle of justice. Triage algorithms utilized in emergency departments have the potential to be calibrated in a manner that optimizes the allocation of resources, prioritizing procedures with the greatest therapeutic benefit. This approach effectively preserves resources for procedures that have a greater probability of success. Additionally, algorithms are developed by human programmers that possess intrinsic fallibility [115]. If the programmer of an AI fails to account for regional or statewide inequalities in the kind of emergency department situations that are most common in a certain area (for example, frostbite or heatstroke), there is a possibility that errors will be produced [116]. The impact of bias, specifically unconscious bias, such as sexism, racism, heterosexism, and similar forms of bias, might potentially shape the code that individuals create [117]. In specific situations, individuals may lack awareness of their own biases, leading them to create programs that accidentally propagate their insensitivities. In cases of "deep learning," wherein AI assimilates data from its processes and adapts appropriately, the presence of biased inputs during the initial stages of the algorithm has the potential to influence the entire sequence, producing biased outputs [118].

The effectiveness of an AI system, even if it is precisely constructed, is dependent upon human implementation, particularly about health, justice, and sustainability. This phenomenon presents benefits as well as drawbacks. Healthcare practitioners have a moral responsibility for their interpretation and application of AI algorithms [119]. However, the ineffective or abnormal execution of AI algorithms has the potential to cause greater harm, appearing in various ways such as medical errors, the development of societal inequalities, or the generation of excessive medical waste [120]. Therefore, the integration of sustainable development principles and the responsible utilization of AI in the field of healthcare occurs frequently. The sustainable utilization of AI in the field of healthcare encompasses various practices. One such practice is the examination of extensive textual data to identify new outbreaks characterized by unique symptom patterns.

Additionally, AI can be employed to recognize and analyze infection patterns within healthcare facilities [121, 122]. The utilization of predictive analytics has the potential to reduce the occurrence of epidemics, thus enhancing the overall well-being of people. The use of current AI technology has the potential to utilize data regarding infection patterns to provide prompt therapies, thus ensuring equitable access to timely healthcare for patients. The analysis of symptom patterns has the potential to reveal previously undiagnosed diseases and prevent the environmental effect associated with delayed medical care. This, in turn, contributes to the conservation of resources. The detection of infection patterns may not serve as reliable indicators of real infection. Due to its dependence on monitoring infected individuals, rich data text is a reactive rather than proactive approach. The presence of both false positives and false negatives has the potential to undermine the reliability of a given outcome [123]. To ensure that the infected individual receives treatment promptly, it is vital to have local medical facilities that are not only supplied with the appropriate types of medical therapy but also have staff that are trained to provide care. Despite the presence of these established mechanisms and the availability of rich text data to locate a local clinic offering suitable care, it cannot be assumed that an unwell individual would be able to physically reach the clinic or afford the expenses associated with it, due to limited mobility or financial limitations [124].

4.7 AI for ambitious climate-action goals (SDG 13)

Sustainable Development Goal 13 (SDG 13) requires significant climate change mitigation and adaptation, and AI has quickly emerged as a potent instrument for achieving this goal. Use of enormous databases to better understand climate change developments and their implications, plays a crucial role in climate modeling, prediction, and mitigation methods. Technologies powered by AI improve the accuracy of weather forecasts, reduce energy waste, and open the way for the expansion of renewable energy. In addition, by analyzing satellite data and automating management processes, AI helps keep an eye on and care for natural resources like forests and seas. By offering early warning systems for natural catastrophes and facilitating the effective allocation of resources during emergencies, this technology also contributes to climate adaptation initiatives. In conclusion, AI allows governments, businesses, and citizens to take the initiative in the fight against climate change and toward the ambitious objectives of SDG 13. The evidence base and analysis on the contribution of AI to SDG13 on climate action and other more general goals connected to the climate problem, as well as to the fulfilment of the Agreement of Paris, will be expanded in this area. Figure 4.3 provides a review of how AI may help in the achievement of SDG 13.

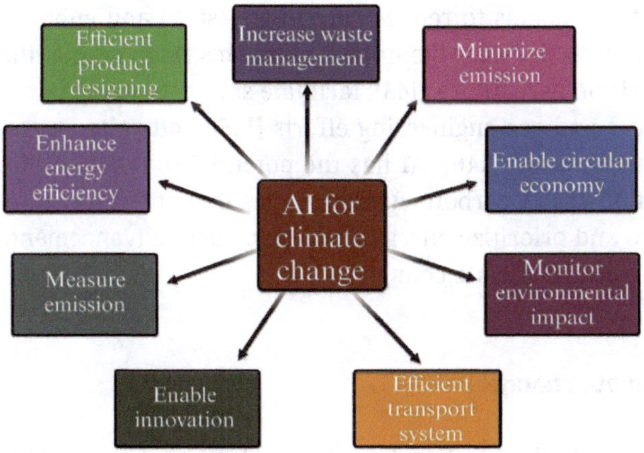

Figure 4.3: AI applications that may assist in achieving SDG 13 (climate action).

4.7.1 The possible function of AI in climate change mitigation

Climate change may hinder the accomplishment of 72 SDG objectives, including those for healthy and sustainable societies [125, 126]. The frequency and severity of natural disasters such as hurricanes, droughts, fires, and floods have all increased [127]. Global ecosystems, which include agricultural products and the natural materials that people depend on, are unstable. According to the 2018 Intergovernmental Panel on Climate Change assessment, unless global greenhouse gas (GHG) emissions are reduced over the next 30 years, the world will face catastrophic repercussions [128]. However, these pollutants continue increasing year after year. The comprehensive approach to climate change encompasses both mitigation, which involves the reduction of emissions, and adaptation, which involves preparing for the unavoidable effects.

Both have several complex dimensions. The reduction of GHG emissions involves enhancements in several sectors, including electrical systems, transportation, buildings, industry, and land use. To effectively address resistance and disaster management, it is essential to engage in strategic planning for adaptation, which includes an extensive understanding of climate patterns and the occurrence of severe events. The use of (AI holds promise in developing worldwide activities to minimize GHG emissions and facilitate appropriate planning [129, 130]. AI developments will probably help in understanding and modeling the effects of climate change. AI has proven to be advantageous in the implementation of many strategies aimed at mitigating climate change. For example, AI has the potential to analyze patterns and process data related to temperature changes and carbon emissions [131, 132]. It can also be used to forecast extreme weather events resulting from climate change [133], identify the effects of climate on human health [134], understand energy requirements and manage energy consumption [135, 136], monitor the impact of climate change on biodiversity

[137, 138], optimize transportation systems to reduce carbon emissions and enhance energy efficiency and routing [139–141], assess the impact on oceans [142], predict the consequences for enabling precision agriculture [143], facilitate smart recycling initiatives [144], aid in carbon capture and geo-engineering efforts [145], and raise awareness about the impact of climate change [146]. AI has the potential to facilitate the development and implementation of low-carbon energy systems that effectively integrate renewable energy sources and prioritize energy efficiency. These advancements are crucial in mitigating the impacts of climate change.

4.7.2 AI for understanding climate change

Climate change poses significant challenges to a broad range of social areas. This needs significant adaptation to address future weather changes. AI has experienced significant advancements, leading to significant developments in several study regions. Additionally, AI has been suggested as a potential tool for assisting in climate studies [147, 148]. The integration of AI into the Earth System Model (ESM) enables the identification of climatic associations, thus enhancing the accuracy of weather predictions, particularly for extreme weather events. While the creation of ESMs is of primary significance, it is recommended that equal attention be given to the integration of AI techniques to enhance the understanding of current models and simulations [149]. The progression of AI will contribute to the comprehension of climate change and facilitate the development of models to assess its potential consequences. Hence, the provision of assistance to enhance the ability to adapt to climate change is advocated [150]. AI methods are employed to analyze large quantities of unorganized and diverse data. These techniques are capable of uncovering and extracting intricate and advanced relationships within the data, without the need for an explicit analytical model of these relationships. This aids in comprehending climate anomalies [151, 152] and [153].

AI is improving our understanding of the effects of climate change on biomass [154], hydrology [155], severe occurrences such as droughts [156], and support for mitigation actions [157, 158]. The utilization of AI methods for extracting significant patterns and datasets from the continuously growing influx of data is crucial for addressing the issues associated with weather forecasting, climate monitoring, and long-term prediction. Various AI approaches are used to detect inter-seasonal interactions, therefore establishing possible links between climate-induced risks. These techniques are valuable in facilitating adaptation planning, such as determining optimal timing for crop planting and implementing measures to minimize disasters. Recent advancements in technology, such as the utilization of drones and the integration of AI in the Internet of Things (IoT), have contributed to the enhancement of efficiency in many systems. These improvements have provided opportunities to expand the range of missions by delivering improved spatial and temporal resolution. One recent development in the field involves the utilization of hybrid frameworks that incorporate deep learning techniques to classify photographs to assist in the management of natural catastrophes, such as avalanches, cyclones, and fires [159, 160]. AI also assumes a crucial part in many approaches aimed at addressing the climate catastrophe, mostly centered on the mitigation of current emissions. Recent research has emphasized the significance of employing AI to reduce environmental emissions generated by companies and urban areas [81, 88]. Additionally, AI has been shown to promote the concept of a circular economy [161, 162].

4.7.3 Use of AI to deal with climate change

AI has the potential to decrease expenses, increase efficiency, improve resource utilization, and optimize the delivery of public services [163]. AI has been suggested as an accelerator for ambitious policy initiatives aimed at managing climate change. A typical suggestion involves utilizing AI for the efficient administration of personal carbon allowances [164]. However, it is important to acknowledge the existence of risks and disadvantages connected with AI to effectively mitigate any potential negative consequences in the short or long term [165, 166]. AI has the potential to exert a substantial influence on the worldwide energy demand. The utilization of AI technologies, research, and product creation sometimes necessitates substantial computational resources, which can only be accessed via advanced computing centers. Recent research has examined the energy consumption and emissions linked to the training and development of AI models, demonstrating the larger effects of this fast advancement [167].

There is now increasing evidence of the significant climatic effect associated with the development of AI [168]. The carbon footprints mostly originate from the rapid advancement and instruction of AI algorithms, with less regard for their broader consequences on the Earth system. According to certain estimations, it is projected that the collective power consumption attributed to information and communication technology (ICT) may increase to almost 20% of the worldwide electricity demand by the year 2030, in contrast to the current level of approximately 1% [169]. The role of AI is expected to experience gradual growth due to the growing volume of data derived from various sources. AI is expected to have a significant impact on the present conversation surrounding the creation of green, low-carbon power and the optimization of energy storage strategies. Numerous efforts have been made to mitigate the carbon emissions associated with data centers through the implementation of energy-efficient infrastructure and the use of renewable energy sources [169]. Given the present condition of the wide range of dependencies on AI and its associated services (such as data collection and storage, hardware requirements, and global shipments, as well as the training of AI/machine learning models), there remains a lack of certainty in accurately determining the complete carbon footprint of AI. With the rising need for AI infrastructure comes the question of whether or not the environmental benefits of AI's efficiency improvements can be achieved on a global scale.

4.8 Conclusion

While technological advancements benefit society and the environment, abuse may lead to destruction of humanity. The potential for AI to revolutionize the delivery of healthcare services is becoming increasingly apparent today. Currently, there have been several advancements in AI technology that have significantly revolutionized and restructured the provision and implementation of healthcare services. The clinical and operational effectiveness of healthcare will improve as a result of the continued and successful application of AI in this sector. The accuracy of AI forecasts will be maintained and improved by ongoing investments in the study and development of AI algorithms and machine learning, which will also lead to optimum levels of data privacy usage and protection and network security. Therefore, it is vital to regulate its expansion for optimal exploitation of these developments. When AI is used in healthcare, the effects of ethical issues can be lessened, but they cannot be completely removed. The utilization of AI in healthcare is associated with advantages and disadvantages, as well as ethical considerations, rendering it impossible to completely eradicate the ethical issues related to this technology. However, constant monitoring can reduce the impact of these moral concerns to an acceptable level. Ethical problems will be easier to overcome if there is a healthy balance between AI and human activity. AI should only be used for complicated jobs and diagnosis, not for everything. AI in the field of healthcare is limited to the information it has been designed to analyze, whereas an experienced medical practitioner possesses the ability to observe and diagnose conditions that may not be immediately observable.

This study aimed to review current evidence on AI ethics, identify ethical issues in healthcare, identify gaps, and propose steps toward an evidence-based approach to ethical issues. It also aimed to collect evidence related to the ethical principles emphasized in the Ethics Guidelines for trustworthy AI; additionally, gather data on practical methods for adhering to these ethical standards and obtain perspectives from various stakeholders. AI has extensive use in various fields of medical research and clinical practice. In addition to its inherent advantages and prospective applications in the healthcare sector, AI gives rise to different ethical concerns and presents difficulties to ethical governance. These concerns are especially prevalent in emerging AI domains, where there is higher potential for harmful effects on the environment and implications for the future of humanity. The integration of AI is imperative within the field of healthcare, particularly in the administration and supervision of health services. This is mostly due to its ability to facilitate medical decision-making, particularly in the area of predictive analysis, as it pertains to the diagnosis and treatment of patients. The challenges associated with the adoption of AI in the public health sector include the facilitation of early adoption, ensuring sustainable implementation within the health system, and the lack of consideration for the user's perspective. Additionally, it is observed that technology is not being efficiently utilized, despite its necessity for the successful adoption of AI in the public health sector. Ethical support for medical treatment can be achieved via the utilization of the criteria of health, justice, and resource conservation. Also, the integration of sustainability into healthcare is actualized.

References

[1] Higgins D, Madai V. From bit to bedside: A practical framework for artificial intelligence product development in healthcare. Advanced Intelligent Systems. 2020 Oct;2(10):2000052.

[2] Tobore I, Li J, Yuhang L, et al. Deep learning intervention for health care challenges: Some biomedical domain considerations. JMIR mHealth uHealth. 2019;7:e11966,.

[3] Pee LG, Pan SL, Cui L. Artificial intelligence in healthcare robots: A social informatics study of knowledge embodiment. Journal of the Association for Information Science and Technology. 2019;70:351–69.

[4] Coeckelbergh M. Health care, capabilities, and AI assistive technologies. Ethical Theory and Moral Practice. 2010;13:181–90.

[5] Ilan Y. Improving global healthcare and reducing costs using second-generation artificial intelligence-based digital pills: A market disruptor. International Journal of Environmental Research and Public Health. 2021;18(2):811.

[6] Vrontis D, Christofi M, Pereira V, Tarba S, Makrides A, Trichina E. Artificial intelligence, robotics, advanced technologies and human resource management: A systematic review. The International Journal of Human Resource Management. 2022;33(6):1237–66.

[7] Jiang L, Wu Z, Xu X, Zhan Y, Jin X, Wang L, Qiu Y. Opportunities and challenges of artificial intelligence in the medical field: Current application, emerging problems, and problem-solving strategies. Journal of International Medical Research. 2021;49(3):03000605211000157.

[8] Obermeyer Z, Topol EJ. Artificial intelligence, bias, and patients' perspectives. The Lancet. 2021;397(10289):2038.

[9] Bates DW, Levine D, Syrowatka A, Kuznetsova M, Craig KJT, Rui A, Jackson GP, Rhee K. The potential of artificial intelligence to improve patient safety: A scoping review. NPJ Digital Medicine. 2021;4(1):1–8.

[10] Morley J, Machado CC, Burr C, Cowls J, Joshi I, Taddeo M, Floridi L. The ethics of AI in health care: A mapping review. Social Science & Medicine. 2020;260:113172.

[11] Rajpurkar P, Chen E, Banerjee O, Topol EJ. AI in health and medicine. Nature Medicine. 2022;28(1):31–38.

[12] Wang F, Preininger A. AI in health: State of the art, challenges, and future directions. Yearbook of Medical Informatics. 2019;28(01):016–026.

[13] Panch T, Szolovits P, Atun R. Artificial intelligence, machine learning and health systems. Journal of Global Health. 2018;8(2).

[14] Kaushik S, Choudhury A, Sheron PK, Dasgupta N, Natarajan S, Pickett LA, Dutt V. AI in healthcare: Time-series forecasting using statistical, neural, and ensemble architectures. Frontiers in Big Data. 2020;3:4.

[15] Halawa F, Madathil SC, Gittler A, Khasawneh MT. Advancing evidence-based healthcare facility design: A systematic literature review. Health Care Management Science. 2020;23(3):453–80.

[16] Glaize A, Duenas A, Di Martinelly C, Fagnot I. Healthcare decision-making applications using multicriteria decision analysis: A scoping review. Journal of Multi-Criteria Decision Analysis. 2019;26(1–2):62–83.
[17] Mohapatra S, Swarnkar T. Artificial intelligence for smart healthcare management: Brief study. Intelligent and Cloud Computing. 2021:365–3.
[18] Gupta S, Kar AK, Baabdullah A, Al-Khowaiter WAA. Big data with cognitive computing: A review for the future. International Journal of Information Management. 2018;42:78–89.
[19] Dogru AK, Keskin BB. Ai in operations management: Applications, challenges and opportunities. Journal of Data, Information and Management. 2020;2(2):67–74.
[20] Kar AK, Dwivedi YK. Theory building with big data-driven research–moving away from the æwhatg towards the æwhyg. International Journal of Information Management. 2020;54:102205.
[21] Whicher D, Rapp T. The value of artificial intelligence for healthcare decision making lessons learned. Value in Health. 2022;25(3):328–30.
[22] Yadav H, Kar AK, Kashiramka S. How does entrepreneurial orientation and SDG orientation of CEOs evolve before and during a pandemic. Journal of Enterprise Information Management. 2022 Feb 18;35(1):160–78.
[23] Pencheon D, Rissel CE, Hadfield G, Madden DL. Health sector leadership in mitigating climate change: Experience from the UK and NSW. New South Wales Public Health Bulletin. 2010 Feb 4;20(12):173–76.
[24] Eckelman MJ, Sherman J. Environmental impacts of the U.S. health care system and effects on public health. PLoS One. 2016;1(6):0157014.
[25] Mill E, Garn W, Ryman-Tubb N. Managing sustainability tensions in artificial intelligence: Insights from paradox theory. In: Proceedings of the 2022 AAAI/ACM Conference on AI, Ethics, and Society. 2022 Jul 26 (pp. 491–98).
[26] Farjana SH, Huda N, Parvez Mahmud MA. Life cycle assessment of cobalt extraction process. Journal of Sustainable Mining. 2019;18:150–61.
[27] AlAhmad YM, Haggeer DM, Alsayed AY, Haik MY, AbuAfifeh LM, Aljaber MH, Mohamed AA, Balideh M, Almutawa N, Mahmoud MH. The effect of telemedicine on patients' compliance in family medicine follow-ups in Qatar. Avicenna. 2022;2021(1):3.
[28] Dignum V. The role and challenges of education for responsible AI. London Review of Education. 2021 Jan 13;19(1):1–1.
[29] Barman M, Hussain T, Abuswiril H, Illahi MN, Sharif M, Saman HT, Hassan M, Gaafar M, Abu J, Ahmad MKK, Kamar MA. Embracing healthcare delivery challenges during a pandemic. Review from a nodal designated COVID-19 center in Qatar. Avicenna. 2021;2:8.
[30] Dilsizian SE, Siegel EL. Artificial intelligence in medicine and cardiac imaging: Harnessing big data and advanced computing to provide personalized medical diagnosis and treatment. Current Cardiology Reports. 2014;16(1):1–8.
[31] Kooli C. COVID-19: Public health issues and ethical dilemmas. Ethics, Medicine and Public Health. 2021;17:100635, ISSN 2352-5525.
[32] Zandi D, Reis A, Vayena E, Goodman K. New ethical challenges of digital technologies, machine learning, and artificial intelligence in public health: A call for papers. Bulletin of the World Health Organization. 2019;97(1):2–2.
[33] Shiraishi J, Li Q, Appelbaum D, Doi K. Computer-aided diagnosis and artificial intelligence in clinical imaging. Seminars in Nuclear Medicine, WB Saunders. 2011;41(6):449–62.
[34] Jiang F, Jiang Y, Zhi H, Dong Y, Li H, Ma S, Wang Y, Dong Q, Shen H, Wang Y. Artificial intelligence in healthcare: Past, present and future. Stroke and Vascular Neurology. 2017;2:4.
[35] Davenport T, Kalakota R. The potential for artificial intelligence in healthcare. Future Healthcare Journal. 2019;6(2):94.

[36] McCall B. COVID-19 and artificial intelligence: Protecting healthcare workers and curbing the spread. The Lancet Digital Health. 2020;2(4):e166–e167.
[37] Rong G, Mendez A, Assi EB, Zhao B, Sawan M. Artificial intelligence in healthcare: Review and prediction case studies. Engineering. 2020;6(3):291–301.
[38] Hadley TD, Pettit RW, Malik T, Khoei AA, Salihu HM. Artificial intelligence in global health – A framework and strategy for adoption and sustainability. International Journal of Maternal and Child Health and AIDS. 2020;9:121–7.
[39] Nakata N. Recent technical development of artificial intelligence for diagnostic medical imaging. Japanese Journal of Radiology. 2019;37:103–08.
[40] Salmon JW, Thompson SL, Salmon JW, Thompson SL. Big data: Information technology as control over the profession of medicine. The Corporatization of American Health Care: The Rise of Corporate Hegemony and the Loss of Professional Autonomy. 2021:181–254.
[41] Sadasivam RS, Borglund EM, Adams R, Marlin BM, Houston TK. Impact of a collective intelligence tailored messaging system on smoking cessation: The perspect randomized experiment. Journal of Med Internet Research. 2016;18:e285.
[42] Judson TJ, Odisho AY, Young JJ, Bigazzi O, Steuer D, Gonzales R, Neinstein AB. Implementation of a digital chatbot to screen health system employees during the COVID-19 pandemic. Journal of the American Medical Informatics Association. 2020 Sep;27(9):1450–55.
[43] Semantha FH, Azam S, Yeo KC, Shanmugam B. A systematic literature review on privacy by design in the healthcare sector. Electronics. 2020;9:452.
[44] Calancea CG, Alboaie L, Panu A. A SwarmESB-based architecture for an European healthcare insurance system in compliance with GDPR. Parallel and Distributed Computing, Applications and Technologies. 2019:261–67.
[45] Wolff J, Pauling J, Keck A, Baumbach J. Success factors of artificial intelligence implementation in healthcare. Frontiers in Digital Health. 2021;3:594971.
[46] Asan O, Bayrak AE, Choudhury A. Artificial intelligence and human trust in healthcare: Focus on clinicians. Journal of Med Internet Research. 2020 Jun 19;22(6).
[47] Veronese A, Silveira A, Lemos AN. Artificial intelligence, Digital Single Market and the proposal of a right to fair and reasonable inferences: A legal issue between ethics and techniques. UNIO–EU Law Journal. 2019 Jul 2;5(2):75–91.
[48] Chen H, Li L, Chen Y. Explore success factors that impact artificial intelligence adoption on telecom industry in China. Journal of Management Analytics. 2021 Jan 2;8(1):36–68.
[49] Kularatna S, Hettiarachchi R, Senanayake S, Murphy C, Donovan C, March S. Cost-effectiveness analysis of paediatric mental health interventions: A systematic review of model-based economic evaluations. BMC Health Services Research. 2022 Dec;22(1):1–7.
[50] Sidhom O. Physical and mental health aspects in COVID-19: Two sides of a coin. Avicenna. 2021;2:6.
[51] Noorbakhsh-Sabet N, Zand R, Zhang Y, Abedi V. Artificial intelligence transforms the future of health care. The American Journal of Medicine. 2019 Jul 1;132(7):795–801.
[52] Belisle-Pipon JC, Couture V, Roy MC, Ganache I, Goetghebeur M, Cohen IG. What makes artificial intelligence exceptional in health technology assessment?. Frontiers in Artificial Intelligence. 2021;4:736697.
[53] Spatharou A, Hieronimus S, Jenkins J. Transforming Healthcare with AI: The Impact on the Workforce and Organizations. McKinsey & Company; 2020, 10.
[54] Lee D, Yoon SN. Application of artificial intelligence-based technologies in the healthcare industry: Opportunities and challenges. International Journal of Environmental Research and Public Health. 2021;18(1):271.
[55] Cook MJ, O'Brien TJ, Berkovic SF, Murphy M, Morokoff A, Fabinyi G, D'Souza W, Yerra R, Archer J, Litewka L, Hosking S, Lightfoot P, Ruedebusch V, Sheffield WD, Snyder D, Leyde K, Himes

D. Prediction of seizure likelihood with a long-term, implanted seizure advisory system in patients with drug-resistant epilepsy: A first-in-man study. The Lancet Neurology. 2013;12(6):563–71.
[56] Villar JR, Gonzalez S, Sedano J, Chira C, Trejo-Gabriel-Galan JM. Improving human activity recognition and its application in early stroke diagnosis. International Journal of Neural Systems. 2015;25(04):1450036.
[57] Somashekhar SP, Kumar R, Rauthan A, Arun KR, Patil P, Ramya YE. Abstract S6–07: double-blinded validation study to assess performance of IBM artificial intelligence platform, Watson for oncology in comparison with Manipal multidisciplinary tumour board – first study of 638 breast cancer cases, 2017.
[58] Hassabis D, Kumaran D, Summerfield C, Botvinick M. Neuroscience-inspired artificial intelligence. Neuron. 2017;95(2):245–58.
[59] Siddique S, Chow JC. Artificial intelligence in radiotherapy. Reports of Practical Oncology & Radiotherapy. 2020;25(4):656–66.
[60] Challen R, Denny J, Pitt M, Gompels L, Edwards T, Tsaneva-Atanasova K. Artificial intelligence, bias and clinical safety. BMJ Quality and Safety. 2019;28(3):231–37.
[61] Choudhury A, Asan O. Role of artificial intelligence in patient safety outcomes: Systematic literature review. JMIR Medical Informatics. 2020;8(7):e18599.
[62] Hashimoto DA, Rosman G, Rus D, Meireles OR. Artificial intelligence in surgery: Promises and perils. Annals of Surgery. 2018;268(1):70.
[63] Gerke S, Minssen T, Cohen G. Ethical and Legal Challenges of Artificial Intelligence-Driven Healthcare. In: Artificial Intelligence in Healthcare (pp. 295–336). Academic Press; 2020 Jan 1.
[64] Kantarjian H, Yu PP. Artificial intelligence, big data, and cancer. JAMA Oncology. 2015 Aug 1;1(5):573–74.
[65] Chen IY, Pierson E, Rose S, Joshi S, Ferryman K, Ghassemi M. Ethical machine learning in healthcare. Annual Review of Biomedical Data Science. 2021;4:123–44.
[66] Madaio MA, Stark L, Wortman Vaughan J, Wallach H. Co-designing checklists to understand organizational challenges and opportunities around fairness in AI. In: Proceedings of the 2020 CHI Conference on Human Factors in Computing Systems, 2020, 1–14.
[67] Landi H. Healthcare data breaches hit all-time high in 2021, impacting 45M people. Health Tech. 2022.
[68] Kooli C, Al Muftah H. Artificial intelligence in healthcare: A comprehensive review of its ethical concerns. Technological Sustainability. 2022 Mar 4;1(2):121–31.
[69] Goggin B. Inside Facebook's suicide algorithm: Here's how the company uses artificial intelligence to predict your mental state from your posts. Business Insider. 2019 Jan 18.
[70] Klugman CM, Dunn LB, Schwartz J, Cohen IG. The ethics of smart pills and self-acting devices: Autonomy, truth-telling, and trust at the dawn of digital medicine. The American Journal of Bioethics. 2018;18(9):38–47.
[71] Friedman B, Felten E, Millett LI. Informed consent online: A conceptual model and design principles. University of Washington Computer Science and Engineering Technical Report, 2000, 8.
[72] Pavlova T, Zarutska E, Pavlov R, Kolomoichenko O. Ethics and law in Kant's views: The principle of complementarity. International Journal of Ethics and Systems. 2019;35(4):651–64.
[73] Voigt P, Von Dem Bussche A. The EU General Data Protection Regulation (GDPR). A Practical Guide, 1st ed. Cham: Springer International Publishing; 2017 Aug 10;vol. 10(3152676):10–5555.
[74] Shah H, Roberts F. Not born of woman: Gendered robots. In: Gender in AI and Robotics: The Gender Challenges from an Interdisciplinary Perspective 2023 Feb 26 (pp. 107–27).
[75] Patrick L, Abney K, Bekey GA. Introduction to Robot Ethics. 3–15; 2012.
[76] Guan J. Artificial intelligence in healthcare and medicine: Promises, ethical challenges and governance. Chinese Medical Sciences Journal. 2019 Jun 1;34(2):76–83.

[77] Petrini C. Bioethics of clinical applications of stem cells. International Journal of Molecular Sciences. 2017;18(4):814.
[78] Poggio T, Shelton CR. Learning in brains and machines. Spat Views. 2000;13(2–3):287–96.
[79] Hochberg LR, Serruya MD, Friehs GM, et al. Neuronal ensemble control of prosthetic devices by a human with tetraplegia. Nature. 2006;442(7099):164–171.
[80] Anderson M, Anderson SL. How should AI be developed, validated, and implemented in patient care?. AMA Journal of Ethics. 2019 Feb 1;21(2):125–30.
[81] Fiske A, Henningsen P, Buyx A. Your robot therapist will see you now: Ethical implications of embodied artificial intelligence in psychiatry, psychology, and psychotherapy. Journal of Medical Internet Research. 2019;21:e13216.
[82] Grote T, Berens P. On the ethics of algorithmic decision-making in healthcare. Journal of Medical Ethics. 2020;46:205–11.
[83] Chen IY, Szolovits P, Ghassemi M. Can AI help reduce disparities in general medical and mental health care?. AMA Journal of Ethics. 2019;21(2):E167–179.
[84] Floridi L, Cowls J, King TC, Taddeo M. How to design AI for social good: Seven essential factors. Science and Engineering Ethics. 2020;26:1771–96.
[85] Braun M, Hummel P, Beck S, Dabrock P. Primer on an ethics of AI-based decision support systems in the clinic. Journal of Medical Ethics. 2020;47:3.
[86] Buruk B, Ekmekci PE, Arda B. A critical perspective on guidelines for responsible and trustworthy artificial intelligence. Medicine, Health Care and Philosophy. 2020;23:387.
[87] Petkus H, Hoogewerf J, Wyatt JC. What do senior physicians think about AI and clinical decision support systems: Quantitative and qualitative analysis of data from specialty societies. Clinical Medicine Journal of the Royal College of Physicians of London. 2020;20:324–28.
[88] Liyanage H, Liaw ST, Jonnagaddala J, Schreiber R, Kuziemsky C, Terry AL, et al. Artifcial intelligence in primary health care: Perceptions, issues, and challenges. Yearbook of Medical Informatics. 2019;28:41–46.
[89] Ma H, Guo X, Ping Y, Wang B, Yang Y, Zhang Z, et al. : PPCD: Privacy-preserving clinical decision with cloud support. PLoS ONE. 2019;14:e0217349.
[90] McCradden MD, Baba A, Saha A, Ahmad S, Boparai K, Fadaiefard P, et al. : Ethical concerns around use of artificial intelligence in health care research from the perspective of patients with meningioma, caregivers and health care providers: A qualitative study. C Open. 2020;8:E90–E95.
[91] Morley J, Floridi L, Kinsey L, Elhalal A. From what to how: An initial review of publicly available AI ethics tools, methods, and research to translate principles into practices. Science and Engineering Ethics. 2020;26(4):2141–68.
[92] Stanfll MH, Marc DT. Health information management: Implications of artifcial intelligence on healthcare data and information management. Yearbook of Medical Informatics. 2019;28:56–64.
[93] Yoon J, Drumright LN, Van Der Schaar M. Anonymization through data synthesis using generative adversarial networks (ADS-GAN). IEEE-EMBS International Conference on Biomedical and Health Informatics. 2020;24:1.
[94] Liaw S-T, Liyanage H, Kuziemsky C, Terry AL, Schreiber R, Jonnagaddala J, et al. Ethical use of electronic health record data and artificial intelligence: Recommendations of the primary care informatics working group of the International Medical Informatics Association. Yearbook of Medical Informatics. 2020;29:051.
[95] Lillywhite A, Wolbring G. Coverage of ethics within the artificial intelligence and machine learning academic literature: The case of disabled people. Assistive Technology. 2021;33(3):129–35.
[96] Harrison C. FDA backs clinician-free AI imaging diagnostic tools. Nature Biotechnology. 2018;36(8):673.

[97] Serag A, Ion-Margineanu A, Qureshi H, McMillan R, Saint Martin M-J, Diamond J, O'Reilly P, Hamilton P. Translational AI and deep learning in diagnostic pathology. Frontiers in Medicine. 2019;6:185.

[98] Tschandl P, Rinner C, Apalla Z, Argenziano G, Codella N, Halpern A, Janda M, Lallas A, Longo C, Malvehy J, Paoli J, Puig S, Rosendahl C, Soyer HP, Zalaudek I, Kittler H. Human–computer collaboration for skin cancer recognition. Nature Medicine. 2020;26(8):1229–34.

[99] Dembrower K, Wahlin E, Liu Y, Salim M, Smith K, Lindholm P, Strand F. Effect of artificial intelligence-based triaging of breast cancer screening mammograms on cancer detection and radiologist workload: A retrospective simulation study. The Lancet Digital Health. 2020;2(9): e468–e474.

[100] Sechopoulos I, Teuwen J, Mann R. Artificial intelligence for breast cancer detection in mammography and digital breast tomosynthesis: State of the art. Seminars in Cancer Biology. 2021;72:214–25.

[101] Tucker TC, Durbin EB, McDowell JK, Huang B. Unlocking the potential of population-based cancer registries. Cancer. 2019;125(21):3729–37.

[102] Majkowska A, Mittal S, Steiner DF, Reicher JJ, McKinney SM, Duggan GE, Eswaran K, Cameron Chen PH, Liu Y, Kalidindi SR, Ding A, Corrado GS, Tse D, Shetty S. Chest radiograph interpretation with deep learning models: Assessment with radiologist adjudicated reference standards and population-adjusted evaluation. Radiology. 2020;294(2):421–31.

[103] Dreyer KJ, Geis JR. When machines think: Radiology's next Frontier. Radiology. 2017;285(3):713–18.

[104] Bukowski M, Farkas R, Beyan O, Moll L, Hahn H, Kiessling F, Schmitz-Rode T. Implementation of eHealth and AI integrated diagnostics with multidisciplinary digitized data: Are we ready from an international perspective?. European Radiology. 2020;30(10):5510–24.

[105] Koushik CSN, Choubey SB, Choubey A. Chapter 7 – Application of Virtual Reality Systems to Psychology and Cognitive Neuroscience Research. In: GR Sinha, JS Suri (eds.), Computer Modelling, and Cognitive Science JSBT-CI (pp. 133–47). Academic Press; 2020.

[106] Kushwaha PK, Kumaresan M. Machine learning algorithm in healthcare system: A review. In: 2021 international conference on technological advancements and innovations (ICTAI) 2021, 478–81.

[107] Ozdemir Z, Barron J, Bandyopadhyay S. An analysis of the adoption of digital health records under switching costs. Information Systems Research. 2011;22(3):491–503.

[108] Murray E, Hekler EB, Andersson G, Collins LM, Doherty A, Hollis C, Rivera DE, West R, Wyatt JC. Evaluating digital health interventions: key questions and approaches, 2016.

[109] Ali O, Abdelbaki W, Shrestha A, Elbasi E, Alryalat MA, Dwivedi YK. A systematic literature review of artificial intelligence in the healthcare sector: Benefits, challenges, methodologies, and functionalities. Journal of Innovation & Knowledge. 2023 Jan 1;8(1):100333.

[110] Verma S, Sharma R, Deb S, Maitra D. Artificial intelligence in marketing: Systematic review and future research direction. International Journal of Information Management Data Insights. 2021;1(1):100002.

[111] Kushwaha AK, Kar AK, Dwivedi YK. Applications of big data in emerging management disciplines: A literature review using text mining. International Journal of Information Management Data Insights. 2021;1(2):100017.

[112] Triantafyllidis AK, Tsanas A. Applications of machine learning in real-life digital health interventions: Review of the literature. Journal of Med Internet Research. 2019;21:e12286.

[113] Bates DW, Saria S, Ohno-Machado L, Shah A, Escobar G. Big Data in health care: Using analytics to identify and manage high-risk and high-cost patients. Health Affairs. 2014;33(7):1123–31.

[114] Osoba OA, Welser W. An Intelligence in Our Image: The Risks of Bias and Errors in Artificial Intelligence. Rand Corporation; 2017.

[115] Yu K-H, Beam AL, Kohane IS. Artificial intelligence in healthcare. Nature Biomedical Engineering. 2018;2(10):719–31.

[116] Simonite T. When it comes to gorillas, google photos remains blind. Wired. 2018 Jan 11:13.
[117] Dastin J. Amazon Scraps Secret AI Recruiting Tool that Showed Bias Against Women. In: Ethic of Data and Analytics (pp. 296–99). Auerbach Publications; 2022.
[118] Shortliffe EH, Sepúlveda MJ. Clinical decision support in the era of artificial intelligence. JAMA. 2018;320(21):2199–200.
[119] Campion N, Thiel CL, Woods NC, Swanzy L, Landis AL, Bilec MM. Sustainable healthcare and environmental life-cycle impacts of disposable supplies: A focus on disposable custom packs. Journal of Cleaner Production. 2015;94:46–55.
[120] Maojo V, Martin-Sanchez F. Bioinformatics: Towards new directions for public health. Methods of Information in Medicine. 2004;43(3):208–14.
[121] Maddox TM, Rumsfeld JS, Payne PRO. Questions for artificial intelligence in health care. JAMA. 2019;321(1):31–32.
[122] Dhar V. Big data and predictive analytics in health care. Big Data. 2014 Sep 1;2(3):113–16.
[123] Bourgois P, Holmes SM, Sue K, Quesada J. Structural vulnerability: Operationalizing the concept to address health disparities in clinical care. Academic Medicine: Journal of the Association of American Medical Colleges. 2017 Mar;92(3):299.
[124] Fuso Nerini F, Sovacool B, Hughes N, Cozzi L, Cosgrave E, Howells M, Tavoni M, Tomei J, Zerriffi H, Milligan B. Connecting climate action with other Sustainable Development Goals. Nature Sustainability. 2019 Aug;2(8):674–80.
[125] Smith J. Climate change: Scientific evidence and the industry of denial. The Missouri Review. 2017;40(3):187–201.
[126] Field CB (ed.). Managing the Risks of Extreme Events and Disasters to Advance Climate Change Adaptation: Special Report of the Intergovernmental Panel on Climate Change. Cambridge University Press; 2012, 28.
[127] Allen MR, Dube OP, Solecki W, Aragón-Durand F, Cramer W, Humphreys S, Kainuma M, Kala J, Mahowald N, Mulugetta Y, Perez R. Framing and context. Global Warming Of. 2018 Aug 2;1(5).
[128] Nishant R, Kennedy M, Corbett J. Artificial intelligence for sustainability: Challenges, opportunities, and a research agenda. International Journal of Information Management. 2020 Aug 1;53:102104.
[129] Vinuesa R, Azizpour H, Leite I, Balaam M, Dignum V, Domisch S, Fell Änder A, Langhans SD, Tegmark M, Fuso Nerini F. The role of artificial intelligence in achieving the Sustainable Development Goals. Nature Communication;11(1):233.
[130] Wu Y, Sharifi A, Yang P, Borjigin H, Murakami D, Yamagata Y. Mapping building carbon emissions within local climate zones in Shanghai. Energy Procedia. 2018 Oct 1;152:815–22.
[131] Barnes EA, Hurrell JW, Ebert-Uphoff I, Anderson C, Anderson D. Viewing forced climate patterns through an AI lens. Geophysical Research Letters. 2019 Nov 28;46(22):13389–98.
[132] Feng P, Wang B, Li Liu D, Waters C, Yu Q. Incorporating machine learning with biophysical model can improve the evaluation of climate extremes impacts on wheat yield in south-eastern Australia. Agricultural and Forest Meteorology. 2019 Sep 15;275:100–13.
[133] Berrang-Ford L, Sietsma AJ, Callaghan M, Minx JC, Scheelbeek PF, Haddaway NR, Haines A, Dangour AD. Systematic mapping of global research on climate and health: A machine learning review. The Lancet Planetary Health. 2021 Aug 1;5(8):e514–25.
[134] Aslam S, Khalid A, Javaid N. Towards efficient energy management in smart grids considering microgrids with day-ahead energy forecasting. Electric Power Systems Research. 2020 May 1;182:106232.
[135] Kim TY, Cho SB. Predicting residential energy consumption using CNN-LSTM neural networks. Energy. 2019 Sep 1;182:72–81.
[136] Dujon AM, Schofield G. Importance of machine learning for enhancing ecological studies using information-rich imagery. Endangered Species Research. 2019 Jun 20;39:91–104.

[137] Kulkarni R, Di Minin E. Automated retrieval of information on threatened species from online sources using machine learning. Methods in Ecology and Evolution. 2021 Jul;12(7):1226–39.
[138] Alsrehin NO, Klaib AF, Magableh A. Intelligent transportation and control systems using data mining and machine learning techniques: A comprehensive study. IEEE Access. 2019 Apr 3;7:49830–57.
[139] Jin H, Ma H, Butala MD, Liu EX, Li EP. EMI radiation prediction and structure optimization of packages by deep learning. IEEE Access. 2019 Jul 10;7:93772–80.
[140] Milojevic-Dupont N, Creutzig F. Machine learning for geographically differentiated climate change mitigation in urban areas. Sustainable Cities and Society. 2021 Jan 1;64:102526.
[141] Lou R, Lv Z, Dang S, Su T, Li X. Application of machine learning in ocean data. Multimedia Systems. 2023 Jun;29(3):1815–24.
[142] Sharma A, Jain A, Gupta P, Chowdary V. Machine learning applications for precision agriculture: A comprehensive review. IEEE Access. 2020 Dec 31;9:4843–73.
[143] Rutqvist D, Kleyko D, Blomstedt F. An automated machine learning approach for smart waste management systems. IEEE Transactions on Industrial Informatics. 2019 May 8;16(1):384–92.
[144] Menad NA, Hemmati-Sarapardeh A, Varamesh A, Shamshirband S. Predicting solubility of CO2 in brine by advanced machine learning systems: Application to carbon capture and sequestration. Journal of CO2 Utilization. 2019 Oct 1;33:83–95.
[145] George G, Merrill RK, Schillebeeckx SJ. Digital sustainability and entrepreneurship: How digital innovations are helping tackle climate change and sustainable development. Entrepreneurship Theory and Practice. 2021 Sep;45(5):999–1027.
[146] Schneider T, Lan S, Stuart A, Teixeira J. Earth system modeling 2.0: A blueprint for models that learn from observations and targeted high-resolution simulations. Geophysical Research Letters. 2017 Dec 28;44(24):12–396.
[147] Lin Y, Chiang YY, Franklin M, Eckel SP, Ambite JL. Building autocorrelation-aware representations for fine-scale spatiotemporal prediction. In: 2020 IEEE International Conference on Data Mining (ICDM) 2020 Nov 17 (pp. 352–61).
[148] Huntingford C, Jeffers ES, Bonsall MB, Christensen HM, Lees T, Yang H. Machine learning and artificial intelligence to aid climate change research and preparedness. Environmental Research Letters. 2019 Nov 22;14(12):124007.
[149] Tripathi S, Srinivas VV, Nanjundiah RS. Downscaling of precipitation for climate change scenarios: A support vector machine approach. Journal of Hydrology. 2006 Nov 15;330(3–4).
[150] Dewitte S, Cornelis JP, Müller R, Munteanu A. Artificial intelligence revolutionises weather forecast, climate monitoring and decadal prediction. Remote Sensing. 2021 Aug 13;13(16):3209.
[151] Carlsen H, Nykvist B, Joshi S, Heintz F. Chasing artificial intelligence in shared socioeconomic pathways. One Earth. 2024 Jan 19;7(1):18–22.
[152] Yang H, Piao S, Huntingford C, Peng S, Ciais P, Chen A, Zhou G, Wang X, Gao M, Zscheischler J. Strong but intermittent spatial covariations in tropical land temperature. Geophysical Research Letters. 2019 Jan 16;46(1):356–64.
[153] Wu C, Chen Y, Peng C, Li Z, Hong X. Modeling and estimating aboveground biomass of Dacrydium pierrei in China using machine learning with climate change. Journal of Environmental Management. 2019 Mar 15;234:167–79.
[154] Goyal MK, Bharti B, Quilty J, Adamowski J, Pandey A. Modeling of daily pan evaporation in sub tropical climates using ANN, LS-SVR, Fuzzy Logic, and ANFIS. Expert Systems with Applications. 2014 Sep 1;41(11):5267–76.
[155] Yang Y, Guan H, Batelaan O, McVicar TR, Long D, Piao S, Liang W, Liu B, Jin Z, Simmons CT. Contrasting responses of water use efficiency to drought across global terrestrial ecosystems. Scientific Reports. 2016 Mar 17;6(1):23284.

[156] Buckland C, Bailey R, Thomas D. Using artificial neural networks to predict future dryland responses to human and climate disturbances. Scientific Reports. 2019;9(1):1–13.
[157] Ghiggi G, Humphrey V, Seneviratne SI, Gudmundsson L. GRUN: An observation-based global gridded runoff dataset from 1902 to 2014. Earth System Science Data. 2019 Nov 13;11(4):1655–74.
[158] Hernández D, Cano JC, Silla F, Calafate CT, Cecilia JM. AI-enabled autonomous drones for fast climate change crisis assessment. IEEE Internet of Things Journal. 2021 Jul 19;9(10):7286–97.
[159] Nijhawan R, Rishi M, Tiwari A, Dua R. A novel deep learning framework approach for natural calamities detection. In: Information and Communication Technology for Competitive Strategies: Proceedings of Third International Conference on ICTCS 2017 2019 (pp. 561–69).
[160] Bag S, Pretorius JH, Gupta S, Dwivedi YK. Role of institutional pressures and resources in the adoption of big data analytics powered artificial intelligence, sustainable manufacturing practices and circular economy capabilities. Technological Forecasting and Social Change. 2021 Feb 1;163:120420.
[161] Wilson M, Paschen J, Pitt L. The circular economy meets artificial intelligence (AI): Understanding the opportunities of AI for reverse logistics. Management of Environmental Quality: An International Journal. 2022 Jan 11;33(1):9–25.
[162] Vinuesa R, Azizpour H, Leite I, Balaam M, Dignum V, Domisch S, Felländer A, Langhans SD, Tegmark M, Fuso Nerini F. The role of artificial intelligence in achieving the Sustainable Development Goals. Nature Communications. 2020 Jan 13;11(1):1.
[163] Fuso Nerini F, Fawcett T, Parag Y, Ekins P. Personal carbon allowances revisited. Nature Sustainability. 2021 Dec;4(12):1025–31.
[164] Gupta S, Motlagh M, Rhyner J. The digitalization sustainability matrix: A participatory research tool for investigating digitainability. Sustainability. 2020 Nov 9;12(21):9283.
[165] Gupta S, Langhans SD, Domisch S, Fuso-Nerini F, Felländer A, Battaglini M, Tegmark M, Vinuesa R. Assessing whether artificial intelligence is an enabler or an inhibitor of sustainability at indicator level. Transportation Engineering. 2021 Jun 1;4:100064.
[166] Henderson P, Hu J, Romoff J, Brunskill E, Jurafsky D, Pineau J. Towards the systematic reporting of the energy and carbon footprints of machine learning. The Journal of Machine Learning Research. 2020 Jan 1;21(1):10039–81.
[167] Masanet E, Shehabi A, Lei N, Smith S, Koomey J. Recalibrating global data center energy-use estimates. Science. 2020 Feb 28;367(6481):984–86.
[168] Karnama A, Haghighi EB, Vinuesa R. Organic data centers: A sustainable solution for computing facilities. Results in Engineering. 2019 Dec 1;4:100063.
[169] Hleg AI. High-level expert group on artificial intelligence. Ethics Guidelines for Trustworthy AI. 2019:6.

Chapter 5
Additive manufacturing for advancing healthcare sustainability

Abstract: Additive manufacturing (AM) is recognized as a multifunctional manufacturing technique with significant promise for driving innovation in the production of various medical equipment and instruments, orthoses, prosthetic devices, surgical models, sterile surgical implants, and biological manufacturing. The use of AM technology in the healthcare industry has considerable potential in addressing significant sustainability challenges. The chapter analyzes the many aspects of AM in promoting sustainability in the healthcare sector. It conducts a comprehensive examination of the uses, advantages, and ecological impacts of AM. The study addresses the potential of AM in facilitating the implementation of personalized medicine, improving the efficiency of medical device manufacturing, and optimizing pharmaceutical formulations, the reduction in material waste, energy consumption, and transportation costs associated with healthcare solutions facilitated by AM. The use of AM in the healthcare sector has the potential to address environmental challenges, promote resource efficiency, and improve patient care results, thus promoting sustainability in healthcare. This chapter offers important perspectives for policymakers, healthcare practitioners, and manufacturers who are interested in using AM's revolutionary capabilities to foster a more sustainable healthcare environment. The objective of this chapter is to provide a systematic examination of AM technology and emphasize the most recent advancements and practical implementations in the industry. This chapter explores several aspects of sustainability within the AM process, as well as the economic, social, and environmental benefits that it brings. The study demonstrates the future perspective of AM in terms of technological advancements, application domains, and material considerations. These factors can generate novel concepts and avenues for future studies in the field of AM. The investigation of the environmental sustainability of AM technology is essential. The aim of enhancing environmental impact is becoming more prevalent among organizations, leading to a growing emphasis on sustainability within the field of AM.

5.1 Introduction

The method of additive manufacturing (AM), often known as three-dimensional (3D) printing, fast-developing prototypes, or free-form fabrication, is a manufacturing process that involves the creation of objects by adding material layer by layer. AM utilizes a computer-aided design (CAD) model to fabricate products via a process of 3D printing, where each layer is constructed sequentially [1, 2]. AM is a viable method for producing intricate objects that possess complex geometric characteristics using polymers, metals,

https://doi.org/10.1515/9783111436432-005

and ceramics. Furthermore, AM products exhibit favorable cost-efficiency and can be personalized [3]. The accelerated advancement of manufacturing techniques has had a substantial impact on the world's financial system, in enhancing value via heightened organizational metabolic processes. The production industry plays a significant role in the global economy, contributing 16.96% to the overall value addition. In contrast, the agricultural industry provides 4.61% and business services contribute 78.43%. This contribution is achieved via the conversion of raw materials into completed items that are ready for consumption [4, 5]. Nevertheless, the rapid rate of expansion in the industrial sector not only results in the production of products and services but also leads to the depletion of environmental resources and generation of pollution and waste. These activities have detrimental effects on the surroundings [6]. The AM technique, established in the 1980s, has been evaluated in several medical applications, including as scaffolding for tissue engineering, testing facilities, and systems for delivering drugs [3]. Furthermore, AM is applied for customizing medical equipment and medications as well as extending their functionality to biological tissues to facilitate the recognition of physiological responses [7, 8]. The AM technique is used in the field of medicine for therapeutic modeling purposes. This application of the AM method is particularly beneficial for the preparation and education of physicians, as it enables them to engage in activities that include postoperative and preoperative strategy as well as implantation establishment [9]. AM creates 3D structures using imaging information collected as geometrical parameters by using computer programs for specialized purposes, allowing physicians to generate operating procedures for preparation and surgery simulation [10].

In various aspects, AM exhibits a much higher level of sustainability compared to conventional processes for production, thereby offering a more sustainable approach for the environment and the globe [11]. Consequently, the production industry is acquiring the ability to adapt to AM technology to generate various goods through significant study and development initiatives. Therefore, to optimize the overall ecological footprint on a global scale, companies using AM are in a favorable position to promote sustainability. The concepts of sustainability are used to minimize, reuse, and recycle [12, 13]. It has supplied hospitals as well as healthcare institutions both domestically and internationally with essential personal protective equipment, including customized facial shields, testing swabs, and certain elements for healthcare supplies.

The pharmaceutical sector used AM to deal with coronavirus issues, saving lives and promoting healthcare sustainability [14]. Metals, plastics, ceramics, and composites, in powder, wire, or liquid forms, are only some of the fundamental components now employed in AM. With the use of AM technology, layers are connected to layers. It starts with computer files, like CAD files or the results of a 3D picture of real products [15, 16]. The functional and geometrical features of produced components might vary greatly due to the variety of AM techniques and procedures. It takes some careful planning to determine which AM technology and material would best meet the needs of a given application [17]. AM suggests that generative design in space involves the most intelligent individuals. Due to significant concerns about diversity and changing climates, AM applications are being researched in industrial sectors. Sustainable development promotes environmentally friendly manufacturing due to the growing relevance of sustainability concerns and legislation [18–20]. Sustainability relies on ecological efficiency, and reducing enormous energy and material usage may reduce global warming. 3D printers enable the automated production of customized goods that are ideal for certain applications. These goods have better environmental performance and criteria [21, 22]. Additionally, with their educational value, the models may be used to help patients and their families better understand potentially hazardous and complicated medical procedures. Anatomic customization of prosthetics, orthotics, and splints is another field in which AM has developed extensive applications in the medical field [1]. Most of these programs are meant to be used on the outside of the body. Next, AM technologies will be used to manufacture hardware and devices for healthcare equipment, expanding their functionality and enhancing their effectiveness. The advantages of AM techniques and procedures, their applicability in many industrial sectors, and their effect on manufacturing facilities to allow quick manufacture and mass production have all been documented well by several academics [23–25]. However, additional investigation is necessary to completely actualize the prospective benefits of AM processes to overcome the challenges of expense for material property data, eligibility, and certification, especially for essential elements (aerospace or automotive parts). As indicated in Figure 5.1, the ASTM (American Society for Testing and Materials) International Committee F42 on AM Techniques has organized AM processes into seven classes [26–29].

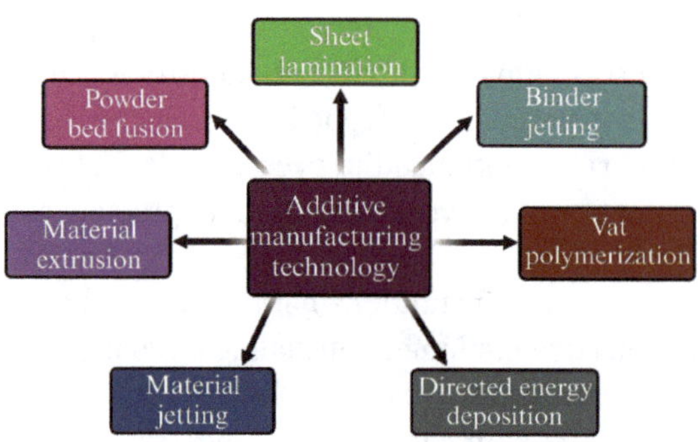

Figure 5.1: Classifications of the additive manufacturing process technologies based on ASTM F42.

5.2 Technologies for additive manufacturing

The establishment of the AM technique started in the 1980s [30]. Considerable advancements have been achieved since that time, and it is expected that AM technology has the potential to significantly change the industrial sector and provide many societal advantages. The advantages include: (1) individualized medical products are projected to considerably promote individual wellness; (2) the reduction of raw material utilization and consumption of energy is a significant factor in promoting healthcare sustainability; and (3) The concept of on-demand manufacturing has the potential to restructure the supply chain for manufacturing to provide cost-effective goods to customers in a shorter time frame, all while minimizing resource use.

The AM methodology is employed for the production of different structures and complicated elements. The use of such a method was first implemented by Charles Hull in 1986 for the stereolithography (SLA) technique [31]. Other printing techniques were found throughout time, and in less than three decades, the use of AM technology was greatly expanded, which changed production and logistical procedures. The investment in AM technology has seen a substantial increase, rising from $4 billion in 2014 to more than $21 billion by the year 2020 [32]. The recent development may be related to the advancements in AM technology and materials, which have promoted more investment in different fields like biomedical, aerospace, and automotive [33]. However, the advantages of AM have received significant attention within the manufacturing industry. These benefits include the ability to be involved in mass-customized manufacturing, facilitate prototyping, promote sustainable production practices, and reduce lead time and costs [34]. In recent times, advancements in the AM process have rendered it more appealing. These advancements include bioprinting, four-dimensional (4D) printing, nanoscale printing, and metamaterial printing [35].

Additionally, one of the benefits of AM techniques is their ability to facilitate the development of creative ideas and products by smaller enterprises and consumers, enabling them to serve as self-designers and manufacturers [36]. The environmental effect of AM is significant, leading to more sustainable manufacturing processes. The advantages of AM in terms of sustainability may be effectively described as including great resource efficiency, extended manufacturing life, and a redesigned value chain [37–39]. However, the implementation of AM techniques has presented several limitations, including substantial expenses associated with machinery, materials, and energy. These factors are hindering the extensive use of this technology [40]. Another issue with traditional processes is the excessive waste of materials and the absence of control mechanisms to adapt to current circumstances. Computer-controlled machinery has helped with the latter issue, although material waste remains a difficulty [41]. Since all AM operations are system-controlled, a single computer may manage an infinite amount of equipment. In all AM techniques, the first step is to deposit a thin layer of material and then repeat this cycle until the final 3D item is made [42]. There are three basic steps to AM technology:

i. A computerized 3D solid model is created and transformed into an appropriate AM file format, such as the classic patterning language [32] or the newer AM file format [34].
ii. The file is delivered to an AM system for manipulation, such as adjusting its position, orientation, or scale.
iii. The AM technology builds the components layer by layer.

Various AM mechanisms construct and stabilize layers in specific ways. Metal or plastic powder may be melted or sintered (formed into a coherent structure without melting) using heat energy from a laser or an electron beam that is focused using optics. In other methods, powdered ceramic or polymer is sprayed with a binder or solvent using printing heads similar to inkjet printers (Table 5.1). The following is a brief overview of several important AM processes.

Table 5.1: An overview of the most frequently used additive manufacturing (AM) techniques.

S. no.	Method	ASTM (American Society for Testing and Materials) standards	Initial materials (steel, aluminum, and alloy)	Applied energy (mechanical energy)	References
1.	Inkjet printing (IJP)	Droplet-based printing	Liquid materials and ink	Nozzle piezoelectric	[43–45]
2.	Fused deposition modeling (FDM)	Material extrusion	Nylon, ABS, ceramics, investments casting –wax, and alloy	Heating	[43, 45–47]
3.	Laminated object manufacturing (LOM)	–	Paper, polymer, ceramics, and metal	Heated laser beam and high power	[48, 49]
4.	Stereolithography (SLA)	Vat polymerization	Photopolymer solution	UV laser's rays	[43, 44, 46, 47]
5.	Selective laser sintering (SLS)	Material extrusion	Paraffin wax, nylon, metal, and ceramic	Lasers of extreme intensity	[43, 44, 46]
6.	Three-dimensional printing (3DP)	Material extrusion	Metal, ceramic, and binder	piezoelectric nozzle and heat	[43, 44]

5.2.1 Inkjet printing (IJP):

The inkjet technology is a nonimpact dot-matrix technique designed for producing 2D pictures. It was first used in the late 1800s, and the first acceptable inkjet device was given a patent (US Patent 2566443) on September 4, 1951 [50]. It utilizes inks, which are liquid-phase compounds with a solute distributed in a solvent. A nozzle ejects an exact amount of ink from an opening through a rapid, quasi-adiabatic drop in chamber volume through piezoelectric action (Figure 5.2). The small droplet settles to the surface and dries by evaporation of the solvent due to gravity. IJP has been used to

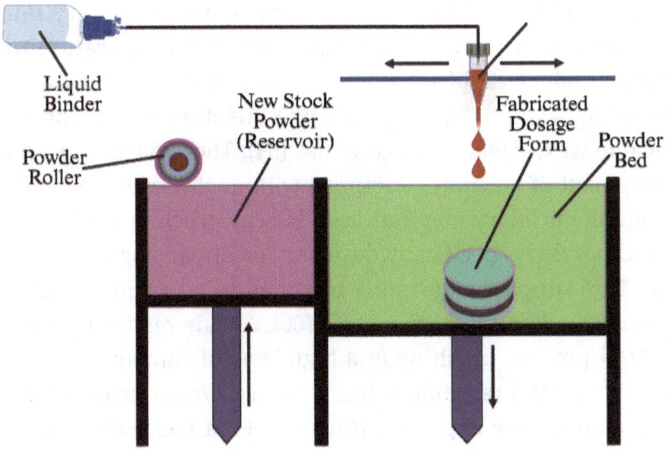

Figure 5.2: Inkjet printing (IJP) technology.

create several different products, such as solar energy cells, detectors, and transistors with thin films [51], all of which employ different kinds of substances. IJP allows for rapid reaction and individualized service. It has a few problems, like print heads that are easily broken and ink refills that cost quite a lot.

To print a 3D item, pre-patterned substrates are used in many steps of processing [51]. The IJP technology enables fast prototyping by deposition of microscopic photopolymerized ink solutions on substrates for 3D structures and scaffolds, with great precision and efficiency for biomolecules, polymers, and live cells [52]. Typically, the bioink viscosity should be below 10 mPa for optimal printing with this technology. This technology has low cell densities and rapid production speeds in comparison to other 3D printing methods [53]. IJP has two basic modes: drop-on-demand (DOD) and continuous inkjet printing [54]. DOD is a noncontact 3D printing process that employs on-demand ink droplets (25–50 mm) with pressure or pulses of voltage for controlling the material droplets. This approach is often used to create scaffolds for tissue engineering (TE) applications [55]. The IJP technique is used for the fabrication of scaffolding using a combination of ceramic and biopolymer-based biomaterials such as hydroxyapatite, bioactive glass, polylactic acid, polycaprolactone, polyglycolic acid, and others. There is a diverse range of uses for image-based joint processing in various biological fields, such as customized medicine, controlled medication delivery, and prostheses [56, 57].

5.2.2 Fused deposition modeling (FDM)

The patent for fused deposition modeling (FDM), specifically US Patent 5121329, was officially granted on June 9, 1992. However, the basic concept had been previously documented by Crump in an earlier publication [58]. A thermoplastic material in liquid form is extruded using a mobile FDM apparatus and then applied in very thin layers onto a substrate (Figure 5.3). The material is subjected to a temperature increase over its melting point to induce rapid solidification after extrusion, facilitating cold welding with the preceding layers. The range of materials used in this context has been broadened to include wax, metals, and ceramics [59]. There have also been advancements in the development of machines equipped with dual nozzles. One nozzle is dedicated to dispensing the primary material used for constructing the desired component, while the second nozzle is specifically designed for dispensing a more affordable support material. This support material is intended to be easily removed from the part without causing any damage to its surface [60]. A wide range of components may be used in the FDM process, resulting in a high level of component precision with a tolerance of 0.05 mm. FDM technology has a small physical footprint and is associated with minimal maintenance expenses. However, FDM has certain drawbacks, such as the layer seam line, the need for support, the lengthy construction time, and delamination brought on by temperature fluctuations [61].

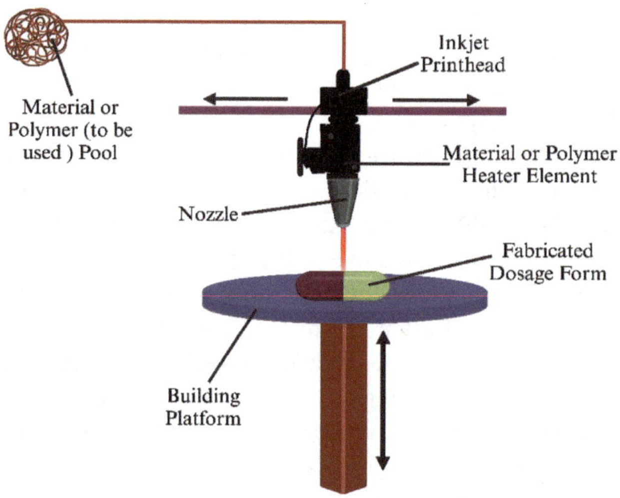

Figure 5.3: Fused deposition modeling (FDM) technique.

In FDM, the production of filament via hot melt extrusion is a crucial step. However, it is important to note that this process exposes the medicines to heat stress, which has the potential to cause discoloration and, in some cases, deterioration through oxidation. Once the filament has been produced, the medication is introduced into the 3D printer and subjected to further heating. The lack of available additives and medications that can produce filaments with the necessary physical and mechanical qualities for 3D printing, however, is the main disadvantage [62]. A 3D printing technique was used to fabricate objects directly from powdered materials employing an improved FDM printer. This approach eliminated the need to manufacture filaments loaded with drugs [63]. The printing of the pills was effective when the mixtures were processed at 170 °C using various molecular weights of hydroxypropyl cellulose loaded with 35% of the drug. By using adapted 3D printers, it becomes possible to directly fabricate excipients and pharmaceuticals in a powdered state, leading to a consequent reduction in heat stress exerted on the active pharmaceutical ingredient and thus decreasing the possibility of deterioration [64]. The FDM technique is widely regarded as a very appropriate approach for the fabrication of biopolymers. This method involves the use of thermoplastic filaments that are subjected to heating, causing them to attain a melted or semiliquid condition. Subsequently, such filaments are extruded via an opening onto a printer substrate [65].

5.2.3 Laminate object modeling/manufacturing (LOM)

The LOM patent (US Patent 4752352) was granted on June 21, 1988 [66]. The sticky substance used by LOM enables the sheets to be joined together either by pre-coating it over the substrate or by depositing it before the binding process. The production of 3D components involves a sequential process of lamination as well as cutting 2D cross-sections. The cutting process is conducted with a laser beam, with careful adjustments made to its speed and intensity. This ensures that the depth of the cutting precisely matches the dimension of the layer, preventing any harm to the underlying layers. Many materials may be utilized, such as paper, metals, plastics, textiles, synthetics, and composites. The LOM technique is cost-effective and does not produce any hazardous emissions. Furthermore, this process has the potential to be automated with little human intervention. However, there are a few concerns with the *Z*-axis precision of LOM that lead to complications with dimension consistency (Figure 5.4). The occurrence of interior cavities has the potential to adversely impact the quality of the product. Furthermore, the allocation of postproduction time is necessary to effectively reduce any waste that may be present. Additionally, in some instances, supplementary operations must be undertaken to produce components that possess precise functionality [67]. The approach involves feeding adhesive-coated thin-film material, integrating cutting and lamination operations, and resulting in a finished product [68]. The process of layer manufacturing begins by attaching a sheet of paper, metal [69], or an artificial polymer to the substrates using a laminated roller. Following the deposition of the layer onto the construction platform, the blades [70] or the carbon dioxide laser [71], which are affixed to the print head, proceed to delineate the outlines of the layer by the instructions provided by the CAD file [68–71]. After the adhesive-coated films are attached between layers, the film is then cut with a laser beam into the desired design [68]. Additionally, the platform is subsequently lowered, simplifying the distribution of a new layer of manufacturing material (Figure 5.4) [68]. After covering the prior layer using film, the subsequent cross-sectional design is cut and laminated [72]. Subsequently, the cutting procedure is repeatedly performed in a

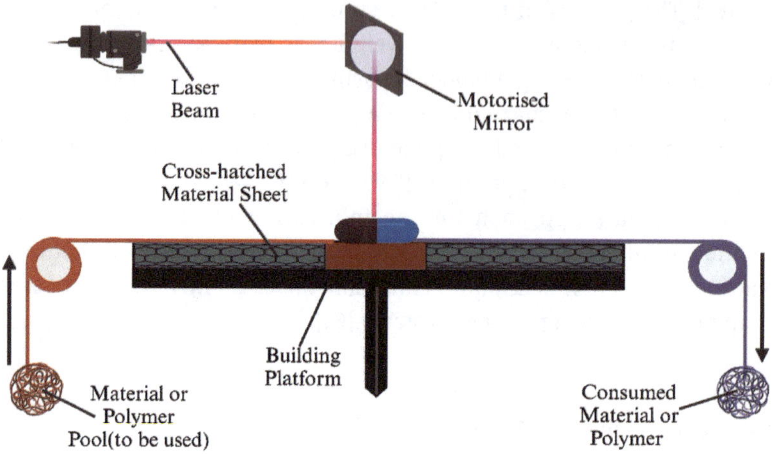

Figure 5.4: Laminate object modeling/manufacturing (LOM) technique.

layer-by-layer manner, along with the required quantity of sheets in the finished item or dose form [68].

In comparison to other AM techniques, Laminated Object Manufacturing (LOM) provides a rapid method for efficiently producing larger-sized goods [73]. However, it is essential to ensure the consistency of the material composition inside each layer [68]. In addition to quicker prototyping [68, 73–75], the LOM provides a high tensile strength in the lamination dimension [76] with adequate quality [77–79]. Furthermore, using prefabricated film sheets as a base material makes this approach flexible, adaptive, and suitable for many materials, involving the lengthy fibers in reinforcement composites [72, 80]. Supporting material is not needed during laminating since the construction material remains in situ and cannot move beyond the model's shape [68]. Additionally, the LOM approach has significant anisotropy efficiency for elaborate designs and complicated 3D geometries [80, 81]. LOM techniques are useful since they do not need a heat phase during production to ensure the layers' adhesive adhesion. Compared to other techniques like FDM, this phase has a low influence on nonuniformity in created parts [82]. The limitation of LOM is that the substances used in the technique are restricted by their ability to be turned into sheets and combined with adhesive [83]. Sheet bonding issues arise in the LOM process, leading to insufficient adhesion, process failure, and frustrating separation from the frame [68]. Another drawback of LOM is that improperly regulated phase or roller temperatures can cause adhesive delamination or structural damage [84].

5.2.4 SLA method or stereolithography

The patent for SLA apparatus, as documented in US Patent 4575330, was officially granted on March 11, 1986. The process was subsequently disclosed to the public via the publication authored by Hull [85]. An intriguing method of 3D printing, known as selective laser sintering (SLS), creates patterned structures by using materials that have been precisely cross-linked under the influence of ultraviolet (UV) or visible light (Figure 5.5) [86]. These materials include elastic materials, thermosetting polymers, ceramic-based resins, and bioink materials. This method is often used to create biopolymer-based scaffolds and complicated structures for both soft and hard regeneration of tissue [87]. This approach enables the achievement of high printing resolution, allowing for values as small as 20 μm. It is widely regarded as one of the most precise 3D printing methods available [88].

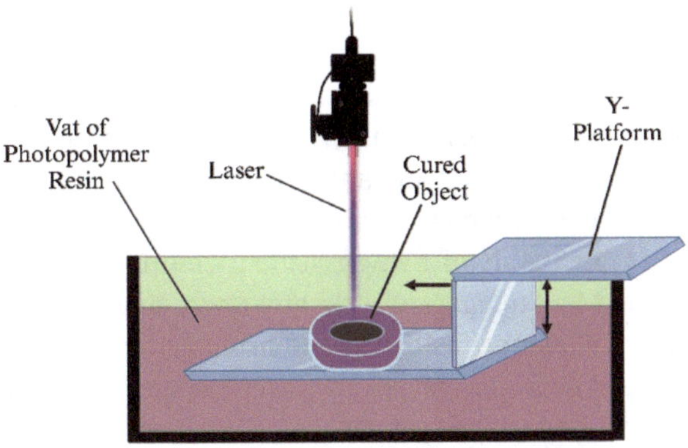

Figure 5.5: Stereolithography (SLA) technique.

The SLA process involves the use of a photogenic monomer resin in conjunction with an UV laser to construct components in a sequential layer-by-layer manner (Figure 5.5). It needs support structures to connect the component to the construction platform. The beam of laser energy solidifies the design by tracing a cross-section of the component on the liquid resin substrate at every single layer. Subsequently, the construction stage is lowered to provide a comprehensive coating of the product. The stage is then elevated to a point where a blade is used to remove the resin, resulting in the preservation of a single layer of resin over the component. Subsequently, the component is gradually descended by a single stratum and allowed to rest until the liquid has reached a state of equilibrium, guaranteeing a uniform surface before the construction of the subsequent stratum [89].

After the completion of the portion, the support frameworks may be carefully removed. SLA is ideally suited for use in the manufacturing industry because it reduces the time required for producing prototype parts and provides a high level of surface finish. The primary drawback of SLA is its modest product size, often no more than a 2-foot cube. Another drawback is the financial burden. Photopolymer costs $300–500, without including the equipment. In comparison to other AM methods, the selection of materials available for use in SLA has been limited [90]. The thermoset resins used in SLA technology have little degradation under light source activity. To create tissue-degradable scaffolds, a mixture of biodegradable materials, including poly(trimethylene carbonate), poly(propylene) fumarate, and poly(D, L-lactic acid), was used [91]. SLA is not appropriate for printing live cells simultaneously owing to photoinitiators insoluble in water resulting in toxic cells. In addition, cells suffer damage as a consequence of the activity of the UV radiation during the curing process. Cells are integrated into scaffolds after their development because of this particular reason [92].

5.2.5 Selective laser sintering (SLS)

The SLS patent (US Patent 4863538) was granted on September 5, 1989; however, the technique was previously described by Deckard and Beaman [93, 94]. The process of SLS (Figure 5.6) involves the use of a laser with significant power to merge minute particles of the chosen construction material, which may include polymers, metals, ceramics, glass, or any other substance capable of being crushed. The powder bed used in the fabrication process is heated to a temperature slightly lower than the material's melting point. It is done to reduce heating distortion and promote the bonding of the current layer with the previously deposited layer. The laser's beam is used to sinter the material, enabling the drawing of each layer on the powder bed. The sintered material is responsible for the formation of the component, while the unsin-

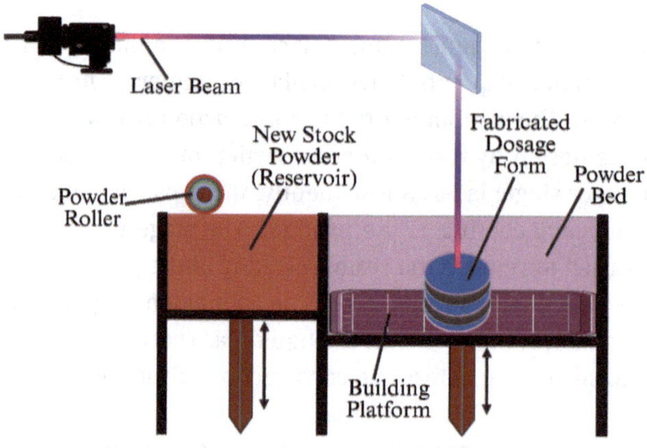

Figure 5.6: Selective laser sintering (SLS) technique.

tered powder serves as a supportive element for the structure. After the completion of the construction, the unsintered powder may be removed and reused.

SLS provides the advantage of expeditiously constructing intricate components that possess enhanced durability and superior usefulness compared to alternative AM techniques. Post-curing is not necessary, and the construction process is characterized by rapid build time. Nevertheless, the operation of SLS is intricate due to the need to determine numerous build factors. The attainable surface smoothness is comparatively inferior to that achieved using SLA, and the process of changing materials poses challenges [67]. The SLS process requires managing process parameters to achieve the quality of the product under specified ranges. Extensive research has been conducted in the pharmaceutical sector to examine the many attributes of the SLS process. In this context, the engineering department's knowledge of relevant procedural parameters has been utilized to facilitate the production of medications and medical devices [95–100]. The SLS process has been optimized to enhance the mechanical properties, dimensional precision, subsurface and surface appearance, and other important quality attributes of the manufactured dosage forms. Furthermore, the critical quality attributes (CQAs) are mostly influenced by the accuracy of the STL (stereolithography) file conversion process, which originates from the CAD program. However, it is important to consider other variables that may significantly impact the CQAs of the process [100–104]. Additionally, laser beam density significantly affects 3DP dosage form densities and mechanical durability [101, 102, 104–108].

The SLS printing method has undergone continual refinement, resulting in the widespread use of the SLS printer for the production of pharmaceutical formulations. This printer uses a diode laser as its primary mechanism [102, 106–108]. The energy density is influenced by many factors, including the velocity of the laser as it moves over the powder bed, the amount of laser energy used, and the distance between scan lines, also known as the hatch distance [103, 109, 110]. However, the efficacy of laser sintering is significantly influenced by the particle size of the powder. The impact of particle size on content homogeneity, chemical and physical stability, solubility, and bioavailability of dose formulations is influenced by its effect on their surface area [103, 110]. Paracetamol and copovidone SODFs were successfully manufactured by Gueche et al. using the SLS and a CO_2 laser for the first time. However, it was also suggested by the authors that the CO_2 laser may have an impact on a wider range of thermolabile medications, and thus, more investigations should be conducted to thoroughly assess their deterioration [6, 150]. Furthermore, multiple oral formulations [101, 111–114], implants [115, 116], and controlled-release formulations [117, 137] have been produced and analyzed in various regions worldwide.

5.2.6 Three-dimensional (3D) bioprinting

The patent for 3DP (US Patent 5204055) was granted on April 20, 1993; however, the work was previously published by Sachs et al. [118]. The 3DP (three-dimensional printing) process operates by the application of powdered substance onto a substrate, which is then selectively fused using a binder that is dispensed through a nozzle. To prevent significant disruption upon contact with the binder, the substance is first stabilized by spraying it with water droplets. After the layer-by-layer application procedure has been completed, the loose powder is swept away. Further processing may improve bonding by firing the item at high temperatures. This method may be used in the manufacturing of metal, ceramic, and composite items that are composed of both metal and ceramic. Faster manufacturing and lower material costs are two benefits of 3D printing [119]. Indeed, it is perhaps the most expeditious of all AM techniques. There are several constraints, including rough surface quality, size restrictions, and high cost. The progress of 3D printing technology has led to the creation of commercial bioprinters such as BioBots, Aether, Regenhu, and Cellink [120–123]. 3D bioprinting is a new and innovative technology that is based on AM technology. It combines living cells with useful materials to make biological parts that have changed the TE, bone healing, and pharmaceutical industries [124–127].

Bioinks serve as the raw materials for 3D bioprinting and contribute to the creation of complex and diverse structures, such as vasculatures. They also facilitate improved cell adhesion, development, and differentiation when combined with natural tissues [128]. When comparing ordinary 3D printing technologies to the field of 3D bioprinting, the process of developing artificial tissues poses more challenges. These challenges arise from the need to carefully pick cell growth, kinds, and differentiation factors, as well as the architecture and functionality of the tissues [129]. In contemporary times, the field of 3D bioprinting is effectively addressing the needs arising from traumas, tumors, tooth extraction, and accidents. This is achieved by the manipulation of porosity and the even distribution of materials, thus enhancing their compatibility with the human body. However, obstacles such as cell inclusion, structural activities, and feedstock needs must be addressed in this technique [130]. The stages involved in 3D bioprinting are shown in a simplified format in (Figure 5.7). Bioinks are composed of cells that are combined with a carrier, often composed mostly of hydrogel materials. The carrier materials used in this context may consist of synthetic substances, natural substances, or a combination of both synthetic and natural components. The presence of viable cells inside bioprinted structures has led to the hypothesis that these structures have the potential to self-generate vasculature, and growth factors, and exhibit improved integration with host tissues [131].

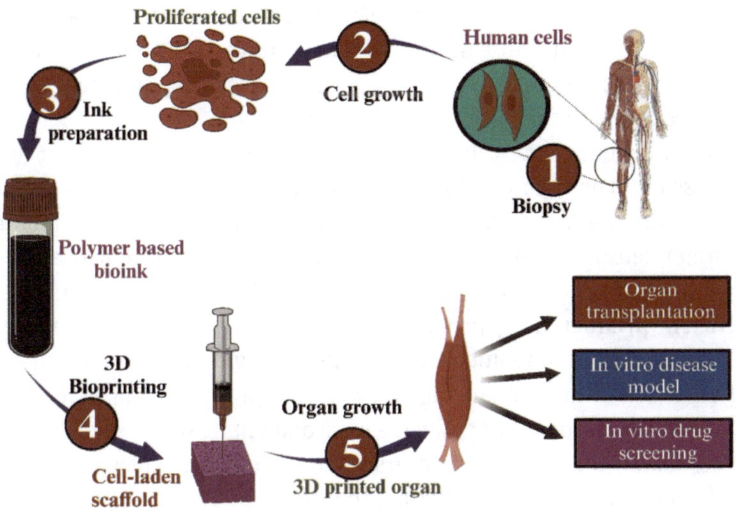

Figure 5.7: The diagrammatic representation of the three-dimensional bioprinting process.

5.3 Environmental sustainability in additive manufacturing

Sustainable manufacturing stimulates newly developed sustainable innovations and supports environmentally friendly practices to minimize pollution. The automobile industry is characterized by high levels of competition and a significant need for energy. Various industries are now implementing strategies to enhance production and reduce waste [132, 133]. The importance of sustainable production for many reasons cannot be overstated. Sustainable manufacturing is not simply environmentally responsible. Furthermore, this will significantly enhance the safeguarding of the infrastructure, labor force, customers, and the global population as a whole. The economic viability of sustainable manufacturing is also noteworthy. The use of this approach results in a reduction in energy consumption, a decrease in processing time, and a minimization of waste and superfluous resources. The main advantage is in enhancing the performance of the facilities and operations. The reliability and automation of the system provide enhanced protection against possible risks.

Implementing sustainable manufacturing techniques in operations may enhance the industry's image, boost consumer loyalty, and promote innovative management. Customers are increasingly looking for businesses that use sustainable production methods [134], according to the international trend toward sustainable development. Promoting waste-efficient production with less environmental impact is crucial.

Planning and maintaining updates on ecological waste by detecting, quantifying, measuring, and managing it, as well as using best practices, will lessen the effects on the climate and make the surroundings more sustainable. Waste, raw materials, power, paper, packing fuel, and other resources need to be reduced, and sustainable manufacturing is the only method that can do this. This unique technique provides waste-reduction potential. This reduces waste by recycling resources. In some operations, it is possible to coordinate the cycles of metal presses.

AM has the potential to reduce the reliance on machines, jigs, and fixtures in the environment of large-scale production facilities. In the future, a manufacturing system needs to adopt sustainable practices that effectively mitigate all environmental concerns. It is recommended that industries adopt more sustainable practices by reducing their energy use and resource utilization [135]. In a highly competitive market, firms may attain efficient production by providing items that are of superior quality, cost-effective, and long-lasting, accompanied by exceptional service and support.

The concept of sustainable manufacturing refers to the production of manufactured goods using methods that aim to limit adverse environmental effects, save energy and natural resources, ensure the safety of workers, communities, and customers, and maintain economic viability [136]. When evaluating the sustainability of AM processes or any manufacturing process, it is essential to take into account the whole life cycle to determine the overall sustainability effect. The manufacturing process itself is just one of several environmental consequences that are linked to the life cycle of the product. Currently, there is a limited amount of empirical evidence available about the comparison between AM items and traditionally made products in terms of energy consumption, supply chain dynamics, pollution levels, and other possible environmental consequences during their lifespan [137]. The manufacturing sector is responsible for around 33% of energy use in the United States [138, 139]. Enhancing energy efficiency methods aids in fortifying the economic sustainability of American manufacturing, while concurrently safeguarding the environment, by diminishing energy use and its corresponding expenditures. Therefore, energy efficiency, together with the price of energy and its availability, has a significant influence on the degree to which US industries can compete internationally and their overall economic well-being [136].

There are significant AM benefits that are relevant to sustainable manufacturing:

- The additive method reduces waste compared to traditional manufacturing methods such as stamping or sculpting, due to its inherent nature.
- There is no need for the use of specific tools or fixtures of AM.
- The capacity to construct structurally efficient components, while simultaneously preserving their mechanical integrity.
- The use of this technology decreases the need for substantial quantities of raw materials in the supply chain and transport processes.
- In comparison to standard machining and casting methods, this technology demonstrates significant material efficiency.
- This technology can fabricate optimized geometries that exhibit strength-to-weight ratios that are almost flawless when compared to wrought materials.

- The component exhibits reduced environmental effects throughout its life cycle, leading to a decreased carbon footprint, lower embodied energy, and an improved economic model.
- The capacity to produce replacement parts as needed enables the reduction or elimination of inventory.

5.4 Advances in additive manufacturing for reducing environmental deterioration

Additive manufacturing is an innovative technique with promised environmental benefits due to its resource-efficient and ecologic manufacturing processes. Resource efficiency is one of its main benefits; in contrast to typical subtraction manufacturing techniques, which produce a lot of waste, additive manufacturing creates products layer by layer, using just the necessary material. This decrease in waste output is critical in reducing the environmental impact of resource extraction, transportation, and trash disposal. Furthermore, by removing the need for energy-intensive gear and lengthy production lines, additive manufacturing may significantly cut energy usage, especially when driven by renewable energy sources. The ability of localized production to reduce emissions and energy consumption associated with transportation is important since it allows for manufacturing things in closer proximity to their intended destination. Additionally, the technology enables the creation of intricate, lightweight structures that were previously unattainable using conventional techniques, thereby contributing to fuel efficiency in industries such as aerospace and transportation. Sustainability requires optimal designs that use less materials and energy, which customization allows. The development of biodegradable and recyclable materials through AM shows its promise for ecologically friendly production. Its use in repair and maintenance improves product lifespans and reduces replacements, minimizing environmental impact. AM may improve resource efficiency, waste reduction, and sustainable production, making it a possible environmental solution.

AM involves layer-by-layer addition of raw materials using a CAD model. The method was first tested on polymers and metal powders; however, it has since been expanded to include latex, cement, graphite, and graphene. The use of fast prototyping facilitates efficient production processes, increases productivity, and simplifies the development of distinctive goods. Additionally, it has the effect of decreasing manufacturing expenses and carbon dioxide emissions associated with the production process, while also diminishing the energy requirements throughout the life cycle of the product.

Numerous samples and variations are conducted to achieve the finalization of a perfect idea throughout the manufacturing process. Metals can undergo recycling in the form of powder, and the mechanical characteristics of the final product may be enhanced by the use of several post-processing procedures [135, 140]. Support structures play a crucial role in the effective production of 3D-printed components. The process of fabricating components with complex design features such as branches, apertures, and spans is inherently complex. The implementation of support structures is crucial in mitigating the risk of collapse during the printing process since these features are inherently susceptible to collapse in the absence of adequate support. Supports may also function as heat dissipaters in processes that entail elevated temperatures, such as metal AM [141, 142].

The sustainability of 3D printing relies heavily on the content used. Recycled plastic material for 3D printers has been a topic of research and development in recent years. The popularity of 3D printing has seen a substantial rise in recent years. This technology revolutionizes every industry it affects and makes product design and production easier than before [143, 144]. Consequently, it has significant potential in several aspects, including environmental concerns. 3D printing eliminates waste and enables reuse, making it desirable to the fashion sector. It is used to create brand-new shoes with the necessary style. The substantial quantity of plastic pollution present in the atmosphere significantly contributes to the process of climate change. Plastic emits greenhouse gases throughout its life cycle, contributing to the phenomenon of global warming. Additionally, microplastics contain detrimental compounds that have the potential to pollute both the water supply and the atmosphere. This method enables printing on almost any kind of plastic as well as environmentally friendly material. The material, in contrast to other synthetic substances, may be 3D printed, responds well to normal finishing methods, and decomposes without the need for a commercial composting facility. The material has a low weight and cost, making it a highly suitable option for AM as a substitute for plastics [145].

The continuous advancement of AM has the potential to assist plastics producers in reducing their environmental impact by minimizing their carbon footprint. Topology optimization and generative design techniques are used to minimize the quantity of plastic utilized in a particular product. Despite the apparent advantages, such as AM being appropriate for complicated and tiny components, there is a lack of comprehensive study on its sustainability. Moreover, since the quality of the component is determined by the process and not by the expertise of the operator, production can be precisely customized to customer requirements. All the benefits are associated with one or two of the primary aspects, wherein the advantages either relate to the reduction of time and/or cost in the procedure or are derived from a natural adaptability of AM techniques. Despite the many benefits associated with AM technology, understanding the relationship between the process, characteristics, and geometry of AM components is a complex matter that needs further investigation. While AM has the potential to catalyze enhancing industrial sustainability, its implementation may have consequences on the industrial system that could lead to a different scenario. This scenario involves a decrease in eco-efficient localized production, an increase in customer demands for customized goods, and a high rate of product obsolescence, all of which contribute to an overall increase in resource expenditure.

5.4.1 Additive manufacturing to reduce resource needs

The AM technique requires fewer resources and fewer personnel to produce a desired item. It reduces the manufacturing cost of any component. Additionally, 3D printing reduces delivery costs and saves time. The use of this method during the design stage may significantly contribute to the enhancement of project planning. It facilitates the improvement of product design and ensures that all criteria specified in the CAD plans are met. Based on the drawings, a 3D model of the part/product may be built to assist in giving the best design options [146, 147]. It is essential to satisfy the needs of the consumer and deliver an appropriate response. AM provides us with more design liberty, enabling us to generate novel approaches and solutions to meet the requirements. The investigation examines the advantages of its implementation in the industrial sector and offers enhanced prospects for resolving diverse issues [146]. Furthermore, the capacity of 3D printing to produce lightweight and structurally optimized components leads to a reduction in material use, particularly in sectors such as aerospace and automotive, thus promoting resource conservation. Moreover, the ability of this technology to produce goods as needed and in specific locations has the potential to greatly decrease the need for maintaining extensive stockpiles, hence resulting in a significant reduction in resource demands. In general, additive manufacturing is a viable and sustainable solution to address the issue of resource depletion and promote the use of environmentally conscious industrial methods.

5.4.2 Reducing waste by using additive manufacturing

There are numerous debates regarding whether the use of AM is a more sustainable alternative to conventional manufacturing. It can reduce plastic waste by utilizing only the material necessary for manufacturing final components and minor support structures. Technological advancements, such as the use of binder jet/powder bed techniques, have the potential to reduce waste generation by facilitating the recycling or reutilization of surplus powder materials. Sustainable production is of greatest significance in large industries, including manufacturing, automotive, and aerospace. Several sectors have already adopted the use of 3D printing technology, with a focus on using recycled materials and exploring opportunities for additional growth [148, 149]. Therefore, the technology of additive manufacturing is used to accelerate the prototype process and provide assistance in reverse engineering parts. The use of this methodology has become a prevailing practice in several customized metal and plastic manufacturing companies, as designers accumulate additional experience in the process of additive manufacturing.

This methodology maximizes the use of materials, ensuring that just essential materials are utilized, hence efficiently reducing waste. The decrease in waste generation has a consequential impact on environmental deterioration, mitigating the resource-intensive procedures linked to the extraction of raw materials, their transportation, and the disposal of garbage. AM plays a crucial role in fostering a sustainable and environmentally conscious industrial ecosystem by effectively reducing material waste. This methodology maximizes the use of materials, ensuring that just essential materials are utilized, hence efficiently reducing waste. The decrease in waste generation has a consequential impact on environmental deterioration, mitigating the resource-intensive procedures linked to the extraction of raw materials, their transportation, and the disposal of garbage. AM plays a crucial role in fostering a sustainable and environmentally conscious industrial ecosystem by effectively reducing material waste. Additive manufacturing may also create complicated shapes with less material. Traditional manufacturing methods are frequently limited by mold, tool, or machining process restrictions, resulting in designs that need more material to attain structural integrity. While traditional manufacturing methods need substantial amounts of material, sectors like aerospace and automotive benefit greatly from 3D printing because of the ability to create lightweight but durable structures. Further reducing resource needs and supply chain waste are the on-demand and localized manufacturing capabilities of 3D printing, which eliminate the need for massive stocks and long-distance transportation. AM is an important prospect for waste reduction in the manufacturing industry since it is aligned with sustainable practices and contributes to environmental conservation.

5.4.3 Additive manufacturing for supply chain cost reduction

This flexible infrastructure provides value to supply lines and their everyday tasks. AM has the potential to decrease supply chain costs by enabling direct delivery of items to customers. One of the primary benefits of separate components is their lightweight nature. 3D printers can produce whole objects without requiring any further assembly, hence reducing the length of the supply chain. Thus, companies may observe significant cost savings in production and minimize the accumulation of undesirable behaviors. Additional benefits of 3D printing are its adaptability and the possibility of personalized design. The method of research and development may be enhanced by incorporating various design modifications [150, 151]. The process of customizing items reduces expenses and results in decreased overall costs. AM has been widely adopted by enterprises as a means of increasing productivity. The traditional manufacturing method provides limited opportunities for organizations seeking to develop new, creative, and customized goods. The positive effects of AM in the supply chain are evident in several sectors, including medical, defense, aerospace, dentistry, and electronics. The advancement of this technology has significant prospects for the production of individualized, superior-quality goods at a reduced cost. AM has emerged as an appealing solution for several industrial businesses [152].

As regards the environmental aspect of sustainable development, the implementation of industry results in a reduction in the generation of waste via the optimization of production processes and the promotion of efficient recycling and remanufacturing operations. The integration of diverse sensor technologies is implemented inside industrial processes. These sensors can give valuable information on the performance of the product, including its functionalities, use patterns, defect models, performance indicators, environmental impact, and stress-induced performance, among other factors. To enhance goods and processes without harming competitiveness, information is employed in simulation systems to lessen environmental impact [153, 154]. It helps with the management and prevention of manufacturing and use mistakes throughout the product's life cycle. To attain sustainability and responsibility, manufacturers will create innovative, competitive, and ecologically friendly items. Manufacturers are actively searching for methods to reduce production operating expenses. Through the use of AM technology, businesses have effectively optimized several operations within their supply chain. This proposed system would enable manufacturers to customize components, staff, and services to optimize processes, decrease manufacturing operating expenses, and enhance overall efficiency. Manufacturers may minimize waste by using 3D printing [155].

5.5 Advantages of additive manufacturing

Various AM approaches aim to expand the process possibilities for producers and consumers, rather than replacing existing production methods. Each method has benefits and drawbacks of its own, and the selection of which to use depends on the application. To evaluate the benefits of AM processes, it is necessary to identify their distinguishing characteristics. The evaluation of the benefits of AM may be based on three essential features: time, cost, and flexibility (Figure 5.8). One of the primary objectives of using AM is to optimize manufacturing time and enhance production velocity. This helps speed up prototyping and minimizes manufacturing time for spares and alternative components [156]. Altering the location of production facilities and decentralizing production at several regional sites adjacent to principal markets might enhance the supply chain performance of spare parts [157]. Distributed spare part manufacturing offers advantages such as decreased downtime, costs, capacity utilization, inventory management, resilience, and adaptability to supply chain fluctuations [158]. On-site manufacturing allows for faster and more cost-effective creation of bespoke components by reducing transportation time. While conventional production requires extensive material removal, AM efficiently reuses waste materials for subsequent construction. Case studies have indicated that the amount of material waste produced by AM is decreased by 40% when compared to trash produced by conventional techniques and that 95–98% of the materials that are discarded may be recycled [159].

The cost-effectiveness of AM products may be attributed to the decrease in labor costs and the avoidance of expensive storage. Furthermore, AM does not need the use

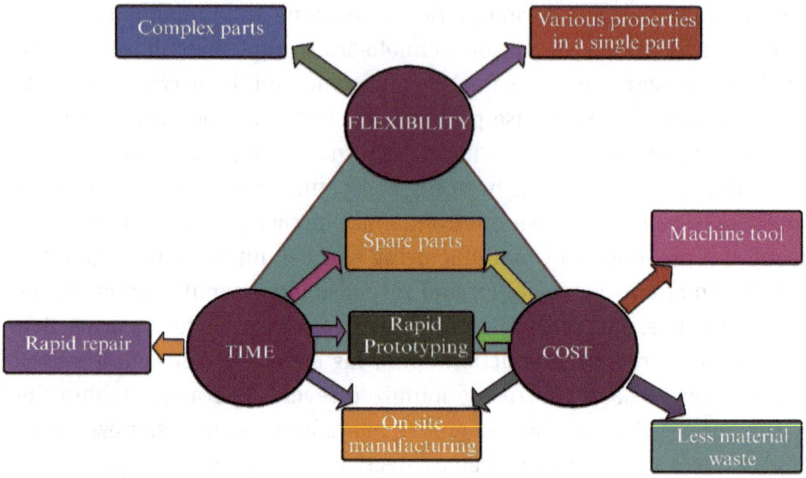

Figure 5.8: The relationship between the three fundamental qualities and the benefits of additive manufacturing (AM).

of supplementary resources such as fixtures, cutters, jigs, and cooling agents. Also, making things only when they are ordered lowers the risk of having unwanted finished goods in stock. Flexibility may be described as an integral component of the process. This implies that designers possess a higher degree of adaptability in designing intricate components and possess the autonomy to readily modify process parameters to their requirements. Due to the absence or minimal presence of tooling restrictions in AM, it becomes possible to produce components with intricate geometries. Consequently, the manufacturing limitations do not hinder the operation of the parts. Additionally, a single component might have variable qualities, such as strength and ductility [160, 161]. Moreover, given that the quality of the parts is dependent upon the process rather than the abilities of the operator, it is possible to achieve output that precisely aligns with client demand. Figure 5.8 demonstrates the interrelationships of the three aforementioned fundamental characteristics along with the benefits offered by AM methodologies. All the benefits are associated with one or two primary elements, wherein the advantages either pertain to the reduction of time and/or cost in the process or arise from the inherent flexibility of the AM methods. Even with all of these benefits, the relationship between the process, qualities, and shape of AM components is very difficult and needs more research.

The pharmaceutical dosage forms produced from AM have been shown to enhance the quality of life for patients, leading to long-term health benefits. This is particularly true when these dosage forms are used in conjunction with pharmacometabolomics and pharmacogenomics [162, 163]. The utilization of pharmaceutical AM technology enables the development of personalized treatment programs. Polypills, in particular, can enhance patient adherence, provide flexibility in medication dose and modification, and contribute to the advancement of precision medicine [164, 165]. The AM also provides patients and healthcare practitioners with a range of choices for dose forms, including the production of tablets that are easily dispersible or have controlled-release properties, specifically designed for individuals with specific requirements [164–167].

Additionally, allowing community and hospital pharmacies to customize therapeutic regimens using appropriate AM techniques improves patient health and reduces the financial burden of serious illnesses on the healthcare system. Furthermore, the incorporation of AM methods in community and hospital pharmacies allows for the advancement of compounding pharmacy practices, particularly in the field of precision medicine, on a broader and more readily available scale [164, 166]. Table 5.2 presents a comprehensive overview of the merits and demerits associated with the many AM methodologies used within the pharmaceutical industry.

Table 5.2: An overview of the merits and demerits associated with different additive manufacturing (AM) techniques used within the pharmaceutical industry.

Techniques	Merit	Demerit	References
Inkjet printing	A variety of fluids may be used. The resolve is excellent. Microarray printing has found several uses in various fields, including genomics, proteomics, and drug discovery. This technology allows for the simultaneous capacity to effectively merge or blend different types of polymers.	Continuous inkjet printing (CIP) requires a substantial quantity of water. Inefficient use of resources and the gradual pace of production. The issue of high cost is significant. It is important to use caution to ensure proper handling of resins.	[168]
Fused disposition modeling	Affordable and user-friendly. A wide variety of materials may be used.	The surface has a damaged and uneven texture. The fragility of the object is seen at a low resolution in the *Z*-axis.	[169]
Laminated object modeling	The surface finish is characterized by a high level of smoothness and precision. Affordable costs; presence of a support structure is unnecessary. The formation of a three-dimensional structure is achieved by the processes of cutting, laminating, and bonding.	Worsening a problem; delays in delamination.	[170, 171]

Table 5.2 (continued)

Techniques	Merit	Demerit	References
Stereolithography (SLA)	High-resolution layer connection issues are avoided and nozzles are not required. Immediate effects: a well-polished surface.	Homogeneous DNA amplification; raw material prices are quite high. Prints have a low tendency to change shape. Resins need careful treatment to ensure optimal performance.	[172]
Selective laser sintering	Binds together the granules of powder; a large variety of individual elements creates medications with extended-release potential; superior preciseness.	Objects that are weaker and more permeable may be created. Expense, complexity, and risk of laser damage increase with increasing temperature.	[105]

5.6 Current challenges and prospects in healthcare enabled by AM

Recent challenges in healthcare enabled by AM include regulatory challenges, standardization issues, and cost-effectiveness concerns. Regulatory entities have challenges in effectively adapting to the fast advancements of AM technology, resulting in prolonged approval processes and generating a sense of ambiguity among healthcare practitioners. The implementation of standardization initiatives plays a pivotal role in guaranteeing the excellence and compatibility of 3D-printed medical equipment and implants. Furthermore, while AM has potential in the field of customized medicine, its extensive use is hindered by the excessive expenses linked to advanced 3D printing machinery and materials, thus limiting its general implementation. Future efforts should concentrate on developing strong industry standards, simplifying regulatory procedures, and reducing costs via economies of scale and innovation. Furthermore, developments in tissue engineering and bioprinting have enormous potential to produce functioning organs and tissues, transform transplant processes, and enhance patient outcomes over a long period.

5.6.1 Tissue engineering and regenerative medicine

The progress in tissue engineering and regenerative medicine (TERM) is closely linked to the advancements in AM technology for the production of enhanced biomimetic goods. In light of the encouraging outcomes seen in the development of 3D-printed biomedical structures using various biomaterials, there has been a shift in research focus toward the production of intelligent materials. Smart structures are often triggered by the application of external stimuli or the occurrence of physiological changes in the surrounding environment. Additionally, the incorporation of intelligent materials is revolutionizing current designs of bioinks, entering a new age that challenges the conventional notion of requisite materials and technologies. The programming of materials to exhibit dynamic reactions has been extensively explored in the field. Two prominent processes that have been suggested are the use of osmotic pressure, which relies on the swelling property of hydrogels, and thermal actuation, which is based on the shape-memory characteristics of bioinks [173, 174]. These techniques have been identified as crucial and feasible for allowing programmable materials. One of the most essential elements in the future direction of TERM involves the development of useful smart biomaterials specifically designed for scaffolding, surgical implants, and other products within the TERM field.

5.6.2 Pharmaceuticals/drug delivery

Pharmaceutical research and development is progressing ahead of bulk manufacturing and toward individualized dosing for groups of individuals with similar metabolic profiles. The scientific literature suggests that the pharmaceutical sector can make use of all of the ASTM categories listed in Table 5.1, except those dealing with direct thermal deposition and sheet lamination deposition. In the fields of pharmaceutics and medication delivery, FDM and inkjet printing are the preferred 3D printing technologies. These technologies have several advantages over conventional drug manufacturing processes, including high precision at low cost, rapid drug release, individualized dosing, and high dose loading, although they are less productive overall. Furthermore, 4D printing is a novel idea in the pharmaceutical business, with just a few technologies, such as direct ink writing for intravenous drug delivery systems, being deployed [175].

In the future, the pharmacy business may start using other 3D printing technologies. Limited biomaterials are another barrier to AM's widespread use in the pharmaceutical business. Sterilization processes for polymer-based pharmaceuticals, for instance, cannot be too intense or employ temperatures greater than the biopolymer's glass transition point. But shape-recovery hydrogels showed that they could be used for more medicinal purposes in the future. The biggest challenge for AM technology in the mass manufacture of customized pharmaceuticals is scaling up the technologies. As a result, further efforts will be necessary in the future to solve the problems that were discussed before.

5.6.3 Synthetic organs

The major problem and most important thing about the bioprinting of organoids is the precision of the bioprinted tissues. Extensive research initiatives were undertaken to enhance AM technologies within this field, to produce products that closely resemble real tissues. While IJP methods can attain 20–100 µm resolution, the products' mechanical stability is not satisfactory. Similarly, current extrusion-based systems that use microfilaments can achieve a resolution ranging from 200 to 400 µm. However, it is important to note that the attempt at higher resolutions in these systems results in the generation of elevated shear stress, which may have detrimental effects on the cells present in the bioinks. The problem of resolution in inkjet and extrusion-based printing systems often occurs when the size exceeds 50 µm as a result of nozzle blockage [176].

Furthermore, advancements in the formulation of bioinks are crucial in enhancing both resolution capability and printability. There exists a scarcity of biomaterials that possess the necessary characteristics to function as suitable cell carriers, namely enough mechanical stability and minimal cytotoxicity. In addition, the duration of the printing process in bioprinted tissue manufacture is a significant consideration, as prolonged printing time might result in reduced functioning of the bioinks and potential harm to partly printed tissues. The use of multiheaded and parallel bioprinters, together with the implementation of continuous liquid interface production (CLIP), may effectively mitigate the challenges associated with temporal constraints. Hence, the next trajectory in the field of bioprinting organoids will include the advancement of innovative bioinks and the use of more advanced AM methods.

5.7 Conclusion

Research conducted in recent years indicates the significance of AM in facilitating the advancement of healthcare in the future. AM technology has emerged as a novel strategy in the production and design of personalized medical devices, hence enhancing the effectiveness of clinical goods. While there are several 3D printing technologies available, not all of them are suitable for use in the pharmaceutical production industry. However, these technologies organize great potential for future applications and can be utilized in specific manufacturing processes, such as the production of intrauterine devices, for example, implants, microparticles, orodispersible films, and other similar products that require exceptional standards of quality, purity, and precision. AM has significantly contributed to the advancement of several sectors, particularly in the fields of medicine, aerospace, and automotive. AM is a very effective method that contributes to cost and time savings, complexity reduction, quick prototyping, and the facilitation of highly decentralized production. Still, in addition to the many benefits associated with AM technology, there exist some obstacles that impede its rapid expansion. These barriers include restrictions in terms of size, production time, material constraints, as well as the expenses associated with machinery and production. In summary, AM has emerged as a significant catalyst in advancing the progress of healthcare sustainability. AM can produce medical equipment and implants that can be customized to individual patients, possess high efficiency, and exhibit biocompatibility. This technology has the advantage of reducing material waste, energy consumption, and transportation expenses often associated with conventional manufacturing methods. The potential of this intervention to improve patient outcomes, reduce resource use, and support local manufacturing highlights its significance in cultivating a more sustainable healthcare environment. However, the realization of these advantages requires continuous study, active cooperation among relevant parties, and the advancement of environmentally sustainable printing materials. As AM continues to undergo advancements, it has the potential to enhance healthcare practices and make substantial contributions toward achieving a more environmentally friendly and sustainable future within the healthcare industry.

References

[1] Hale L, Linley E, Kalaskar DM. A digital workflow for design and fabrication of bespoke orthoses using 3D scanning and 3D printing, a patient-based case study. Scientific Reports. 2020;10(1):7028.

[2] Panwisawas C, Tang YT, Reed RC. Metal 3D printing as a disruptive technology for superalloys. Nature Communications. 2020;11(1):2327.

[3] Eshkalak SK, Ghomi ER, Dai Y, Choudhury D, Ramakrishna S. The role of three-dimensional printing in healthcare and medicine. Materials and Design. 2020;194:108940.

[4] Santacreu AM, Zhu H. Manufacturing and service sector roles in the evolution of innovation and productivity. Economic Synopses. 2018;2:1–3.

[5] Halstenberg FA, Lindow K, Stark R. Utilization of product lifecycle data from PLM systems in platforms for industrial symbiosis. Procedia Manufacturing. 2017;8:369–76.

[6] Peng T, Kellens K, Tang R, Chen C, Chen G. Sustainability of additive manufacturing: An overview on its energy demand and environmental impact. Additive Manufacturing. 2018;21:694–704.

[7] Awad A, Trenfield SJ, Goyanes A, Gaisford S, Basit AW. Reshaping drug development using 3D printing. Drug Discovery Today: Technologies. 2018;23(8):1547–55.

[8] Bigham A, Foroughi F, Rezvani Ghomi E, Rafienia M, Neisiany RE, Ramakrishna S. The journey of multifunctional bone scaffolds fabricated from traditional toward modern techniques. Bio-Design and Manufacturing. 2020;3:281–306.

[9] Shi B, Huang H. Computational technology for nasal cartilage-related clinical research and application. International Journal of Oral Science. 2020;12(1):21.

[10] Salmi M. Additive manufacturing processes in medical applications. Materials. 2021 Jan 3;14(1):191.

[11] Ford S, Despeisse M. Additive manufacturing and sustainability: An exploratory study of the advantages and challenges. Journal of Cleaner Production. 2016;137:1573–87.

[12] Mani M, Lyons KW, Gupta SK. Sustainability characterization for additive manufacturing. Journal of Research of the National Institute of Standards and Technology. 2014;119:419.

[13] Sreenivasan R, Goel A, Bourell DL. Sustainability issues in laser-based Despeisse additive manufacturing. Physics Procedia. 2010;5:81–90.

[14] Despeisse M, Ford S. The role of additive manufacturing in improving resource efficiency and sustainability. In: Advances in production management systems: Innovative production management towards sustainable growth: IFIP WG 5.7 International Conference, APMS 2015, Tokyo, Japan, September 7–9, 2015, Proceedings, Part II 0 2015 (pp. 129–36).
[15] Jiang J, Newman ST, Zhong RY. A review of multiple degrees of freedom for additive manufacturing machines. International Journal of Computer Integrated Manufacturing. 2021;34(2):195–211.
[16] Javaid M, Haleem A. Additive manufacturing applications in orthopedics: A review. Journal of Clinical Orthopedics and Trauma. 2018;9(3):202–06.
[17] Huang Y, Leu MC, Mazumder J, Donmez A. Additive manufacturing: Current state, future potential, gaps and needs, and recommendations. Journal of Manufacturing Science and Engineering. 2015 Feb 1;137(1):014001.
[18] Baumers M, Tuck C, Bourell DL, Sreenivasan R, Hague R. Sustainability of additive manufacturing: Measuring the energy consumption of the laser sintering process. Proceedings of the Institution of Mechanical Engineers, Part B: Journal of Engineering Manufacture. 2011, 225(12), 2228–39.
[19] Khorram, Niaki M, Nonino F, Palombi G, Torabi SA. Economic sustainability of additive manufacturing: Contextual factors driving its performance in rapid prototyping. Journal of Manufacturing Technology Management. 2019;30(2):353–65.
[20] Yang S, Min W, Ghibaudo J, Zhao YF. Understanding the sustainability potential of part consolidation design supported by additive manufacturing. Journal of Cleaner Production. 2019;232:722–38.
[21] Gebisa AW, Lemu HG. Design for manufacturing to design for Additive Manufacturing: Analysis of implications for design optimality and product sustainability. Procedia Manufacturing. 2017;13:724–31.
[22] Ghobadian A, Talavera I, Bhattacharya A, Kumar V, Garza-Reyes AJ, O'regan N. Examining legitimatization of additive manufacturing in the interplay between innovation, lean manufacturing, and sustainability. International Journal of Production Economics. 2020;219:457e468.
[23] Gardan J. Additive manufacturing technologies: State of the art and trends. Additive Manufacturing Handbook. 2017 May 19:149–68.
[24] Harris ID. Additive manufacturing: A transformational advanced manufacturing technology: Additive manufacturing represents a new paradigm and offers a range of opportunities for design, functionality, and cost. Advanced Materials & Processes. 2012;170(5):25–30.
[25] Wong KV, Hernandez A. A review of additive manufacturing. ISRN Mechanical Engineering. 2012;2012.
[26] Hartke K. Manufacturing Technology Support (MATES). Task Order 0021: Air Force Technology and Industrial Base Research and Analysis, Subtask Order 06: Direct Digital Manufacturing. Air Force Research Laboratory Materials and Manufacturing Directorate. 2011:1–94.
[27] Gibson I, Rosen DW, Stucker B, Khorasani M, Rosen D, Stucker B, Khorasani M. Additive Manufacturing Technologies. Cham, Switzerland: Springer; 2021.
[28] Hopkinson N. Additive Manufacturing: Technology and Applications; 2010.
[29] Levy GN, Schindel R, Kruth JP. Rapid manufacturing and rapid tooling with layer manufacturing (LM) technologies, state of the art and future perspectives. CIRP Annals. 2003;52(2):589–609.
[30] Kruth JP. Material increase manufacturing by rapid prototyping techniques. CIRP Annals. 1991;40 (2):603–14.
[31] Ngo TD, Kashani A, Imbalzano G, Nguyen KT, Hui D. Additive manufacturing (3D printing): A review of materials, methods, applications, and challenges. Composites Part B: Engineering. 2018;143:172–96.
[32] Thompson MK, Moroni G, Vaneker T, Fadel G, Campbell RI, Gibson I, Bernard A, Schulz J, Graf P, Ahuja B, Martina F. Design for additive manufacturing: Trends, opportunities, considerations, and constraints. CIRP Annals. 2016;65(2):737–60.

[33] Dilberoglu UM, Gharehpapagh B, Yaman U, Dolen M. The role of additive manufacturing in the era of industry 4.0. Procedia Manufacturing. 2017;11:545–54.
[34] Berman B. 3-D printing: The new industrial revolution. Business Horizons. 2012;55(2):155–62.
[35] Chang J, He J, Mao M, Zhou W, Lei Q, Li X, Li D, Chua CK, Zhao X. Advanced material strategies for next-generation additive manufacturing. Materials. 2018;11(1):166.
[36] Gao W, Zhang Y, Ramanujan D, Ramani K, Chen Y, Williams CB, Wang CC, Shin YC, Zhang S, Zavattieri PD. The status, challenges, and future of additive manufacturing in engineering. Computer-Aided Design. 2015;69:65–89.
[37] Kohtala C. Addressing sustainability in research on distributed production: An integrated literature review. Journal of Cleaner Production. 2015;106:654–68.
[38] Huang SH, Liu P, Mokasdar A, Hou L. Additive manufacturing and its societal impact: A literature review. The International Journal of Advanced Manufacturing Technology. 2013 Jul 67:1191–203.
[39] Stock T, Seliger G. Opportunities of sustainable manufacturing in industry 4.0. Procedure CIRP. 2016;40:56–41.
[40] Huang R, Riddle M, Graziano D, Warren J, Das S, Nimbalkar S, Cresko J, Masanet E. Energy and emissions saving potential of additive manufacturing: The case of lightweight aircraft components. Journal of Cleaner Production. 2016;135:1559–70.
[41] Newman ST, Nassehi A, Imani-Asrai R, Dhokia V. Energy efficient process planning for CNC machining. Cirp Journal of Manufacturing Science & Technology. 2012;5(2):127–36.
[42] Fotovvati B, Namdari N, Dehghanghadikolaei A. Fatigue performance of selective laser melted Ti6Al4V components: State of the art. Materials Research Express. 2018;6(1):012002.
[43] Bandyopadhyay A, Bose S, Das S. 3D printing of biomaterials. MRS Bulletin. 2015;40(2):108–15.
[44] Trenfield SJ, Awad A, Madla CM, Hatton GB, Firth J, Goyanes A, Gaisford S, Basit AW. Shaping the future: Recent advances of 3D printing in drug delivery and healthcare. Expert Opinion on Drug Delivery. 2019;10:1081–94.
[45] Jakus AE. An introduction to 3D printing – Past, present, and future promise. 3D Printing in Orthopaedic Surgery. 2019:1–15.
[46] Ligon SC, Liska R, Stampfl J, Gurr M, Mülhaupt R. Polymers for 3D printing and customized additive manufacturing. Chemical Reviews. 2017;117(15):10212–90.
[47] Liaw CY, Guvendiren M. Current and emerging applications of 3D printing in medicine. Biofabrication. 2017;9(2):024102.
[48] Klosterman D, Chartoff R, Graves G, Osborne N, Priore B. Interfacial characteristics of composites fabricated by laminated object manufacturing. Composites: Part A Applied Science and Manufacturing. 1998;29:9–10, 1165–74.
[49] Wring T. Review of the applications possible with a helisys LOM model. In: Proceedings of the 3rd European conference on rapid prototyping & manufacturing. University of Nottingham, UK 1994, (Vol. 87).
[50] Le HP. Progress and trends in ink-jet printing technology. Journal of Imaging Science and Technology. 1998;42(1):49–62.
[51] Singh M, Haverinen HM, Dhagat P, Jabbour GE. Inkjet printing – Process and its applications. Advances in Materials. 2010;22(6):673–85.
[52] Hassan M, Dave K, Chandrawati R, Dehghani F, Gomes VG. 3D printing of biopolymer nanocomposites for tissue engineering: Nanomaterials, processing and structure-function relation. European Polymer Journal. 2019;121:109340.
[53] Jiao T, Lian Q, Zhao T, Wang H, Li D. Preparation, mechanical and biological properties of inkjet printed alginate/gelatin hydrogel. Journal of Bionic Engineering. 2021;18:574–83.
[54] Zhang H, Moon SK, Ngo TH. 3D printed electronics of non-contact ink writing techniques: Status and promise. International Journal of Precision Engineering and Manufacturing-Green Technology. 2020 Mar 7:511–24.

[55] Bisht B, Hope A, Mukherjee A, Paul MK. Advances in the fabrication of scaffold and 3D printing of biomimetic bone graft. Annals of Biomedical Engineering. 2021;49(4):1128–50.
[56] Wang S, Zhao S, Yu J, Gu Z, Zhang Y. Advances in translational 3D printing for cartilage, bone, and osteochondral tissue engineering. Small. 2022;18(36):2201869.
[57] Chinnappan A, Jayathilaka WA, Khatibzadeh M, Kowsari E, Ramakrishna S. A review on inkjet printing of CNT composites for smart applications. Applied Materials Today. 2017;9:372–86.
[58] Crump SS. Fast, precise, safe prototypes with FDM. ASME PED. 1991;50:53–60.
[59] Kruth JP, Leu MC, Nakagawa T. Progress in additive manufacturing and rapid prototyping. Cirp Annals. 1998;47(2):525–40.
[60] Pham DT, Gault RS. A comparison of rapid prototyping technologies. International Journal of Machine Tools and Manufacture. 1998;38:10–11, 1257–87.
[61] Hazrat Ali M, Abilgaziyev A. Fused deposition modeling based 3D printing: Design, ideas, simulations. Fused Deposition Modeling Based 3D Printing. 2021:23–42.
[62] Beg S, Almalki WH, Malik A, Farhan M, Aatif M, Rahman Z, et al. 3D printing for drug delivery and biomedical applications. Drug Discovery Today. 2020;25:9.
[63] Kiran ASK, Veluru JB, Merum S, Radhamani AV, Doble M, Kumar TSS, et al. Additive manufacturing technologies: An overview of challenges and perspective of using electrospraying. Nanocomposites [Internet]. 2018;4(4):190–214.
[64] Repka MA, Majumdar S, Kumar Battu S, Srirangam R, Upadhye SB. Applications of hot–melt extrusion for drug delivery. Expert Opinion on Drug Delivery. 2008;5(12):1357–76.
[65] Waheed S, Cabot JM, Macdonald NP, Lewis T, Guijt RM, Paull B, et al. 3D printed microfluidic devices: Enablers and barriers. Lab Chip. 2016;16:1993–2013.
[66] Feygin M, Hsieh B. Laminated object manufacturing (LOM): A simpler process. In: The 2nd Solid Freeform Fabrication Symposium, Austin, TX, 1991, 123–30.
[67] Kamrani AK, Nasr EA. Engineering Design and Rapid Prototyping. Springer Science & Business Media; 2010.
[68] Varghese R, Salvi S, Sood P, Karsiya J, Kumar D. Recent advancements in Additive Manufacturing Techniques Employed in the Pharmaceutical Industry: A Bird's Eye View. Annals of 3D Printed Medicine. 2022 Oct 20:100081.
[69] Vermeeren CA. An historic overview of the development of fibre metal laminates. Applied Composite Materials. 2003 Jul 10:189–205.
[70] Kechagias J, Anagnostopoulos V, Zervos S, Chryssolouris G. Estimation of build times in rapid prototyping processes. In: Proceedings of the 6th European conference on rapid prototyping & manufacturing. Nottingham: University of Nottingham; 1997, 137–48.
[71] Kechagias J, Maropoulos S, Karagiannis S. Process build–time estimator algorithm for laminated object manufacturing. Rapid Prototyping Journal. 2004;10(5):297–304.
[72] Chryssolouris G, Kechagias J, Moustakas P, Koutras E. An experimental investigation of the tensile strength of parts produced by laminated object manufacturing (LOM) process. CIRP Journal of Manufacturing Science and Technology. 2003;32(5):319–22.
[73] Wang Y, Zhou Y, Lin L, Corker J, Fan M. Overview of 3D additive manufacturing (AM) and corresponding AM composites. Composites Part A: Applied Science and Manufacturing. 2020;139:106114.
[74] Gomes C, Travitzky N, Greil P, Acchar W, Birol H, Pedro Novaes de Oliveira A, et al. Laminated object manufacturing of LZSA glass–ceramics. Rapid Prototyping Journal. 2011;17(6):424–28.
[75] Mueller B, Kochan D. Laminated object manufacturing for rapid tooling and patternmaking in foundry industry. Computing Science and Computer Engineering. 1999;39(1):47–53.
[76] Gross BC, Erkal JL, Lockwood SY, Chen C, Spence DM. Evaluation of 3D printing and its potential impact on biotechnology and the chemical sciences. Analytical Chemistry. 2014;86(7):3240–53.

[77] Jacobs PF. Rapid Prototyping & Manufacturing: Fundamentals of Stereolithography. Society of Manufacturing Engineers; 1992.
[78] Curti C, Kirby DJ, Russell CA. Current formulation approaches in the design and development of solid oral dosage forms through three–dimensional printing. Additive manufacturing. 2020;5 (2):111–23.
[79] Fastø MM, Genina N, Kaae S, Kalvemark Sporrong S. Perceptions, preferences € and acceptability of patient designed 3D printed medicine by polypharmacy patients: A pilot study. International Journal of Clinical Pharmacology. 2019;41(5):1290–98.
[80] Park J, Tari MJ, Hahn HT. Characterization of the laminated object manufacturing (LOM) process. Rapid Prototyping Journal. 2000 Mar 1;6(1):36–50.
[81] Sonmez FO, Hahn HT. Thermomechanical analysis of the laminated object manufacturing (LOM) process. Rapid Prototyping Journal. 1998;4(1):26–36.
[82] Pak SS, Nisnevich G. Interlaminate strength and processing efficiency improvements in laminated object manufacturing. In: 5th international conference on rapid prototyping; 1994, 171–80.
[83] Reece SR, Styger LJ. Laminated object manufacturing: Process practice and research experience. In: First national conference on rapid prototyping and tooling research. UK: Buckinghamshire College; 1995, 101–10.
[84] Reeves PE, Cobb RC. Surface deviation modeling of LMT Processes–a comparative analysis. In: Proceedings of fifth European conference on rapid prototyping and manufacturing; 1995.
[85] Hull C. Stereolithography: Plastic prototype from CAD data without tooling. Mod Cast. 1988;8:38.
[86] Kumar P, Rajak DK, Abubakar M, Ali SGM, Hussain M. 3D printing technology for biomedical practice: A review. Journal of Materials Engineering and Performance. 2021;30:5342–55.
[87] Manapat JZ, Chen Q, Ye P, Advincula RC. 3D printing of polymer nanocomposites via stereolithography. Macromolecular Materials and Engineering. 2017;302:1600553.
[88] Kang JH, Sakthiabirami K, Jang KJ, Jang JG, Oh GJ, Park C, et al. Mechanical and biological evaluation of lattice structured hydroxyapatite scaffolds produced via stereolithography additive manufacturing. Materials and Design. 2022;214:110372.
[89] Renap K, Kruth JP. Recoating issues in stereolithography. Rapid Prototyping Journal. 1995;1:4–16.
[90] Low ZX, Chua YT, Ray BM, Mattia D, Metcalfe IS, Patterson DA. Perspective on 3D printing of separation membranes and comparison to related unconventional fabrication techniques. Journal of Membrane Science and Technology. 2017 Feb 1;523:596–613.
[91] Bahati D, Bricha M, El, Mabrouk K. Vat photopolymerization additive manufacturing technology for bone tissue engineering applications. Advanced Engineering Materials. 2023;25:1, 2200859.
[92] Palucci RR, Rosace G. Nanomaterials for 3D printing of polymers via stereolithography: Concept, technologies, and applications. Macromolecular Materials and Engineering. 2021;306:2100345.
[93] Guo N, Leu MC. Additive manufacturing: Technology, applications and research needs. Frontiers of Mechanical Engineering. 2013 Sep 8:215–43.
[94] Deckard C, Beaman JJ. Process and control issues in selective laser sintering. ASME PED. 1988;33:191–97.
[95] Awad A, Fina F, Trenfield S, Patel P, Goyanes A, Gaisford S, et al. 3D printed pellets (miniprintlets): A novel, multi–drug, controlled release platform technology. Pharmaceutics. 2019;11(4):148.
[96] Trenfield SJ, Goyanes A, Telford R, Wilsdon D, Rowland M, Gaisford S, et al. 3D printed drug products: Non–destructive dose verification using a rapid point –and–shoot approach. International Journal of Pharmaceutical Sciences. 2018;549(1–2):283–92.
[97] Fina F, Goyanes A, Gaisford S, Basit AW. Selective laser sintering (SLS) 3D printing of medicines. International Journal of Pharmacy and Pharmaceutical Sciences. 2017;529(1–2):285–93.
[98] Trenfield SJ, Tan HX, Goyanes A, Wilsdon D, Rowland M, Gaisford S, Basit AW. Non-destructive dose verification of two drugs within 3D printed polyprintlets. International Journal of Pharmaceutics. 2020 Mar 15;577:119066.

[99] Caulfield B, McHugh PE, Lohfeld S. Dependence of mechanical properties of polyamide components on build parameters in the SLS process. Journal of Materials Processing Technology. 2007;182(1–3):477–88.
[100] Gueche YA, Sanchez–Ballester NM, Bataille B, Aubert A, Leclercq L, Rossi J–C, et al. Selective laser sintering of solid oral dosage forms with copovidone and paracetamol using a CO2 laser. Pharmaceutics. 2021;13(2):160.
[101] Allahham N, Fina F, Marcuta C, Kraschew L, Mohr W, Gaisford S, et al. Selective laser sintering 3D printing of orally disintegrating printlets containing ondansetron. Pharmaceutics. 2020;12(2):110.
[102] Mohamed EM, Barakh ASF, Rahman Z, Dharani S, Ozkan T, Kuttolamadom MA, et al. Formulation optimization of selective laser sintering 3D–printed tablets of clindamycin palmitate hydrochloride by response surface methodology. AAPS PharmSciTech. 2020;21:6.
[103] Schmid M, Amado A, Wegener K. Polymer powders for selective laser sintering (SLS). In: AIP Conference proceedings, 2015 May 22 (Vol. 1664, No. 1).
[104] Barakh A, Mohamed SF, Ozkan EM, Kuttolamadom T, Khan MA, Asadi A, et al. Understanding the effects of formulation and process variables on the printlets quality manufactured by selective laser sintering 3D printing. International Journal of Pharmaceutical Sciences. 2019;5700:118651.
[105] Hamed R, Mohamed EM, Rahman Z, Khan MA. 3D–printing of lopinavir printlets by selective laser sintering and quantification of crystalline fraction by XRPD-chemometric models. International Journal of Pharmaceutical Sciences. 2021;592:120059.
[106] Kulinowski P, Malczewski P, Pesta E, °aszcz M, Mendyk A, Polak S, et al. Selective laser sintering (SLS) technique for pharmaceutical applications – Development of high dose controlled-release printlets. Additive manufacturing. 2021;38:101761.
[107] Charoo NA, Barakh Ali SF, Mohamed EM, Kuttolamadom MA, Ozkan T, Khan MA, Rahman Z. Selective laser sintering 3D printing–an overview of the technology and pharmaceutical applications. Drug Development and Industrial Pharmacy. 2020 Jun 2;46(6):869–77.
[108] Zhang H, LeBlanc S. Processing parameters for selective laser sintering or melting of oxide ceramics additive manufacturing of high–performance metals and alloys – modeling and optimization [internet]. InTech. 2018.
[109] Salmoria GV, Vieira FE, Ghizoni GB, Marques MS, Kanis LA. 3D printing of PCL/Fluorouracil tablets by selective laser sintering: Properties of implantable drug delivery for cartilage cancer treatment. drugs. 2017;4(6).
[110] Pingali KC, Saranteas K, Foroughi R, Muzzio FJ. Practical methods for improving flow properties of active pharmaceutical ingredients. Drug Development and Industrial Pharmacy. 2009;35(12):1460–69.
[111] Yang Y, Xu Y, Wei S, Shan W. Oral preparations with tunable dissolution behavior based on selective laser sintering technique. International Journal of Pharmaceutical Sciences. 2021;593:120127.
[112] Leong KF, Chua CK, Gui WS, Verani. Building porous biopolymeric microstructures for controlled drug delivery devices using selective laser sintering. International Journal of Advanced Manufacturing Technology. 2006;31(5–6):483–89.
[113] Goole J, Amighi K. 3D printing in pharmaceutics: A new tool for designing customized drug delivery systems. International Journal of Pharmaceutical Sciences. 2016;499(1 –2):376–94.
[114] Hamburg MA, Collins FS. The path to personalized medicine. New England Journal of Medicine. 2010;363:4, 30.
[115] Tremblay J, Hamet P. Role of genomics on the path to personalized medicine. Metabolism. 2013;62: S2–5.
[116] El Aita I, Breitkreutz J, Quodbach J. On–demand manufacturing of immediate release levetiracetam tablets using pressure–assisted microsyringe printing. European Journal of Pharmaceutical Research. 2019;134:29–36.
[117] Khaled SA, Burley JC, Alexander MR, Yang J, Roberts CJ. 3D printing of tablets containing multiple drugs with defined release profiles. International Journal of Pharmacy and Pharmaceutical Sciences. 2015;494(2):643–50.

[118] Sachs E, Cima M, Cornie J. Three-dimensional printing: Rapid tooling andprototypes directly from a CAD model. CIRP Annals - Manufacturing Technology. 1990;39:201–04.
[119] Lembo J. Three-dimensional printing: A unique solid freeform fabrication technology that fabricates tools and parts from a variety of materials. Advanced Materials & Processes. 2003 Jan 1;161 (1):52–54.
[120] Zhang S, Bilal M, Zdarta J, Cui J, Kumar A, Franco M, et al. Biopolymers and nanostructured materials to develop pectinases-based immobilized nano-biocatalytic systems for biotechnological applications. Food Research International. 2021;140:109979.
[121] Montero Y, Souza AG, Oliveira ÉR, Rosa D, dos S. Nanocellulose functionalized with cinnamon essential oil: A potential application in active biodegradable packaging for strawberry. Sustainable Materials and Technologies. 2021;29:00289.
[122] Yang GH, Kim W, Kim J, Kim G. A skeleton muscle model using GelMA-based cell-aligned bioink processed with an electric-field assisted 3D/4D bioprinting. Theranostics. 2021;11:48–63.
[123] Tong A, Pham QL, Abatemarco P, Mathew A, Gupta D, Iyer S, et al. Review of low-cost 3D bioprinters: State of the market and observed future trends. SLAS Technology (Translating Life Sciences Innovation). 2021;26:333–66.
[124] Cano-Vicent A, Tambuwala MM, Hassan SS, Barh D, Aljabali AAA, Birkett M, et al. Fused deposition modelling: Current status, methodology, applications and future prospects. Additive manufacturing. 2021;47:102378.
[125] Bertana V, Catania F, Cocuzza M, Ferrero S, Scaltrito L, Pirri CF. Medical and biomedical applications of 3D and 32. Asian Journal of Pharmaceutical Sciences. 2023;18:100812.
[126] Singh S, Prakash C, Ramakrishna S. 3D printing of polyether-ether-ketone for biomedical applications. European Polymer Journal. 2019;114:234–48.
[127] Patil N, Andasubramanian B. Functionalized polylysine biomaterials for advanced medical applications: A review. European Polymer Journal. 2021;146:110248.
[128] Fazal F, Raghav S, Callanan A, Koutsos V, Radacsi N. Recent advancements in the bioprinting of vascular grafts. Biofabrication. 2021;13:32003.
[129] Higgins M, Leung S, Radacsi N. 3D printing surgical phantoms and their role in the visualization of medical procedures. Annals of 3D Printed Medicine. 2022;6:100057.
[130] Rezvani Ghomi E, Khosravi F, Neisiany RE, Singh S, Ramakrishna S. Future of additive manufacturing in healthcare. Current Opinion in Biomedical Engineering. 2021;17:100255.
[131] Jakus AE. An introduction to 3D printing past, present, and future promise. 3D Printing in Orthopaedic Surgery. 2018:1.
[132] Calza F, Parmentola A, Tutore I. Types of green innovations: Ways of implementation in a non-green industry. Sustainability. 2017 Jul 26;9(8):1301.
[133] Niaki MK, Torabi SA, Nonino F. Why manufacturers adopt additive manufacturing technologies: The role of sustainability. Journal of Cleaner Production. 2019;222:381–92.
[134] Villamil C, Nylander J, Hallstedt SI, Schulte J, Watz M. Additive manufacturing from a strategic sustainability perspective. In: DS 92: Proceedings of the DESIGN 2018 15th International Design Conference, 2018, 1381–92.
[135] Jiang J, Xiong Y, Zhang Z, Rosen DW. Machine learning integrated design for additive manufacturing. Journal of Intelligent Manufacturing. 2022;33(4):1073–86.
[136] Moldavska A, Welo T. The concept of sustainable manufacturing and its definitions: A content-analysis based literature review. Journal of Cleaner Production. 2017 Nov 10;166:744–55.
[137] Grünbaum D, Okubo A. Modelling social animal aggregations. Frontiers in Mathematical Biology. 1994:296–325.
[138] Hens L, Cabello-Eras JJ, Sagastume-Gutiérez A, Garcia-Lorenzo D, Cogollos-Martinez JB, Vandecasteele C. University–industry interaction on cleaner production. The case of the Cleaner

Production Center at the University of Cienfuegos in Cuba, a country in transition. Journal of Cleaner Production. 2017 Jan 20;142:63–68.
[139] Mani M, Lyons KW, Gupta SK. Sustainability characterization for additive manufacturing. Journal of Research of the National Institute of Standards and Technology. 2014;119:419.
[140] Colorado HA, Velásquez EI, Monteiro SN. Sustainability of additive manufacturing: The circular economy of materials and environmental perspectives. Journal of Materials Research and Technology. 2020;9(4):8221–34.
[141] Jiang J, Xu X, Stringer J. Support structures for additive manufacturing: A review. Journal of Manufacturing and Materials Processing. 2018;2(4):64.
[142] Ahn DG. Direct metal additive manufacturing processes and their sustainable applications for green technology: A review. International Journal of Precision Engineering and Manufacturing-Green Technology. 2016;3:381–95.
[143] Khosravani MR, Reinicke T. On the environmental impacts of 3D printing technology. Applied Materials Today. 2020 Sep 1;20:100689.
[144] Machado CG, Despeisse M, Winroth M, da Silva EH. Additive manufacturing from the sustainability perspective: Proposal for a self-assessment tool. Procedia CIRP. 2019;81:482–87.
[145] Freitas D, Almeida HA, Bártolo H, Bártolo PJ. Sustainability in extrusion-based additive manufacturing technologies. Progress in Additive Manufacturing. 2016:65–78.
[146] Oyesola M, Mathe N, Mpofu K, Fatoba S. Sustainability of Additive Manufacturing for the South African aerospace industry: A business model for laser technology production, commercialization and market prospects. Procedia CIRP. 2018;72:1530–35.
[147] Den Boer J, Lambrechts W, Krikke H. Additive manufacturing in military and humanitarian missions: Advantages and challenges in the spare parts supply chain. Journal of Cleaner Production. 2020;257:120301.
[148] Meteyer S, Xu X, Perry N, Zhao YF. Energy and material flow analysis of binder-jetting additive manufacturing processes. Procedia Cirp. 2014;15:19–25.
[149] Chen D, Heyer S, Ibbotson S, Salonitis K, Steingrímsson JG, Thiede S. Direct digital manufacturing: Definition, evolution, and sustainability implications. Journal of Cleaner Production. 2015;107:615–25.
[150] Faludi J, Bayley C, Bhogal S, Iribarne M. Comparing environmental impacts of additive manufacturing vs traditional machining via life-cycle assessment. Rapid Prototyping Journal. 2015;21(1):14–33.
[151] Rejeski D, Zhao F, Huang Y. Research needs and recommendations on environmental implications of additive manufacturing. Additive Manufacturing. 2018;19:21–28.
[152] Gu W, Styger E, Warner DH. Assessment of additive manufacturing for increasing sustainability and productivity of smallholder agriculture. 3D Printing and Additive Manufacturing. 2020;7(6):300–10.
[153] Haleem A, Javaid M. Role of CT and MRI in the design and development of orthopaedic model using additive manufacturing. Journal of Clinical Orthopaedics & Trauma. 2018;9(3):213–17.
[154] Strano G, Hao L, Everson RM, Evans KE. A new approach to the design and optimisation of support structures in additive manufacturing. The International Journal of Advanced Manufacturing Technology. 2013;66:1247–54.
[155] Despeisse M, Yang M, Evans S, Ford S, Minshall T. Sustainable value road mapping framework for additive manufacturing. Procedia CIRP. 2017;61:594–99.
[156] Holmström J, Partanen J, Tuomi J, Walter M. Rapid manufacturing in the spare parts supply chain: Alternative approaches to capacity deployment. Journal of Manufacturing Technology Management. 2010;21:687–97.
[157] Hill C. International business: Competing in the global marketplace. Strategic Plannings. 2008:24.
[158] Sutherland W, Aveling R, Brooks T, Clout M, Dicks L, Fellman L, Fleishman E, Gibbons D, Keim B, Lickorish F. A horizon scan of global conservation issues for 2014. Trends in Ecology & Evolution. 2014;29:15–22.

[159] Petrovic V, Vicente Haro Gonzalez J, Jordá Ferrando O, Delgado Gordillo J, Ramón Blasco Puchades J, Portolés Griñan L. Additive layered manufacturing: Sectors of industrial application shown through case studies. International Journal of Production Research. 2011;49:1061–79.

[160] Bobbio LD, Otis RA, Borgonia JP, Dillon RP, Shapiro AA, Liu Z-K, Beese AM. Additive manufacturing of a functionally graded material from ti-6al-4v to invar: Experimental characterization and thermodynamic calculations. Acta Material. 2017;127:133–42.

[161] Leu MC, Deuser BK, Tang L, Landers RG, Hilmas GE, Watts JL. Freeze-form extrusion fabrication of functionally graded materials. CIRP Annals. 2012:61, 223–226.

[162] Wang J, Goyanes A, Gaisford S, Basit AW. Stereolithographic (SLA) 3D printing of oral modified –release dosage forms. International Journal of Pharmacy and Pharmaceutical Sciences. 2016;503(1–2):207–12.

[163] Karakurt I, Aydoğdu A, Çıkrıkcı S, Orozco J, Lin L. Stereolithography (SLA) 3D printing of ascorbic acid loaded hydrogels: A controlled release study. International Journal of Pharmaceutics. 2020;584:119428.

[164] Holländer J, Genina N, Jukarainen H, Khajeheian M, Rosling A, Mäkilä E, Sandler N. Three-dimensional printed PCL-based implantable prototypes of medical devices for controlled drug delivery. Journal of Pharmaceutical Sciences. 2016;105(9):2665–76.

[165] Goyanes A, Madla CM, Umerji A, Montero. JM, Lamas MJ, et al. Automated therapy preparation of isoleucine formulations using 3D printing for the treatment of MSUD: First single-center, prospective, crossover study in patients. International Journal of Pharmacy and Pharmaceutical Sciences. 2019;567:118497.

[166] Ragelle H, Rahimian S, Guzzi EA, Westenskow PD, Tibbitt MW, Schwach G, Langer R. Additive manufacturing in drug delivery: Innovative drug product design and opportunities for industrial application. Advanced Drug Delivery Reviews. 2021 Nov 1;178:113990.

[167] Martinez PR, Goyanes A, Basit AW, Gaisford S. Fabrication of drug–loaded hydrogels with stereolithographic 3D printing. International Journal of Pharmaceutical Sciences. 2017;532(1):313.

[168] Sandler N, Määttänen A, Ihalainen P, Kronberg L, Meierjohann A, Viitala T, Peltonen J. Inkjet printing of drug substances and use of porous substrates-towards individualized dosing. Journal of Pharmaceutical Sciences. 2011;100(8):3386–95.

[169] Saldivar J–S, Taylor D, Sugarman E, Cullors A, Garces J, Oades K, et al. Initial assessment of the benefits of implementing pharmacogenetics into the medical management of patients in a long –term care facility. Pharmacogenomics. 2016;1.

[170] Huang W, Zheng Q, Sun W, Xu H, Yang X. Levofloxacin implants with predefined microstructure fabricated by three–dimensional printing technique. International Journal of Pharmacy and Pharmaceutical Sciences. 2007;339(1–2):33–38.

[171] Guideline IH. Pharmaceutical development Q8. Current Step. 2005 Nov 4:11.

[172] Trivedi M, Jee J, Silva S, Blomgren C, Pontinha VM, Dixon DL, et al. Additive manufacturing of pharmaceuticals for precision medicine applications: A review of the promises and perils in implementation. Additive Manufacturing. 2018;23:319–28.

[173] Sydney A, Matsumoto EA, Nuzzo RG, Mahadevan L, Lewis JA. Biomimetic 4D printing. Nature Mater. 2016;15:413–18.

[174] Politi S, Carotenuto F, Rinaldi A, Di, Nardo P, Manzari V, Albertini MC, Araneo R, Ramakrishna S, Teodori L. Smart ECM-Based Electrospun Biomaterials for Skeletal Muscle Regeneration. Nanomaterials. 2020:10.

[175] Durga P, Reddy R, Sharma V. Additive manufacturing in drug delivery applications: A review. International Journal of Pharmaceutics. 2020;589:119820.

[176] Xie Z, Gao M, Lobo AO, Webster TJ. 3D bioprinting in tissue engineering for medical applications: The classic and the hybrid. Polymers. 2020:12.

Chapter 6
Revolutionizing healthcare sustainability through mHealth applications

Abstract: Health is essential for enhancing the quality of life. The provision of efficient healthcare services is an important factor in promoting the social and economic advancement of a country. A significant proportion of individuals living in developing nations, particularly those in rural and isolated regions, lack sufficient access to healthcare facilities. The recent development of technologies for telecommunication has the potential to enhance healthcare services. The utilization of mobile health (mHealth) holds significant promise as a supportive system within the healthcare sector for addressing the disparities in providing healthcare by the urban and rural hospitals. Health concerns are often considered to be the primary obstacle to achieving sustainable global development. The integration of information and communication technologies into the healthcare sector, particularly the use of mobile communications, is revolutionizing the provision of healthcare services by enhancing accessibility, affordability, and availability. The chapter analyzes the incorporation of mHealth apps into healthcare systems, proving their capacity to improve sustainability. mHealth enables individuals to take an active role in managing their health, enhances the efficiency of healthcare delivery, and minimizes the environmental impact associated with conventional healthcare procedures. In conclusion, mHealth apps provide a great opportunity to enhance the efficiency, cost-effectiveness, and environmental sustainability of healthcare systems. The purpose of the study is to assess the potential, difficulties, and obstacles involved in developing a mobile healthcare system in poor countries. A prospective mHealth model has been developed, which is based on mobile cellular networks. It presents a comprehensive set of recommendations to facilitate the effective utilization of a mobile healthcare system.

6.1 Introduction

The implementation of mobile health (mHealth) applications is a significant advancement in the field of global healthcare and has received substantial attention [1]. mHealth refers to the application of mobile devices and wireless equipment in the field of medicine and public health [2]. mHealth applications perform a range of tasks related to individual healthcare and medical data, thereby playing a significant role in the efficient performance of public health management [3]. The utilization of digital health, particularly mHealth, has been observed to facilitate beneficial modifications in health behaviors aimed at preventing the occurrence of chronic and acute diseases [4]. Online healthcare is more adaptable and convenient compared to conventional treatment methods because it eliminates geographic restrictions and shortens the wait time

https://doi.org/10.1515/9783111436432-006

for doctor appointments [5]. In 2021, there will be more than 3.5 million mHealth applications available on the market [6], indicating widespread adoption and a substantial user base. The provision of healthcare services is a critical factor in ensuring the long-term financial sustainability and development of a nation. The status of health has a crucial role in the national growth of the economy and society. The recent development of smart mobile phones has presented an exceptional chance to supply healthcare services to underprivileged communities in a prompt, interactive, and personalized approach. The field of mHealth, often referred to as mHealth, is revolutionizing the provision of healthcare by enhancing its accessibility, affordability, and availability [7]. The use of mobile devices has had a substantial impact on the accessibility of information, the streamlining of workflow processes, and the advancement of evidence-based practice, hence enabling individuals to make well-informed and efficient decisions at the location where care is provided [8]. The use of mHealth has been implemented to gather health data remotely, monitor patients from a distance, and enhance the quality of life for individuals through the use of mobile devices [9].

The use of mHealth has become more prevalent in developing nations due to the significant increase in mobile phone usage. Mobile phones are considered superior platforms for delivering health care in underdeveloped nations due to their small size, mobility, low energy consumption, and capacity to function with inadequate infrastructure. The use of mHealth in poor countries aims to increase access to excellent healthcare and improve the ability of health systems to serve bigger populations [10–12]. The introduction of mHealth apps has improved the delivery of healthcare facilities, benefiting both healthcare professionals and patients [10]. mHealth applications are especially useful in remote regions where there is limited availability of modern medical services [11]. The impact of digitalization on society was evident in the use of mHealth applications for delivering medical treatments during the COVID-19 pandemic [12]. mHealth applications enabled users to decrease the frequency of clinic visits and minimize travel to areas with a higher risk of health complications [13]. The WHO also promotes the progress and broad acceptance of mHealth technology, to enhance healthcare facilities in different countries to attain complete health coverage, meet the health-related Sustainable Development Goals, and accomplish other health objectives described in its Global Action Plan [14]. The WHO has developed a toolbox to promote national eHealth systems [15]. For a health plan to effectively yield its advantages to the public, it must be sustained and maintained over an extended period [16]. The use of such technology has significantly enhanced the efficiency of decision-making and production processes within the health and healthcare sectors. This occurs via the efficient transmission of timely and accurate data to the appropriate individuals [17]. It has also substantially improved information availability, and workflow, and encouraged the use of evidence-based techniques to make educated and effective choices immediately at the point of care [18–24].

Mobile communications have already been used in the field of healthcare, specifically in the area of mHealth. The technology has been employed for remote data collecting, remote monitoring, and enhancing the overall quality of life for patients [25].

Mobile communications can significantly enhance healthcare in rural and resource-poor situations, as shown by increased interest and evidence [25]. The potential influence of mobile communications-based healthcare, also known as mHealth, is anticipated to have a broader scope compared to previous advancements. In the rapidly evolving healthcare industry, consumers are demanding providers change from volume-based treatment (pay for service) to a value-based system (value price), known as a "population health" strategy [26]. It provides advantages not only to the patient but also serves as a motivating factor for healthcare providers to give high-quality treatment at a reduced cost. Additionally, it provides patients with a lifestyle management strategy that has significant promise for enhancing their health. mHealth technologies are effective in therapy, and further study on their long-term sustainability is required [27]. Most mHealth initiatives that are first implemented in a community do not include long-term action plans or follow-ups [28]. The financial, economic, and social aspects of mHealth practices are important areas of concern that have received little consideration in terms of study and in the development of mHealth initiatives.

6.2 mHealth definition

The term "mHealth" refers to the use of mobile communication devices, such as Personal Digital Assistants and telephones, for the provision of health services and transmission of health-related information. Improving the safety of patients and the quality of treatment is the primary purpose of the adoption of mobile devices designed to record, save, retrieve, and transfer data in real time between patients and providers [25]. According to Istepanian [29], the term "mHealth" refers to the use of mobile communications and network technology in the field of healthcare. Table 6.1 added to this description, by focusing on any wireless technologies (Bluetooth, GPRS/3G, Wi-Fi, and WiMAX) that can frequently send different health-related information and support to mobile devices like cell phones, smartphones, PDAs, computers, and tablet personal computer (PC). Several mHealth services are being utilized by both patients and health professionals, such as mobile online healthcare services in India, Bangladesh, and Mexico [30, 31]. In their study on mHealth platforms, Motamarri et al. [7] (Table 6.1) characterized them as personalized, interactive services that provide 24/7 access to medical advice and information on mobile platforms.

mHealth is a part of eHealth that involves the delivery of medical services to patients via the use of mobile devices [33]. Prevalent healthcare methods include the provision of timely health treatments to individuals, irrespective of their location or time limitations. It aims to enhance both the availability and quality of healthcare [34, 35]. mHealth identifies as a particular development, shifting from the current perspective. The authors propose that the mobile platform is a modern type of mass media, characterized by particular features that may be used to enhance patient empowerment and improve the delivery of healthcare services. According to the study, mHealth can sim-

Table 6.1: Definition of mHealth provided in different studies.

S. no.	mHealth definition	Authors
1.	The term "mHealth" is related to the developing mobile telephone and network technologies that are used in the healthcare sector.	Istepanian et al. [29]
2.	mHealth is a personalized and engaging service that gives everyone, everywhere, and at any time access to medical advice and information through mobile devices.	Motamarri et al. [7]
3.	The term "mHealth" deals with the use of integrated wireless devices for the purpose of monitoring health-related indications.	Bardram et al. [32]
4.	Mobile health refers to the use of mobile communication devices such as personal digital assistants (PDAs) and mobile phones for the provision of health services and dissemination of health-related information.	UN foundation and Vodafone foundation [25]
5.	mHealth refers to an area of study within the broader field of eHealth, wherein mobile devices are used as a means of providing healthcare services to people.	Mechael [33]

plify and improve healthcare delivery processes, save expenses, simplify service provision, establish closer relationships with patients, and provide them with more convenience and attractiveness. They consider it as a revolutionary change toward better global healthcare for all people. Overall, according to the study, mHealth and eHealth are closely related, their technologies complement one another, and both may effectively enhance global health outcomes [25]. mHealth identifies as a particular development, shifting from the current perspective [36] The mobile phone has gained significant importance in everyday life, especially in the domain of mHealth, as a result of advances in technology [37]. Enhancing the availability of healthcare and encouraging people to enhance their health situations is of significant importance [38].

According to a study [39], mHealth apps may have a major influence on patient self-management and positively influence high-need patients, improving their physical and mental well-being. The use of mHealth applications is suggested as a means to decrease the time spent on visits, reduce expenses related to mobility, minimize waiting times, and decrease the number of healthcare providers, all the while taking into consideration the preferences of patients [40]. However, these digital treatments are not optional [41] but rather deemed necessary to support future healthcare delivery [42]. The field of mHealth applications has its foundations in an extensive range of studies focused on assessing quality. The existing body of research focuses on the analysis of the structure and development of mHealth applications [43], as well as the assessment of the effectiveness and reliability of current technology applications [44]. Concerns about users' privacy and safety were raised by a mHealth study that looked at how people felt about and thought about online healthcare tools [45]. A study investigating the utilization of mHealth technology to assist the aging population in adopting and maintaining healthy lifestyle choices showed positive outcomes in terms of providing health education, self-tracking, and developing user-provider interactions [46]. The study examined many factors that either facilitate or hinder the use of mHealth for interventions in maternal healthcare. These factors include the attitudes of consumers, community involvement, the understanding of participants, and the perceived effectiveness of the interventions. Research on mHealth treatments with individuals diagnosed with significant mental disorders yielded results indicating that such interventions are effective in evaluating, monitoring, and minimizing challenges related to daily functioning in this population [47, 48]. The most recent innovations in healthcare that have been implemented in mobile medical treatment facilities have been developed to promote health and well-being, although challenges remain to be overcome [49]. Current mHealth applications are evaluated in terms of the level of healthcare or the influence on health and cost variables, which mostly concentrate on clinical efficacy and ignore other elements of long-term sustainability, such as socio-environmental challenges [50].

The sustainable integration of mHealth treatments into complicated healthcare systems is a major barrier, despite their substantial growth and long-term benefit. To achieve sustainability, it is recommended to use the 4A scheme proposed by the World Health Organization (WHO), which includes accessibility, availability, suitableness, and affordability. This scheme acts as a guiding framework for the implementation of mHealth devices. Existing mHealth practices contributing to the sustainability of healthcare systems have been a significant area of study. Despite this, the elements that contribute to the implementation of sustainable mHealth programs have yet to be fully realized [51, 52]. New evaluation procedures for mHealth projects should address sustainability. This will allow healthcare authorities to establish precise guidelines for the advancement of technology and examine case studies that contribute to sustainability [53].

6.3 The significant role of mHealth in patient-centered care

The patient-centric mHealth model refers to a healthcare strategy that emphasizes focus on the patient's requirements, preferences, and active participation in their healthcare by using mHealth technology. In this particular model, individuals are provided with the means to assume control over their health via the use of mobile devices, applications, and wearable technology to monitor their overall well-being, access their health data, arrange appointments, and receive personalized health-related information. The use of a patient-centered strategy has been shown to facilitate enhanced patient participation, increased compliance with treatment regimens, enhanced levels of satisfaction, and ultimately, improved health outcomes. Additionally, it facilitates improved communication between patients and healthcare professionals, mitigates healthcare inequities, and fosters cost-effective and efficient healthcare delivery, fundamentally transforming the healthcare experience and delivery [54]. Although concerns about the potential for the elderly to experience a lack of personal care due to a growing dependence on technology rather than human interaction [55], the general assessment of mHealth shows a prevailing sense of optimism. Increased efficiency may be achieved via resource optimization, cost reduction, and better staff utilization, as well as indirect advantages through better care [56].

The use of mHealth has the potential to significantly change the field of healthcare by promoting a patient-centric approach. Patient-centric care focuses a strong emphasis on the active engagement of individuals in making healthcare choices and managing their health. mHealth plays an important role in facilitating this conceptual shift by providing patients with potent tools. The use of mobile apps and wearable devices enables people to conveniently recover their health information, effectively monitor chronic illnesses, and carefully maintain vital signs. Thus, this enables individuals to make more educated healthcare decisions. Moreover, mHealth enables personalized treatment plans and promotes a stronger patient-provider connection by facilitating continuous contact between patients and healthcare professionals via features like virtual consultations. The implementation of mHealth technology has a substantial role in enhancing patient engagement, happiness, and overall health outcomes, hence leading to a transformation of the healthcare landscape toward a more patient-centric approach. Figure 6.1 represents a patient-centric mHealth approach that is sustainable for individuals located in both urban and rural areas of developing nations. This model is made possible by the wide acceptance and extensive use of mobile phones, which have emerged as the primary means of communication in developing countries. The concept has integrated several stakeholders involved in healthcare services.

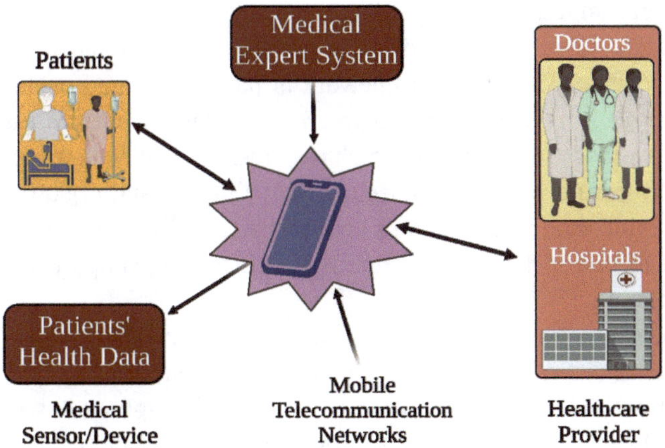

Figure 6.1: A patient-centered mobile health model.

6.4 An integrated model for mHealth lifestyle management (MLM)

When discussing mHealth lifestyle management, investigators typically discuss the behaviors connected with these activities. However, behaviors are only the activities that may be seen from outside. the thinking habits of persons seeking to better their health are crucial [57]. Despite rational thinking, patients may lack the drive to achieve desired actions. Psychology has long understood that focusing on outer behavior is inadequate to understand human decision-making and hence provides a significant viewpoint on engagement and sustainable habits. Human factor psychology depends on the understanding that individuals are part of a larger system that includes the surrounding environment in which they perform their duties [58]. Traditionally, human factor psychology in healthcare has centered on how doctors and nurses think and behave. However, researchers have widened their focus to include the patient as an integral component of the healthcare team lately [58, 59]. Therefore, a "systems approach" emphasizes an environment in which an association exists between the patient's ideas and behaviors [60, 61]. Understanding mental abilities information, and the factors that determine why patients become involved and/or motivated in healthy lifestyle behaviors may provide us with a wealth of information about the difficulties they experience in maintaining such practices. Lippa, Klein, and Shalin [62] emphasize the importance of understanding patient cognition while seeking to examine and change patient behaviors. Poor glucose control was associated with difficulties in self-management, and investigators concluded that this was "not due to a lack of improvement or intelligence, but rather to inadequate understanding of self-care." Participants "did not recognize the significance of physical activities in controlling glucose and were subsequently unable to improve their level of activity," [63]. According to the study conducted by Holden et al., and as supported by the integrated model, individuals play a crucial role in the work system, emphasizing the need to design systems that effectively facilitate human activities [64]. In the framework of patient-centered healthcare, individuals who play a role in managing patient well-being include clinicians, patients, family caregivers, and various groups or teams such as clinicians, family members, faith-based communities, and mental wellness counselors. These people and organizations are responsible for carrying out tasks related to both physical and mental healthcare. In addition to individuals, the sociotechnical environment has been simplified to include four fundamental factors:

- Health tasks include human behaviors performed within a wider work process, irrespective of their level of difficulty, complexity, or range of diversity.
- Health tools and/or technologies refer to the objects individuals employ to assist them in their activities. In the field of healthcare, this encompasses health information technologies, medical devices, and mHealth applications. These tools are designed to enhance usability, accessibility, automation, portability, ubiquity, and functionality.

- Health organizations are external entities or institutions that have an impact on various aspects of healthcare, including work schedules, staff management and training, policies, resource availability, community groups, communication facilities housing arrangements, family roles and responsibilities, interpersonal interactions, culture, social norms, and rules.
- Health ecosystems include both the indoor environment, which relates to the physical aspects such as lighting, noise, temperature, physical spaces, and air quality, as well as the social surroundings. The term "external environment" encompasses ecological and policy elements, along with other external influences that have the potential to influence performance results [65].

Figure 6.2 demonstrates the MLM model, which represents the combination of the Healthcare Information and Management Systems Society (HIMSS) and Systems Engineering Initiative for Patient Safety (SEIPS) 2.0 models. To reflect the cyclical nature of the HLM process, a black and gray spiral and circle are used to display it, complementing the SEIPS 2.0 architecture. As per the HIMSS framework, we incorporated a black centered spiral to represent rising patient involvement levels. (MLM) represents a combination of individual and systemic healthcare factors.

6.5 Multidisciplinary assessment of mHealth sustainability

We discuss the ways in which the four disciplines that have already been defined provide resources to analyze mHealth that extend beyond the medical and economic perspectives. The sustainability factor within each field is summarized to emphasize their complementary nature and improve their capacity to provide a more comprehensive perspective on the sustainability of mHealth. The evaluation of the sustainability of mHealth programs requires the use of a multidisciplinary methodology that encompasses diverse disciplines such as healthcare, technology, environmental science, and ethics. The comprehensive evaluation analyzes the sustainability and influence of mHealth services on healthcare systems over an extended period. It involves various factors like cost-effectiveness, efficient use of resources, and environmental consequences. Ethical considerations, data security, and equitable access also play crucial roles in the assessment. By integrating understanding from several academic fields, we assess the comprehensive sustainability of mHealth implementations and make informed decisions on their integration into healthcare ecosystems.

6.5.1 Economics sustainability

Cost-effectiveness evaluation is the preferred method for allocating healthcare resources. The health-related technological evaluation (health technology assessment, HTA)

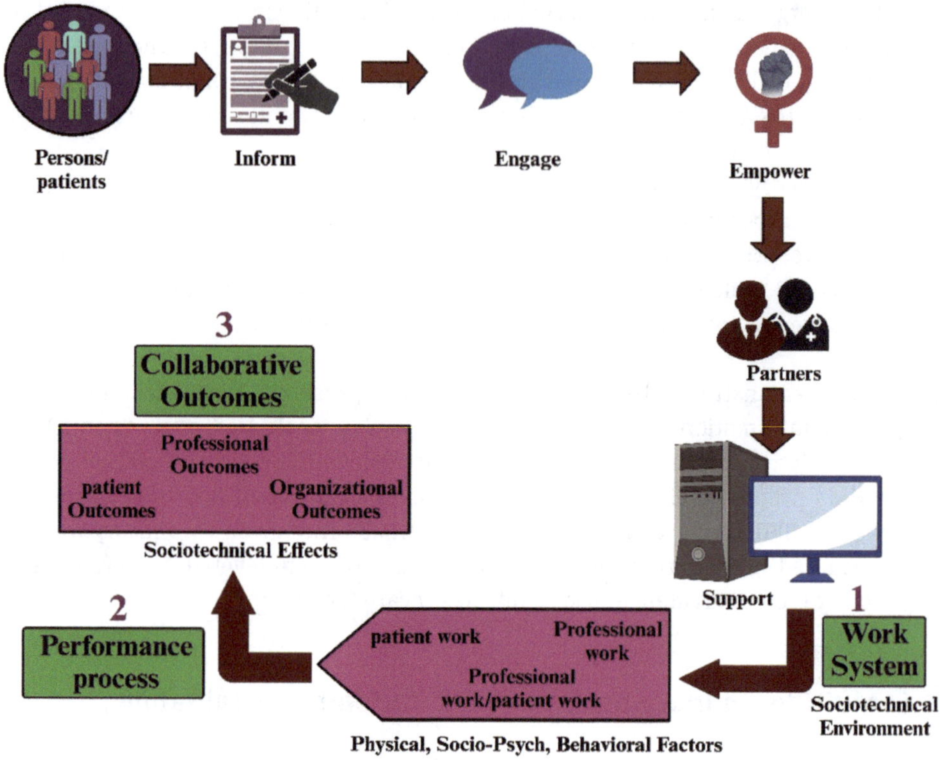

Figure 6.2: mHealthy lifestyle management model.

framework is a widely recognized and comprehensive strategy that assesses the effects, advantages, and impacts of novel healthcare technologies or therapies, while also considering their prices and implications for the healthcare economy. HTA plays an essential part in assisting decision-makers in the healthcare sector to ensure the provision of optimal medical technologies to patients. It is widely used by national health authorities such as the UK National Institute for Health and Care Excellence (NICE), the French Haute Autorité de Santé (HAS), and the Belgian Federal Knowledge Centre for Health Care (KCE), among others.

mHealth treatments may save money and improve hospital efficiency, according to studies. Comparative economic analyses of mHealth techniques and traditional strategies are becoming important. The facilitation of broad acceptance and implementation of these health improvements within healthcare organizations is of greatest significance [66]. However, the best analytic methods for assessing mHealth technologies are unknown, resulting in unsatisfactory evaluation [67, 68]. The field of economic evaluation differentiates between two types of analyses: cost-effectiveness analyses (CEA) and cost-benefit analyses (CBA) or return investment studies. CEA aims to provide evidence on which programs offer the most value for money, while

CBA is used to demonstrate affordability and estimate the resources needed for program scale-up and sustainability [68].

The selection of an appropriate economic assessment approach is dependent upon many factors, including the desired outcome measures such as survival gains, quality of life improvements, quality-adjusted life years (QALYs), accessibility, prevention of events, or cost savings. Additionally, the selection of the evaluation method is influenced by the specific objectives of mHealth and the current stage of its development. CEAs have emerged as a widely used method for conducting economic evaluations, making them particularly applicable for evaluating the effectiveness of mHealth technologies [66–70]. Various approaches are available for evaluating mHealth technology from a health-economic perspective. An HTA of a medical device will typically evaluate the equipment's efficacy as well as its costs. However, the evaluation of mHealth technology must also include the evaluation of other aspects, such as clinical safety, protecting patient information, security, ease of use and accessibility, connectivity, and technical stability [69, 71]. The assessment of cost-effectiveness for mHealth technologies might vary depending on the perspective being addressed, either individual or organizational, as well as the precise phase of the mHealth service being assessed, such as development or implementation [70]. The evaluation of mHealth's cost-effectiveness often focuses on the financial sustainability and business strategies associated with mHealth, rather than the actual health results [70].

However, mHealth technologies are complicated interventions in an intricate framework, and as such, they might affect factors outside of the users themselves. Some research examined the economic effect of mHealth technology outside the users themselves, such as medical professionals and family members, or beyond the health sector, such as educational or employment sectors [67–72]. However, the existing HTA assessment procedures fail to explicitly address the inclusion of environmental consequences or provide guidance on how to include them in the assessment process. Pollution is often neglected due to its lack of market value and the lack of necessary metrics for sustainability in health-related service delivery [73].

6.5.2 Environmental sustainability

The evaluation of environmental impact is an essential aspect of promoting sustainability; however, the application of such assessments to mHealth devices involves many challenges. Wearable and mHealth technologies are subject to many limitations from a technological perspective, mostly due to their need for mobility. These limitations include requirements with regard to physical characteristics like size, form, appearance, materials, weight, and biocompatibility. It imposes certain design and technical criteria on the product. Most information and communication technology (ICT) studies have attempted to satisfy these conditions. Sustainability was mostly seen from a green perspective. It should be significant that this industry has been weak to accept ideas of sustainability or to implement sustainable practices [74]. Life cycle assessment (LCA) is widely recognized as the most organized technique for evaluating the environmental characteristics of a device across its entire life cycle, making it a valuable tool in sustainable design. The International Organization for Standardization (ISO) 14040 incorporates a standardized framework that assesses the environmental characteristics and possible impact of a product's life cycle, including the whole process – from raw material procurement through manufacture, use, and disposal [75]. Although the LCA technique is not commonly used in the medical field, this approach offers quantitative assessments of the environmental and social implications associated with a certain product configuration, under the condition that reliable data is available. The LCA methodology operates via comparison. It has been previously used within the medical industry to evaluate the impacts of using single-use medical equipment versus reprocessed ones [76]. Additionally, LCA has been utilized to determine the most effective approach for maximizing the advantages of bioplastics as a substitute for petroleum-based medical plastics [77]. The implementation of environmentally aware designing on a larger scale in the medical sector faces several challenges, including greater design time and costs, greater importance being given to other design aspects, a lack of demand, regulatory requirements, and improper business models [74]. Additionally, mHealth device LCA faces the issue of unreliable data inventories from outsourced manufacturers or hidden players in the global supply chain, like other electronic gadgets. Additionally, skin-worn devices need biocompatible materials and may need the use of sophisticated manufacturing processes to make them flexible or extensible [78, 79].

The use of nonconventional electronic materials is a significant challenge due to the complex production processes involved and the potential environmental consequences associated with their disposal. Moreover, the lack of available data for conducting LCAs is further complicated by these materials. Design approaches and methodologies have been offered to assist innovators in adopting a more holistic view of the life cycle of their medical product and so revealing the complicated trade-offs and consequences that occur at various phases of the life cycle. For example, eco-design provides a simpler framework that may assist in the definition of a slow material flow loop that takes into account the scarcity of materials, encourages a decrease in the range of materials and relationships, supports additive manufacturing processes, and avoids energy-intensive processing [74]. It also covers end-of-life aspects such as elimination, recycling, and toxicity of transformation processes. Furthermore, the environmental impact of electronic waste should need particular end-of-life measures to simplify device disposal. Biodegradable electronics [80] have the potential to reduce the amount of electronic waste produced, but they are too new for widespread implementation at the moment. mHealth devices may have environmental implications that are not directly related to the life cycle impacts that the LCA addresses. The Life cycle, Enabling, and Structural framework (LES) [81] explains that indirect effects are allowing effects that are caused by ICT devices. Micro-level impacts may have both beneficial and negative outcomes, such as resource optimization and replacement. Analyzing the indirect benefits of a single invention is challenging since it requires a complete strategy. Studies that have addressed these issues have primarily focused on their positive effects, such as a decrease in emissions associated with telemedicine as a result of lowered transportation [82]. It is important to observe that the LES framework incorporates macro-level structural changes that need examination from a multidisciplinary approach. Many studies address the development of strategies for incorporating sustainability into the planning and development process. The evaluation of environmental impacts using the LCA has two primary components.

On one side, the LCA is a comprehensive and quantitative methodology that provides thorough information. However, it is time-consuming and, as a result, may not be the most advantageous instrument to use during the design process, whereas, design guidelines include a greater degree of qualitative characteristics, making them easier to adhere to throughout the first phases of the design process. Recent mHealth studies aim to quantify the environmental implications of electronics like heart monitoring and smart wearables. However, study on LCAs for wearable electronics is limited. Data availability and knowledge in LCA for electronics are common barriers. A lot of unusual electronics, like electronics that can bend and stretch, are used in mHealth and smart gadgets. However, the existing LCA inadequately addresses this, making future assessments difficult.

6.5.3 Social sustainability

Societal improvements are unique changes arising in social situations using certain new technologies that develop into services or goods. It contributes to social sustainability through addressing issues, increasing democratic control, and promoting work relations and organization [83, 84]. Social innovations (SIs) are being evaluated ecologically to minimize wastage and resource intensity, and transition manufacturing processes towards low-carbon communities [85, 86]. Mehmood and colleagues [87] have established two fundamental concepts that serve as guidance for the development of ecologically friendly SIs. The primary objectives of this study are in two parts: (i) identify socially innovative strategies for addressing environmental challenges, and (ii) adopt a SI framework in the management of sustainable development, thereby promoting inclusive engagement and involvement of marginalized populations that are typically excluded from the process of problem-solving. The European Union regards social enterprises (SEs) as the preferred means for facilitating sustainable transitions [88]. Search engines have the potential to enhance innovations regarding sustainability in society due to their compatibility with social concerns and their commitment to bottom-up dynamics [89].

The organizational structure of SEs allows them to effectively handle the social aspect of sustainable development and engage many stakeholders. This is achieved via a participatory and democratic governance model, which is characteristic of SEs

[87, 89]. Recent discussions of SIs under the perspective of sustainable transition studies (STSs) [90, 91] have helped to better understand the problems of SIs in terms of their transformational potential within larger socio-technical systems. The field of STS focuses on the examination of system innovations, specifically those that exhibit radical characteristics regarding the prevailing socio-technical system. The modifications include multiple fields, including technology, science, market, culture, and politics, among others, and involve many different stakeholders. Transformative SIs refer to innovative interventions that result in significant systemic changes inside established regimes, such as those associated with agro-food, transportation, energy, or health sectors. These changes include alterations in social connections, regulations, and the overall directionality of the respective systems. Transformative SIs can direct systems toward a more sustainable pathway and empower individuals and organizations that are committed to achieving sustainability goals. These approaches address healthcare inequality by addressing end-user demands via a multi-stakeholder and community-engaged approach. In addition to cost-effective technology, collaborative learning initiatives that empower patients [92], social interventions that target health literacy issues via distance learning, peer support, and other strategies [93], as well as the implementation of healthy buildings [94], might yield systemic effects. Sustainable mHealth-related innovations, according to this transformational approach, substitute for adverse economic conditions and inefficiencies in the welfare state, react to expanding social imbalances, and also bring structural improvements in health policy [95, 96].

To achieve significant sustainability, Lehoux et al. [51] propose that the 4A concept developed by the World Health Organization (WHO), which includes availability, accessibility, appropriateness, and affordability, might serve as a guiding principle for future developments. The encouragement of principles such as subsidiarity, self-care, and reduced labor intensity is also recommended. Subsidiarity refers to innovations that promote the patient or primary caregiver, who represents the most independent unit within the system, to deliver the necessary service. Self-care innovations aim to enhance patients' ability to take care of themselves. Additionally, innovations that improve outcomes while simultaneously decreasing labor intensity should be promoted. The level of institutionalized mHealth innovations, including factors such as funding and supporting legislation, as well as their quantity, are regarded as essential for bringing about transformation. It is important to note that neither of these factors alone can provide a solution. However, their collaboration and integration into the sustainable "fabric of society" are what matter [97, 98].

6.5.4 Behavior sustainability

The field of human-computer interaction (HCI) has a significant challenge in addressing sustainability, as it involves the development of a concept of sustainability, specific to HCI, and expanding its impact beyond the limits of HCI. Despite the considerable attention generated by this newly developed discipline, it is currently recognized that there is a lack of a universally accepted framework, methodology, or instrument within the field of HCI for assessing the sustainability of technologies or initiatives. The origins of sustainability-related advancements in the field of HCI may be connected directly to the conferences on human aspects in computer systems in 2007 [99]. According to Blevis, "sustainability can and should be the primary focus on sustainable interaction design (SID)" [100]. In the investigation conducted by Mankoff, a sustainable HCI (SHCI) special category was established with the objective of promoting awareness about sustainability in the field of HCI [101]. These advancements contributed to the expansion of the discipline, despite the rapid emergence of a lack of agreement when the community attempted to define, scope, and expand SHCI [102].

A CHI workshop article on SHCIs recommended engaging in sustainability-focused work outside HCIs and moving beyond compartment models to solve sustainability challenges, which was the nearest the community reached a consensus [103]. While there is ongoing debate over whether SHCI is a process or an outcome [104], the requirement for a comprehensive, interdisciplinary method of evaluating technology has been widely recognized. As an example, Dillahunt et al. developed a list of criteria, including sustainability requirements, with factors such as the incorporation of alternative energy sources, potential for material replacement, reusability of materials, and the recyclability of the device [105]. Additionally, Dillahunt suggested the incorporation of social, economic, and environmental sustainability into technology evaluations [106]. Similarly, Remy et al. recommended that the HCI community use assessment methodologies from other fields to show how and why innovation may be used sustainably in real-world situations [107]. Furthermore, Remy et al. identified five essential components that might provide useful information for conducting evaluations in the field of (S)HCI research. These components are target, mechanics, metrics, methodologies, and scope [102].

The selection of processes, measurements, techniques, and scope for an assessment allows for the achievement of the evaluation's desired aims. The authors did not provide a comprehensive list of particular measurements or approaches. At the same time, the authors emphasized the identification of possible assessment constructs, which include measurable quantities of resources, and lasting implications or modifications in behavior. Additionally, the authors suggested the use of LCA data as an initial foundation for the establishment of a dependable repository or database for SHCI metrics. An LCA offers a comprehensive set of environmental indicators about the whole life cycle of a product. Lundström and Pargman introduced a taxonomy to categorize computer initiatives as either sustainable or unsustainable. This classification is based on an evaluation of three characteristics of sustainability: the credibility of goals, the effect of the project, and the probability of the impact [108]. The evaluation process is dependent upon personal choices, such as determining the limits of the system. They recommended the adoption of a particular approach to examine each dimension.

The environment of an SHCI encompasses a wide range of competing and extensive regions, including compelling technologies and sustainable interface design. Mankoff et al. differentiate between two types of contributions to sustainability in design. The first type involves incorporating environmental sustainability factors, such as energy footprint, into the construction and materials of products. The second type focuses on promoting sustainable lifestyles and facilitating sustainable decision-making through the use of technologies [101]. HCI plays a significant role in enhancing mHealth technologies aimed at behavior modification. This is achieved by the active engagement of healthcare providers and patients in the development stage, as well as by facilitating the addition of accessibility, acceptance of technology, and behavior change considerations throughout the assessment process of these systems. The important success aspects of mHealth innovations have been recognized as usability, technological uptake, and behavior modification [109–112]. Few studies evaluating mHealth usability [113], technological acceptance, and adoption lack consistency [112], and behavior change assessments are typically rigorous [109]. The absence of publications can be explained subsequently. The evaluation of mHealth solutions for behavior modification is a demanding process that involves significant allocation of resources, extends over an extended duration, and necessitates the presence of a fully operational mHealth system [114, 115]. These limits do not meet academic publishing requirements [103] or research project funding eligibility standards. Furthermore, while the significance of behavior modification in the sustainability of mHealth systems is well recognized, there is still ongoing development in terms of its quantification. However, the meaning of "engagement" is established by designers and developers. Designers and developers have a similar issue when attempting to establish the parameters of sustainable and unsustainable behavior [107].

6.6 The global applications of mobile health (mHealth)

Mobile communications are being used by an increasing number of nations to meet various healthcare demands, such as education and awareness-raising, remote data gathering, remote monitoring, communications and training, monitoring illness and pandemics, and helping in diagnosis and treatment [31]. Emerging advancements in mobile devices and integrated systems have enabled the development of new beneficial characteristics. These include technologies like smart houses and environmental assisted living [116]. According to a survey, mHealth projects are now operating in 26 developing nations. Various categories of mHealth services have been implemented. It may serve as a platform for diagnostic and therapeutic assistance in certain situations and as a remote communication tool for stakeholders in others. In general, mHealth applications include many forms of communication such as text, audio, video, and data-driven solutions. The growing popularity of mobile platforms is being utilized to provide health information, particularly regarding sexual health and HIV prevention [25]. In Uganda, a local mobile phone company has partnered with AIC (Aids Information Center) to deliver bulk short message service (SMS) messages to clients, providing them with a comprehensive education on AIDS [30]. This initiative, known as 'Text-to-change,' seeks to enhance awareness of AIDS through voluntarily providing counseling and HIV testing. Additionally, Project Masiluleke is actively addressing the HIV and tuberculosis epidemics in South Africa. The platform enables users to transmit cost-free "Please Call Me" (PCM) messages using their mobile devices to get essential healthcare information. Immediate response is obtained by individuals via consultation with local phone centers specializing in HIV and TB. Projects such as Project Masiluleke in South Africa and Text-to-Change in Uganda, which aim to disseminate health information remotely, have achieved significant success. Comparable services are also accessible in Nigeria with the implementation of the 'Learning about living' initiative, which aims to raise awareness about AIDS by using text messaging as a medium. In the state of Georgia, a comparable awareness initiative is implemented via the use of video content. In India, mobile gaming services are used as a means to disseminate information about HIV/AIDS in an interactive and pleasurable manner [24]. The games have been designed to target users belonging to various socioeconomic and demographic categories. Furthermore, they are compatible with over 100 different brands of mobile phones, spanning from basic devices to advanced smartphones [117].

6.6.1 Collecting information and monitoring disease

The fundamental focus of mobile apps for services remains on mobile data collection. In the present situation, EPiServer, developed by DataDyne.org, introduced an open-source application that facilitates the transmission of information among community health workers (CHWs) via the use of smartphones. CHWs can get electronic patient information sheets via an Internet platform. They may then use their mobile phones to complete these sheets and transmit them to a central clinic computer for real-time analysis. The high level of accessibility and practicality shown by EPiServer facilitates the collection of precise health data by healthcare teams. This program plays a crucial role in promptly evaluating large amounts of data to monitor the occurrence of illnesses. For example, in the African nation of Zambia, a smartphone app has been used to check the progress of the vaccination program, while in other regions, it has been utilized to monitor the inventory of crucial medications in local health clinics.

CHWs exhibit a preference for this particular software due to the convenience it offers in enabling them to carry out their assigned duties via the use of their own mobile devices. In the sub-Saharan African area, community health workers at the regional offices of each nation are using a specific application. This application is backed by the mHealth Alliance, an organization founded via the cooperation between the United Nations Foundation and the Vodafone Foundation. The use of electronic data as a means of storing health records is an example of leapfrog technology, demonstrating the advancement and establishment of a dependable and resilient electronic health records system. Due to its use of open-source software, EPiServer has the advantageous qualities of cost-effectiveness, collaboration capabilities, and suitability for rapid modification and adaption. Nevertheless, the functionality of this system relies on the use of advanced mobile devices such as Blackberry smartphones, as well as a reliable and consistent Internet connection. These resources, however, remain limited in developing regions. Therefore, the implementation of web-independent mHealth solutions using affordable smartphones would result in greater flexibility and sustainability.

6.6.2 Mobile telemedicine to facilitate treatment

We now discuss a telemedicine operation by the largest mobile network provider in a nation. A family instantly contacts a qualified physician at Healthline's call center for desired medical information. The mHealth platform handles 6,000–10,000 calls daily. Study indicates that utilizing the service saves consumers travel time (98%), money (91%), shorter wait times (97%), and lower doctor's fees (83%). Fifty-five percent of Healthline calls come from rural regions, and 77% are from females. About 81% of the rural callers are women, with two-thirds related to their children's health. Approximately 56% of callers had income below the poverty level and were mostly situated in rural regions (63%). About 72% of rural residents reported no access to contemporary healthcare providers within 5 miles of their house [30].

Voice-based mHealth services, mostly call centers, have developed significant popularity in developing nations such as India, Mexico, Pakistan, South Africa, and Bangladesh. These services, alongside text and video, are becoming widely used. Within this platform, individuals have the convenience to access the provided service for nonemergency requirements such as headaches, colds, and coughs, as well as for emergencies including accidents, burns, and severe stomach pain. This may be accomplished by calling a set of different numerical numbers, such as 789 in Bangladesh, directly from their mobile devices. Individuals can rapidly get medical information, guidance, therapies, assessment, diagnosis, referral, and counseling from healthcare professionals, including registered doctors, nurses, and paramedics, at a cost within their means [30]. mHealth technologies are also serving a crucial function in facilitating communications during disaster relief efforts. They assist healthcare professionals and emergency response personnel in enhancing their speed and efficiency in assisting, while also providing essential sustenance and medication to those affected.

The World Food Program (WFP) used mHealth technologies to provide essential food assistance to a total of 100 million individuals affected by disasters across 80 different countries. In this particular example, the UN Foundation and Vodafone Foundation are providing crucial telecommunications assistance to the United Nations and other humanitarian organizations in times of need. In general, the evaluation of mHealth initiatives on a global scale indicates that this platform enhances the provision of healthcare services and effectively contributes to the overall effectiveness of healthcare. There exists compelling data indicating that the provision of timely treatment assistance is benefiting a significant number of individuals in low-income countries. Additionally, the use of SMS reminders has resulted in a significant enhancement in treatment adherence rates, reaching as high as 90%, among individuals diagnosed with HIV and tuberculosis. Several relevant studies done in the United States, Spain, Australia, Finland, and South Korea have shown the comparable advantages of using mobile communications in diabetic self-care, vaccination follow-up, and asthma management [14]. Therefore, mHealth initiatives globally demonstrate a growing trend towards enhanced health outcomes.

6.6.3 Pandemic monitoring with SMS-based mHealth

Frontline SMS is a cost-free software package that is open-source in nature, specifically designed to cater to the needs of nongovernmental organizations (NGOs) operating in underdeveloped regions. Frontline SMS utilizes a PC and a cellular phone to establish a communication hub that operates only via SMS. This system is capable of functioning in any location with cellular network coverage, eliminating the need for an Internet connection. The use of this particular intervention has been seen at several sites within the specified region, with an important example being its implementation in a healthcare facility situated in the rural region. The hospital caters to a population of 250,000 individuals, even with having only two physicians. This limited staffing arrangement enables the hospital to effectively address emergency medical needs, monitor patients, document adherence to HIV and TB medications, stay informed about patient conditions, engage remote communities for communication testing, offer immediate drug dosage and usage information, and facilitate connections among members of HIV/AIDS support groups. The whole operation was very cost-effective since it included recycling mobile phones, receiving a given laptop, and incurring ongoing expenses of less than US $10 each week for SMS texts. Hence, Frontline SMS offers an effective structure for the implementation of mHealth solutions, using open-source technologies. The long-term sustainability of this initiative is rooted in its adaptability, cost-effectiveness, and web-based solutions in settings with limited resources [117].

The implementation of SMS-based mHealth solutions has emerged as a significant method with extensive ramifications in the field of pandemic management. The approach described in this study utilizes the widespread use of mobile phones to establish a system for collecting and monitoring data in real-time. This system enables prompt reactions to developing epidemics and facilitates the tracking of disease patterns. In addition to the aforementioned benefits, the use of SMS-based mHealth initiatives enhances the communication channels between healthcare authorities and the general public, hence promoting more transparency and encouraging confidence. Moreover, it facilitates efficient communication monitoring and timely identification of cases, hence enhancing efforts in pandemic preparation and response. This potent instrument assumes particular significance in settings with limited resources, as it serves as a crucial component within the wider context of public health administration, eventually enhancing the efficacy, adaptability, and durability of pandemic monitoring and control efforts.

6.7 Perspective on the sustainability of mHealth

Assessing the potential of healthcare services such as mHealth extends beyond considerations of safety, functionality, efficacy, and costs, presenting a complex challenge [118]. The development of a multidisciplinary discourse is a crucial first step toward conducting a comprehensive assessment of the sustainability of mHealth systems. This discourse starts by emphasizing the four significant disciplines that have been highlighted as crucial in addressing the sustainability of mHealth developments. This section demonstrates the interdependence of these four disciplines in terms of the gold standard, scope, and time horizon of sustainability evaluation. Additionally, their complementary nature provides insights into unresolved issues associated with assessing the sustainability of mHealth developments. Evaluating the sustainability of mHealth advances requires an extensive perspective that is beyond simply technological viability. The process includes a comprehensive evaluation that involves several dimensions, including financial, environmental, ethical, and sociological aspects. From an economic perspective, mHealth needs to exhibit its cost-effectiveness, thereby ensuring the optimal allocation of resources. Regarding the environment, the primary focus is on the reduction of electronic waste and the conservation of energy. Ethical considerations involve a range of issues, including the protection of data security, safeguarding privacy, and ensuring equal access to mHealth services. Furthermore, it is essential to conduct a thorough analysis of the social implications of mHealth, taking into account its capacity to address healthcare inequalities and enhance general health results. The integration of these several views provides a comprehensive perspective, facilitating well-informed decision-making and the formulation of sustainable mHealth policies that provide advantages for both people and communities. Table 6.2 provides a concise summary of the primary aspects that are taken into account by the four disciplines. The many disciplines differ in their emphasis on desired outcomes, as well as in their assessment periods and perspectives, resulting in distinct criteria for excellence.

Health economics (HE) has established methods for evaluating health technology, but they often ignore environmental concerns and the possibility of long-term health and financial advantages beyond patients' current health state. The strategy for validating mHealth innovations utilizing HTA does not yet address environmental sustainability concerns. The evaluation of the environmental impact associated with the whole life cycle of a mobile device is a key focus in life cycle evaluations.

However, the evaluation fails to provide any understanding of how patients use electronic devices or their level of social agency. Although the LCA primarily addresses

Table 6.2: Sustainability evaluation of mobile health in four different techniques.

S. no.	Discipline	The golden standard	Horizontal timeline	Scope
1.	Health economics (HE)	Improvements in QALYs, a general measure of health improvements that includes both length and quality of life	Short term and long term	Individual
2.	Human-computer interaction (HCI)	Accessibility, social acceptability of technology, and behavioral modification	Short term and long term	A single person or a group of people
3.	Information and communication technology (ICT)	Ecological footprint.	Device	Short term
4.	Transformative social innovations (TSI)	Systemic transformation	Long term	Social

environmental concerns, it fails to delve into the examination of factors related to social sustainability, such as continuing use and fulfillment of unsatisfied social requirements of individuals. The proposed approach in HCI aims to assess mHealth technologies in a manner that goes beyond mere usability and technological adoption. Instead, it emphasizes analyzing users' behavior change. However, it does not specifically address systemic change resulting from innovations. The idea of transformational SI emphasizes how mHealth addresses social concerns and establishes a connection between innovation and the concept of systemic change. The evaluation of the extent to which mHealth effectively supports the overall objectives of a sustainable health system is crucial. This evaluation should include evaluations conducted at two levels: the specific digital health device being implemented, as well as the broader health system as a whole [119]. In other words, intrinsic and systemic sustainability are at risk. The proposed approach is to implement a comprehensive strategy that addresses the underlying issues and minimizes the challenges at hand. It is proposed that integrating the perspectives of the aforementioned four disciplines will provide a more comprehensive assessment of the sustainability of mHealth innovations, including both their design and the resultant benefits and transformations within the healthcare industry.

The comprehensive investigation and assessment of the design, usability, and adaptability of mHealth innovations involve an integrated strategy to understand and evaluate the external factors involved. It is essential to make a meaningful contribution to the development of sustainable health systems. The reason behind the aim of environmental sustainability in health systems arises from the acknowledgment of the interdependence between health and environmental sustainability, and the potential for positive results [120]. The optimal approach for making decisions on the adoption of novel mHealth devices is based on empirical data. A comprehensive range of outcomes across every aspect may be taken into account, including not just performance and efficiency but also improvements in the health and quality of life. The achievement of these overarching health system objectives serves as criteria for evaluating novel digital health services [119].

6.8 Challenges in the implementation of mobile health (mHealth) in developing countries

The use of mHealth has a significant potential for addressing various challenges within existing healthcare systems, hence contributing to the enhancement of healthcare standards in underdeveloped nations. However, the success of this novel health-technical innovation might be influenced by a wide range of problems and limitations. Initially, it is worth noting that although the incorporation of mobility has facilitated the accessibility of mHealth tools in developing regions, it is important to acknowledge the continued existence of network restrictions that may hinder their efficacy. The lack of dependable and affordable internet access in several developing nations is a significant obstacle to the successful implementation of mHealth initiatives. Due to the constraints imposed by restricted internet connection, several nations have difficulties in accessing health-related information. The utilization of voice-recognition mobile phone applications has the potential to serve as a valuable tool in facilitating healthcare services in regions characterized by high illiteracy rates. However, it is important to recognize that the successful implementation of such applications necessitates the availability of a third-generation (3G) network, which is currently lacking in numerous developing countries with limited resources [121, 122].

The use of mHealth has the potential to serve as a beneficial solution for populations characterized by different cultural practices and traditions, particularly in developing regions. However, it is important to recognize that this cultural heterogeneity may pose limitations to the effective implementation of mHealth apps. Individuals living in these nations exhibit various cultural values, beliefs, and practices that have an influence on their health-related behaviors and impede their capacity to assume responsibility for their well-being. One potential approach to addressing diversity is the customization of mHealth apps to align with the specific demands of the community [123, 124]. In addition, the presence of poverty and the prevalence of several spoken languages in developing regions provide additional obstacles that may diminish the efficacy of mHealth technologies, including SMS text messaging [125]. Furthermore, ensuring the confidentiality and privacy of the sent information represents a significant problem [126, 127]. The 1996 Health Information Portability and Accountability Act (HIPAA) brought attention to a significant issue of breaches in electronic communication. This legislation aims to safeguard individuals' sensitive health information [124]. Certain portable devices, such as mobile phones, smartphones, and personal digital assistants include security protections that are restricted in their ability to prevent unwanted access to sensitive information. Additionally, these devices may use complex security mechanisms that prevent the possibility of accessing the device itself [128]. Furthermore, a significant concern among individuals is the potential loss of their electronic devices, which may store sensitive personal medical data. This vulnerability exposes individuals to the risk of exploitation by unauthorized persons [129, 130]. Additionally, an important challenge is the financial implications associated with mHealth devices.

Although mobile devices are increasingly becoming cheaper for individuals in developing countries, they nevertheless remain prohibitively costly for a significant majority, particularly in rural regions [125]. Furthermore, it is interesting to consider that individuals living in developing nations primarily depend on mobile phones that operate on a pay-per-minute basis. It follows that extra minutes are necessary for accessing health consultations or engaging in personal health monitoring. However, it is important to acknowledge that a significant portion of the population in these regions may find the cost of acquiring these additional minutes to be beyond their financial means [124]. In the regions of South Asia and Africa, there exists a significant difference in the ownership of mobile phones among women and men. The difference may be attributed to many factors, including the financial burden associated with mobile phone ownership, limited access to education, resulting in higher illiteracy rates among women, and the lack of reliable energy infrastructure. Individuals often own a cell phone, primarily for professional requirements, while resorting to a temporary phone – borrowing for nonbusiness-related endeavors [131]. Certain poor nations have a limited propensity to embrace mHealth applications due to the challenges associated with attaining a substantial return on investment for such technologies [129, 132].

Furthermore, the absence of mandatory education, the necessary specialization, and the accessibility of pertinent training are significant barriers to the efficient use of mHealth apps. The limited educational attainment and inadequate health literacy skills seen in many underdeveloped nations result in diminished adoption of mHealth apps. Healthcare practitioners have a responsibility to possess a considerable degree of knowledge and ability to use mHealth technology safely and efficiently [132].

In general, the elderly are less quick to accept new technologies than the young. Therefore, older users' acceptability of mHealth technology in low-income regions may be improved by proper education and training [132]. In addition, patients and doctors may face additional difficulties because of the widespread accessibility of mobile devices that may not have enough service support. All of these problems lead to resistance from patients or errors in medical care. The spread of mHealth programs in the poor world is hindered by the present constitutional environment and the lack of rules and standards for mobile phone use [126]. Typically, the components of health systems that influence integration and interoperability nearly never have a single "owner." A uniform mHealth technology architecture and standardized data sharing are necessary to get through such an obstacle. Health-related applications are also often put on hold by developing-world governments until high-speed internet is widely available. As a result, the implementation of mHealth is hindered by the needless investment of resources.

To address this challenge, strategies are required for describing mHealth's intended outcomes while also making use of the readily available commercial wireless infrastructure [129]. To make use of the advanced capabilities of mHealth technology, it is necessary to implement new policies. To create the best mHealth apps for their nations, all stakeholders need to be included in the stages of preparation and implementation. The development of mHealth apps that are customized to local requirements and resources is another major challenge in poor countries. Rural and urban settings in the poor world can have different resources, which might impact the form and content of mHealth. Because of the absence of healthcare professionals and outdated facilities, rural communities need access to cutting-edge medical supplies and technology. However, communities have their own unique set of challenges. Population growth, resource inequity, and poor nutrition all contribute to varying demands on the healthcare system [133]. Furthermore, several poor nations do not have the capability and funding for health research institutes, limiting the ability to design and create mHealth apps based on community requirements [134].

Finally, ensuring high-quality mobile services in emerging markets is a significant obstacle. Many elements, including the mobile device, the mobile network, the information system, the information itself, the dependability, adaptability, and usability of mHealth apps, cost, security, and cultural considerations [127, 129], affect the quality of the mobile service. The service's quality also includes the mHealth devices' battery life and storage capacity. The effectiveness of mHealth services may be improved significantly by modifying these elements. Even though mHealth offers a lot of benefits for people in poor countries, there are still some big problems and obstacles that need to be solved. These challenges need careful analysis and research, as well as efficient participation by governments, NGOs, parties with an interest in this topic, and the health practitioners who will benefit from the much-needed technology.

6.9 Future directions

There are significant and extensive differences between developing-world nations in terms of economic development, availability of technological infrastructure, network generations, cultural behaviors, and educational levels. But mHealth is expected to have a bright future in all of these regions. According to the United Nations Foundation, the primary trends in mobile healthcare will be the same as what has been classified in the technological progress over the previous 40 years, lowering illnesses, increasing network speeds, and decreasing prices. mHealth has made a big difference in the health and lives of many people in the poor world. More test projects will be run in the future, depending on what each country needs. There is a good chance that a project will be accepted and used if it takes into account what the community needs, uses the available technology, makes it easier for people to take care of their health, and works with local groups [135]. Text message apps will still be useful, even as they are used on a big scale in new areas of healthcare. Reports say that the number of people using smartphones will increase quickly. As a result, more apps for smartphones will be made so that they can be used in more healthcare services, like helping people with chronic diseases handle their conditions at home or adopting a healthier lifestyle [135]. Soon, some countries will be able to use 6G/7G technology, which will have extra benefits like higher data transfer speeds and more connectivity, which will help close the gap in medical care [122, 126]. There will be significant development and progress in mHealth in the future. More advanced IoT and wearable devices, more accessible telehealth services, rise of digital therapeutics, increased adoption of decentralized systems for data security, focus on personalized medicine, stronger emphasis on mental health support, global health initiatives utilizing mHealth for data-driven responses, and the proliferation of digital currencies are all developments that are expected to take place shortly.

6.10 Conclusion

The medical industry is recognized for its rapid pace of innovation, leading to gadgets with short lifespans that add significantly to the industry's already substantial environmental burden. Considering sustainability in mHealth solutions as a combination of social, environmental, and technological factors corresponds to the need for a greener healthcare system and sustainable development objectives. mHealth technology has sustainability aspects due to its capacity to spread across geographical boundaries and reach rural places. The initiative can serve a broad population and greatly enhance healthcare quality. Implementing mHealth apps in healthcare delivery gives a sustainable foundation for providing medical services to poor populations in rural areas nationwide. A viable mHealth program requires collaboration between governing authorities and healthcare providers. mHealth applications provide real-time data on patient health, enabling clinical trials, identifying best practices, improving outcomes, and understanding illness development, risk factors, and treatment efficacy. The study analyzes the qualities and talents of medical professionals in adopting and promoting mHealth to enhance decision-making and practices. Mobile communications play a significant role in our daily lives, and as a result, they possess the capacity to revolutionize our well-being and healthcare. Based on our research, it is apparent that the use of mobile platforms for healthcare has significant potential as a transformative application within society.

References

[1] Karageorgos G, Andreadis I, Psychas K, Mourkousis G, Kiourti A, Lazzi G, Nikita KS. The promise of mobile technologies for the health care system in the developing world: A systematic review. IEEE Reviews in Biomedical Engineering. 2018 Sep 5;12:100–22.

[2] Liu Shiyuan mHealth: New Horizons for Health through Mobile Technologies: Based on the Findings of the Second Global Survey on eHealth (Global Observatory for eHealth Series. Healthcare Informatics Research. 2012;18(3):231–33.

[3] Hilty D, Chan S, Torous J, Luo J, Boland R. A framework for competencies for the use of mobile technologies in psychiatry and medicine: Scoping review. JMIR mHealth and uHealth. 2020 Feb 21;8(2):e12229.

[4] Talpur A, Shaikh FK, Baloch N, Felemban E, Khelil A, Alam MM. Validation of wired and wireless interconnected body sensor networks. Sensors. 2019 Aug 26;19(17):3697.

[5] Li Y, Xiang T, Szeto WY. Home health care routing and scheduling problems with the consideration of outpatient services. Transportation Research Part E: Logistics and Transportation Review. 2021 Aug 1;152:102420.

[6] Mohammadzadeh N, Khenarinezhad S, Ghazanfarisavadkoohi E, Safari MS, Pahlevanynejad S. Evaluation of M-Health applications use in epilepsy: A systematic review. Iranian Journal of Public Health. 2021 Mar;50(3):459.

[7] Motamarri S, Akter S, Ray P, Tseng CL. Distinguishing "mHealth" from other healthcare services in a developing country: A study from the service quality perspective. Communications of the Association for Information Systems. 2014;34(1):34.

[8] Kuhn KA, Wurst SH, Bott OJ, Giuse DA. Expanding the scope of health information systems. IMIA Yearbook of Medical Informatics. 2006:43–52.

[9] Jahan S, Chowdhury MM. mHealth: A sustainable healthcare model for developing world. American Journal of Modeling and Optimization. 2014 Jan 17;2(3):73–76.

[10] Zhou L, Bao J, Setiawan IM, Saptono A, Parmanto B. The mHealth app usability questionnaire (MAUQ): Development and validation study. JMIR mHealth and uHealth. 2019 Apr 11;7(4):e11500.

[11] Sonderman KA, Nkurunziza T, Kateera F, Gruendl M, Koch R, Gaju E, Habiyakare C, Matousek A, Nahimana E, Ntakiyiruta G, Riviello R. Using mobile health technology and community health workers to identify and refer caesarean-related surgical site infections in rural Rwanda: A randomized controlled trial protocol. BMJ Open [Internet]. 2018 May 1;8(5):e022214.

[12] Ming LC, Untong N, Aliudin NA, Osili N, Kifli N, Tan CS, Goh KW, Ng PW, Al-Worafi YM, Lee KS, Goh HP. Mobile health apps on COVID-19 launched in the early days of the pandemic: Content analysis and review. JMIR mHealth and uHealth. 2020 Sep 16;8(9).

[13] Kondylakis H, Katehakis DG, Kouroubali A, Logothetidis F, Triantafyllidis A, Kalamaras I, Votis K, Tzovaras D. COVID-19 mobile apps: A systematic review of the literature. Journal of Medical Internet Research. 2020 Dec 9;22(12).

[14] Westin AF. Public attitudes toward electronic health records. AHIP Cover. 2005 Jul 1;46(4):22–25.

[15] Prentice Riazi H, Jafarpour M, Bitaraf E. Towards national eHealth implementation-a comparative study on WHO/ITU national ehealth strategy toolkit In Iran. In: MIE. 2014 Jan 1:246–50.

[16] Lee SM, Lee D. Opportunities and challenges for contactless healthcare services in the post-COVID-19 Era. Technological Forecasting and Social Change. 2021 Jun 1;167:120712.

[17] Geissbuhler A. Access to health information: A key for better health in the knowledge society. Yearbook of Medical Informatics. 2008;17(01):20–21.

[18] Moja L, Kwag KH. Point of care information services: A platform for self-directed continuing medical education for front-line decision makers. Postgraduate Medical Journal. 2015 Feb;91(1072):83–91.

[19] Jen WY, Chao CC, Hung MC, Li YC, Chi YP. Mobile information and communication in the hospital outpatient service. International Journal of Medical Informatics. 2007 Aug 1;76(8):565–74.

[20] Anantharaman V, Han LS. Hospital and emergency ambulance link: Using IT to enhance emergency pre-hospital care. International Journal of Medical Informatics. 2001 May 1;61(2–3):147–61.
[21] Krause A, Hartl D, Theis F, Stangl M, Gerauer KE. Mobile decision support for transplantation patient data. International Journal of Medical Informatics. 2004;73(5):461–64.
[22] Michalowski W, Rubin S, Slowinski R, Wilk S. Mobile clinical support system for pediatric emergencies. Decision Support Systems. 2003;36:161–76.
[23] Chau PYK, Pjh H. Investigating healthcare professionals' decisions to accept telemedicine technology: An empirical test of competing theories. Information and Management. 2002;39(4):297–311.
[24] Andrade R, Wangenheim AV, Bortoluzzi MK. Wireless and PDA: A novel strategy to access DICOM-compliant medical data on mobile devices. International Journal of Medical Informatics. 2003;71(23):157–63.
[25] Akter S, Ray P. mHealth-an ultimate platform to serve the unserved. Yearbook of Medical Informatics. 2010;19(01):94–100.
[26] Dysinger WS. Lifestyle medicine practice: Exploring workable models. American Journal of Lifestyle Medicine. 2016;10:345–47.
[27] Ceptureanu SI, Ceptureanu EG, Luchian CE, Luchian I. Community-based programs sustainability. A multidimensional analysis of sustainability factors. Sustain Times. 2018;10(3):1–15.
[28] Walugembe DR, Sibbald S, le Ber MJ, Kothari A. Sustainability of public health interventions: Where are the gaps? Health Res Policy & Systems. 2019 Jan;17(8):8.
[29] Istepanian R. Introduction to the special section on M-Health: Beyond seamless mobility and global wireless healthcare connectivity. IEEE Transactions on Biomedical Engineering. 2004;8(4):405–13.
[30] Ivatury G, Moore J, Bloch A. A Doctor in your Pocket: Health Hotlines in Developing Countries, Mobilizing Markets: Special Edition of MIT Innovations. In: Journal for the GSMA Mobile World Congress (pp. 119–53). Cambridge: MIT Press; 2009.
[31] Chib A, Van Velthoven MH, Car J. mHealth adoption in low-resource environments: A review of the use of mobile healthcare in developing countries. Journal of Health Communication. 2015 Jan 2;20(1):4–34.
[32] Bardram JE, Baldus H, Favela J. Pervasive Computing in Hospitals. In: JE Bardram, A Mihailidis, D Wan, FL Boca Raton (eds.), Pervasive Computing in Healthcare (pp. 49–77). CRC Press; 2007.
[33] Mechael P. The Case for mHealth in Developing Countries. In: Mobilizing Markets: Special Edition of MIT Innovations Journal for the GSMA Mobile World Congress 2009 (pp. 153–68). Cambridge: MIT Press; 2009.
[34] Varshney U. Pervasive healthcare: Applications, challenges and wireless solutions. Communications of the Association for Information Systems. 2005;16(3):57–72.
[35] Varsheny U. Using wireless technologies in healthcare. International Journal of Mobile Communications. 2006;4(3):354–68.
[36] Schell SF. Public health program capacity for sustainability: A new framework. Implementation Science: IS. 2013 Feb;8(15).
[37] Zhao W, Shahriar H, Clincy V, Bhuiyan ZA. Security and privacy analysis of mhealth application: A case study. In2020 IEEE 19th International Conference on Trust, Security and Privacy in Computing and Communications (TrustCom) 2020 Dec 29 (pp. 1882–87). IEEE.
[38] Sun L, Buijsen M. Mobile health in China: Does it meet availability, accessibility, acceptability and quality standards? Health Policy and Technology. 2022 Sep 1;11(3):100660.
[39] Nouri R, Niakan R, Kalhori S, Ghazisaeedi M, Marchand G, Yasini M. Criteria for assessing the quality of mHealth apps: A systematic review. Journal of the American Medical Informatics Association. 2018 Aug;25(8):1089–98.
[40] Cinar A, Salman FS, Bozkaya B. Prioritized single nurse routing and scheduling for home healthcare services. European Journal of Operational Research. 2021 Mar 16;289(3):867–78.

[41] Suryavanshi K, Gahlot P, Thapa SB, Gandhi A, Raman R. Gamification on OTT platforms: A Behavioural Study for User Engagement. International Journal of Advanced Computer Science and Applications. 2022;13(9).

[42] Zulu JM, Perry HB. Community health workers at the dawn of a new era. Health Research Policy and Systems. 2021 Oct;19(3):1–5.

[43] Chatzipavlou IA, Christoforidou SA, Vlachopoulou M. A recommended guideline for the development of mHealth Apps. mHealth. 2016;2.

[44] Oyebode O, Alqahtani F, Orji R. Using machine learning and thematic analysis methods to evaluate mental health apps based on user reviews. IEEE Access. 2020 Jun 12;8:111141–58.

[45] Zhou W, Jia Y, Peng A, Zhang Y, Liu P. The effect of iot new features on security and privacy: New threats, existing solutions, and challenges yet to be solved. IEEE Internet of Things Journal. 2018 Jun 15;6(2):1606–16.

[46] Faiola A, Papautsky EL, Isola M. Empowering the aging with mobile health: A mHealth framework for supporting sustainable healthy lifestyle behavior. Current Problems in Cardiology. 2019 Aug 1;44(8):232–66.

[47] Udenigwe O, Okonofua FE, Ntoimo LF, Yaya S. Enablers and barriers to the acceptability of mHealth for maternal healthcare in rural Edo, Nigeria. Dialogues in Health. 2022 Dec 1;1:100067.

[48] Jameel L, Valmaggia L, Barnes G, Cella M. mHealth technology to assess, monitor and treat daily functioning difficulties in people with severe mental illness: A systematic review. Journal of Psychiatric Research. 2022 Jan 1;145:35–49.

[49] Yip W, Fu H, Chen AT, Zhai T, Jian W, Xu R, Pan J, Hu M, Zhou Z, Chen Q, Mao W. 10 years of health-care reform in China: Progress and gaps in universal health coverage. The Lancet. 2019 Sep 28;394(10204):1192–204.

[50] Adans-Dester CP, Bamberg S, Bertacchi FP, Caulfield B, Chappie K, Demarchi D, Erb MK, Estrada J, Fabara EE, Freni M, Friedl KE. Can mHealth technology help mitigate the effects of the COVID-19 pandemic? IEEE Open Journal of Engineering in Medicine and Biology. 2020 Aug 7;1:243–48.

[51] Lehoux P, Miller FA, Daudelin G, Urbach DR. How venture capitalists decide which new medical technologies come to exist. Science and Public Policy. 2016 Jun 1;43(3):375–85.

[52] Roncarolo F, Boivin A, Denis JL, Hébert R, Lehoux P. What do we know about the needs and challenges of health systems? A scoping review of the international literature. BMC Health Services Research. 2017 Dec;17:1–8.

[53] Degavre F, Kieffer S, Bol D, Dekimpe R, Desterbecq C, Pirson T, Sandu G, Tubeuf S. Searching for sustainability in health systems: Toward a multidisciplinary evaluation of mobile health innovations. Sustainability. 2022 Apr 27;14(9):5286.

[54] Sharmin J, Mozammel Hoque Chowdhury M. mHealth: A sustainable healthcare model for developing world. American Journal of Modeling and Optimization. 2014;2(3):73–76.

[55] Sharma U, Clarke M. Nurses' and community support workers' experience of telehealth: A longitudinal case study. BMC Health Services Research. 2014 Dec;14:1–7.

[56] De Preux L, Rizmie D. Beyond financial efficiency to support environmental sustainability in economic evaluations. Future Healthcare Journal. 2018 Jun;5(2):103.

[57] Hoffman RR, Feltovich PJ, Ford KM, Woods DD. A rose by any other name . . . would probably be given an acronym [cognitive systems engineering]. IEEE Intelligent Systems. 2002 Jul;17(4):72–80.

[58] Holden RJ, Carayon P, Gurses AP, et al. SEIPS 2.0: A human factors framework for studying and improving the work of healthcare professionals and patients. Ergonomics. 2013;56:1669–86.

[59] Valdez RS, Holden RJ, Novak LL, Veinot TC. Transforming consumer health informatics through a patient work framework: Connecting patients to context. Journal of the American Medical Informatics Association. 2014;22:2–10.

[60] Rutten GM, Meis JJ, Hendriks MR, Hamers FJ, Veenhof C, Kremers SP. The contribution of lifestyle coaching of overweight patients in primary care to more autonomous motivation for physical

activity and healthy dietary behavior: Results of a longitudinal study. International Journal of Behavioral Nutrition and Physical Activity. 2014;11:86.
[61] Critchley CR, Hardie EA, Moore SM. Examining the psychological pathways to behavior change in a group-based lifestyle program to prevent type 2 diabetes. Diabetes Care. 2012;35:699–70.
[62] Lippa KD, Klein HA, Shalin VL. Everyday expertise: Cognitive demands in diabetes self-management. Human Factors. 2008;50:112–20.
[63] Hosseini M, Jones J, Faiola A, Wu H, Vreeman D, Dixon BE. Reconciling disparate information in continuity of care documents: Piloting a system to consolidate structured clinical documents. Journal of Biomedical Science. 2017;74:123–29.
[64] Karsh BT, Holden RJ, Alper SJ, Or CKL. A human factor engineering paradigm for patient safety: Designing to support the performance of the healthcare professional. Quality and Safety in Health Care. 2006;15(Suppl 1).
[65] Lindley SJ, Cook PA, Dennis M, Gilchrist A. Biodiversity, physical health and climate change: A synthesis of recent evidence. Biodiversity and Health in the Face of Climate Change. 2019:17–46.
[66] Bergmo TS. How to measure costs and benefits of eHealth interventions: An overview of methods and frameworks. Journal of Med Internet Research. 2015;17:e254.
[67] Matthew-Maich N, Harris L, Ploeg J, Markle-Reid M, Valaitis R, Ibrahim S, Gafni A, Isaacs S. Designing, implementing, and evaluating mobile health technologies for managing chronic conditions in older adults: A scoping review. JMIR mHealth and uHealth. 2016 Jun 9;4(2):e5127.
[68] LeFevre AE, Shillcutt SD, Broomhead S, Labrique AB, Jones T. Defining a staged-based process for economic and financial evaluations of mHealth programs. Cost Effectiveness and Resource Allocation. 2017 Dec;15:1–6.
[69] Haverinen J, Keränen N, Falkenbach P, Maijala A, Kolehmainen T, Reponen J. Digi-HTA: Health technology assessment framework for digital healthcare services. Finnish Journal of eHealth and eWelfare. 2019 Nov 2;11(4):326–41.
[70] Vis C, Bührmann L, Riper H, Ossebaard HC. Health technology assessment frameworks for eHealth: A systematic review. International Journal of Technology Assessment in Health Care. 2020 Jun;36(3):204–16.
[71] Wilson L, Kim A, Szeto D. The evidence for the economic value of ehealth in the united states today: A systematic review. Journal of the International Society for Telemedicine and eHealth. 2016 Oct 23;4:e21–1.
[72] McNamee P, Murray E, Kelly MP, Bojke L, Chilcott J, Fischer A, West R, Yardley L. Designing and undertaking a health economics study of digital health interventions. American Journal of Preventive Medicine. 2016 Nov 1;51(5):852–60.
[73] Brown C, Harrison D, Burns H, Ziglio E. Governance for Health Equity: Taking Forward the Equity Values and Goals of Health 2020 in the WHO European Region. World Health Organization. Regional Office for Europe; 2014.
[74] Moultrie J, Sutcliffe L, Maier A. A maturity grid assessment tool for environmentally conscious design in the medical device industry. Journal of Cleaner Production. 2016 May 20;122:252–65.
[75] Finkbeiner M, Inaba A, Tan R, Christiansen K, Klüppel HJ. The new international standards for life cycle assessment: ISO 14040 and ISO 14044. The International Journal of Life Cycle Assessment. 2006 Mar;11:80–85.
[76] Brynjarsdottir H, Håkansson M, Pierce J, Baumer E, DiSalvo C, Sengers P. Sustainably unpersuaded: How persuasion narrows our vision of sustainability. In: Proceedings of the sigchi conference on human factors in computing systems 2012 May 5 (pp. 947–56).
[77] Unger SR, Hottle TA, Hobbs SR, Thiel CL, Campion N, Bilec MM, Landis AE. Do single-use medical devices containing biopolymers reduce the environmental impacts of surgical procedures compared with their plastic equivalents? Journal of Health Services Research & Policy. 2017 Oct;22(4):218–25.

[78] García Núñez C, Manjakkal L, Dahiya R. Energy autonomous electronic skin. NPJ Flexible Electronics. 2019 Jan 4;3(1):1.
[79] Nyein HY, Bariya M, Kivimäki L, Uusitalo S, Liaw TS, Jansson E, Ahn CH, Hangasky JA, Zhao J, Lin Y, Happonen T. Regional and correlative sweat analysis using high-throughput microfluidic sensing patches toward decoding sweat. Science Advances. 2019 Aug 16;5(8):eaaw9906.
[80] Wang L, Lou Z, Jiang K, Shen G. Bio-multifunctional smart wearable sensors for medical devices. Advanced Intelligent Systems. 2019 Sep;1(5):1900040.
[81] Hilty LM, Aebischer B. ICT for sustainability: An emerging research field. ICT Innovations for Sustainability. 2015:3–6.
[82] Alshqaqeeq F, Esmaeili MA, Overcash M, Twomey J. Quantifying hospital services by carbon footprint: A systematic literature review of patient care alternatives. Resources, Conservation and Recycling. 2020 Mar 1;154:104560.
[83] Bouchard MJ, Lévesque B. L'innovation et les transformations sociales, une approche théorique plurielle de l'économie sociale. Le cas du Québec. vers une Théorie de l'Économie Sociale et Solidaire. D Hiez, É Lavilunière (eds.), 2013:113–44.
[84] Fossati EC, Degavre F, Lévesque B. L'innovation sociale: Retour sur les marches d'une construction théorique et pratique. Entretien Avec Benoît Lévesque. Revue de la Régulation. Capitalisme, Institutions, Pouvoirs. 2018 May;30(23).
[85] Parra C. 10. Social sustainability: a competing concept to social innovation? 2013.
[86] Loorbach D, Frantzeskaki N, Avelino F. Sustainability transitions research: Transforming science and practice for societal change. Annual Review of Environment and Resources. 2017 Oct 17;42:599–626.
[87] Mehmood A, Parra C. Social Innovation in an Unsustainable World. In: F Moulaert, D MacCallum, A Mehmood, A Hamdouch (eds.), The International Handbook on Social Innovation: Collective Action, Social Learning and Transdisciplinary Research (pp. 53–66). Cheltenham, UK: Edward Elgar; 2013.
[88] Laurent É. The European green deal: From growth-strategy to social-ecological transition? Social Policy in the European Union: State of Play. 2020:97–111.
[89] Hudon M, Huybrechts B. From distant neighbours to bedmates: Exploring the synergies between social economy and sustainable development. Public and Cooperative Economics. 2017;88:141–54.
[90] Kemp R, Schot J, Hoogma R. Regime shifts to sustainability through processes of niche formation: The approach of strategic niche management. Technology Analysis & Strategic Management. 1998 Jan 1;10(2):175–98.
[91] Geels FW. Socio-technical transitions to sustainability: A review of criticisms and elaborations of the Multi-Level Perspective. Current Opinion in Environmental Sustainability. 2019 Aug 1;39:187–201.
[92] Mason C, Barraket J, Friel S, O'Rourke K, Stenta C-P. Social innovation for the promotion of health Equity. Health Promotion International. 2015;30:116–25.
[93] Cangiano S, Romano Z, Loglio M. The growth of digital social innovation in Europe. An Open Design approach to support innovation for the societal good. The Design Journal. 2017 Jul 28;20(sup1): S3546–59.
[94] Smalley KB, Warren JC, Fernández MI. The future of health equity. Health Equity: A Solutions-Focused Approach. 2020 Jul;18:359.
[95] Pel B, Haxeltine A, Avelino F, Dumitru A, Kemp R, Bauler T, Kunze I, Dorland J, Wittmayer J, Jørgensen MS. Towards a theory of transformative social innovation: A relational framework and 12 propositions. Research Policy. 2020 Oct 1;49(8):104080.
[96] Grimm R, Fox C, Baines S, Albertson K. Social innovation, an answer to contemporary societal challenges? Locating the concept in theory and practice. Innovation: The European Journal of Social Science Research. 2013 Dec 1;26(4):436–55.
[97] Macdonnell M, Darzi A. A key to slower health spending growth worldwide will be unlocking innovation to reduce the labor-intensity of care. Health Affairs. 2013 Apr 1;32(4):653–60.

[98] Heals C, Green H. Social innovation in health and social care: Case study results. A Deliverable of the Project: Social Innovation: Driving Force of Social Change (SI-DRIVE). 2016 Dec.
[99] Barkhuus L, Rode JA. From mice to men-24 years of evaluation in CHI. In: Proceedings of the SIGCHI conference on Human factors in computing systems 2007 Apr 28 (Vol. 10, No. 1240624.2180963).
[100] Blevis E. Sustainable interaction design: Invention & disposal, renewal & reuse. In Proceedings of the SIGCHI Conference on Human Factors in Computing Systems, San Jose, CA, USA, 28 April–3 May 2007; pp. 503–12.
[101] Mankoff JC, Blevis E, Borning A, Friedman B, Fussell SR, Hasbrouck J, Woodruff A, Sengers P. Environmental sustainability and interaction. CHI'07 extended abstracts on Human factors in computing systems, 2007 Apr 28, 2121–24.
[102] Remy C, Bates O, Dix A, Thomas V, Hazas M, Friday A, Huang EM. Evaluation beyond usability: Validating sustainable HCI research. In: Proceedings of the 2018 CHI Conference on Human Factors in Computing Systems 2018 Apr 21 (pp. 1–14).
[103] Silberman MS, Nathan L, Knowles B, Bendor R, Clear A, Håkansson M, Dillahunt T, Mankoff J. Next steps for sustainable HCI. Interactions. 2014 Sep 1;21(5):66–69.
[104] Knowles B, Bates O, Håkansson M. This changes sustainable HCI. In: Proceedings of the 2018 CHI Conference on human factors in computing systems 2018 Apr 21 (pp. 1–12).
[105] Dillahunt T, Mankoff J, Forlizzi J. A proposed framework for assessing environmental sustainability in the HCI community. In: Examining Appropriation, Re-Use, and Maintenance of Sustainability workshop at CHI 2010.
[106] Dillahunt TR. Fostering social capital in economically distressed communities. In: Proceedings of the SIGCHI Conference on Human Factors in Computing Systems 2014 Apr 26 (pp. 531–40).
[107] Remy C, Bates O, Thomas V, Huang EM. The limits of evaluating sustainability. In: Proceedings of the 2017 Workshop on Computing Within Limits 2017 Jun 22 (pp. 103–10).
[108] Lundström A, Pargman D. Developing a framework for evaluating the sustainability of computing projects. In: Proceedings of the 2017 Workshop on Computing Within Limits 2017 Jun 22 (pp. 111–17).
[109] Slater H, Campbell JM, Stinson JN, Burley MM, Briggs AM. End user and implementer experiences of mHealth technologies for noncommunicable chronic disease management in young adults: Systematic review. Journal of Medical Internet Research. 2017 Dec 12;19(12).
[110] Brown W, Yen PY, Rojas M, Schnall R. Assessment of the health IT usability evaluation model (Health-ITUEM) for evaluating mobile health (mHealth) technology. Journal of Biomedical Informatics. 2013 Dec 1;46(6):1080–87.
[111] Farao J, Malila B, Conrad N, Mutsvangwa T, Rangaka MX, Douglas TS. A user-centered design framework for mHealth. PLoS ONE. 2020;15:e0237910.
[112] Nadal C, Doherty G, Sas C. Technology acceptability, acceptance and adoption-definitions and measurement. In: 2019 Chi conference on human factors in computing systems, 2019 May 4.
[113] Georgsson M, Staggers N. An evaluation of patients' experienced usability of a diabetes mHealth system using a multi-method approach. Journal of Biomedical Informatics. 2016 Feb 1;59:115–29.
[114] Kasali FA, Awodele O, Kuyoro A, Akinsanya A, Eze M. A conceptual design and evaluation framework for mobile persuasive health technologies (Usability approach).
[115] Merchant G, Weibel N, Patrick K, Fowler JH, Norman GJ, Gupta A, Servetas C, Calfas K, Raste K, Pina L, Donohue M. Click "like" to change your behavior: A mixed methods study of college students' exposure to and engagement with Facebook content designed for weight loss. Journal of Medical Internet Research. 2014 Jun 24;16(6).
[116] Norris T, Stockdale R, Sharma S. Mobile health: Strategy and sustainability. Technology and Health Care. 2008;6(5):326–33.
[117] Li J, Moore N, Akter S, Bleistein S, Ray P. mHealth for influenza pandemic surveillance in developing countries. In: Proceedings of the 43 Hawaii International Conference on System Sciences (HICSS), Hawaii, United States, January 5–8, 2010.

[118] MacNeill AJ, Hopf H, Khanuja A, Alizamir S, Bilec M, Eckelman MJ, Hernandez L, McGain F, Simonsen K, Thiel C, Young S. Transforming the medical device industry: Road map to a circular economy: Study examines a medical device industry transformation. Health Affairs. 2020 Dec 1;39(12):2088–97.

[119] Ricciardi W, Pita Barros P, Bourek A, Brouwer W, Kelsey T, Lehtonen L. How to govern the digital transformation of health services. European Journal of Public Health. 2019 Oct 1;29(Supplement_3).

[120] Podein RJ, Hernke MT. Integrating sustainability and health care. Primary Care: Clinics in Office Practice. 2010 Mar 1;37(1):137–47.

[121] Bolton L. Helpdesk research report. Mobile Telephony for Improved Health Service and Data Management. 2012:1–11.

[122] Mishra S, Singh IP. mHealth: A developing country perspective. Technology. 2008:1–9.

[123] Kahn JG, Yang JS, Kahn JS. 'Mobile' health needs and opportunities in developing countries. Health Affairs. 2010 Feb 1;29(2):252–58.

[124] Waruingi M, Underdahl L. Opportunity in delivery of health care over mobile devices in developing countries. Ajfand. 2009;9(5):1–11.

[125] Kaplan WA. Can the ubiquitous power of mobile phones be used to improve health outcomes in developing countries? Globalization & Health. 2006;2:9.

[126] Rehalia A, Kumar R. A review on M-health system and technologies. International Journal of Current Research and Review. 2012;4(3):53–58.

[127] Hassan MU, Rehmani MH, Chen J. Privacy preservation in blockchain-based IoT systems: Integration issues, prospects, challenges, and future research directions. Future Generation Computer Systems. 2019 Aug 1;97:512–29.

[128] Shin M. Secure remote health monitoring with unreliable mobile devices. Journal of Biomedicine & Biotechnology. 2012;2012:546021.

[129] Patricia M, Hima B, Nadi K, Sarah S, Ada K, Adina G, Lin F, Ossman J. Barriers and gaps affecting mHealth in low- and middle-income countries. Policy White Paper. 2010:1–79.

[130] West D. How mobile devices are transforming healthcare. World Health. 2012;18:1–14.

[131] Noordam AC, Kuepper BM, Stekelenburg J, Milen A. Improvement of maternal health services through the use of mobile phones. Tropical Medicine & International Health: TM & IH. 2011;16(5):622–26.

[132] Yu P, Wu MX, Yu H, Xiao GC. The challenges for the adoption of M-health. International Conference on Service Operations and Logistics and Informatics (SOLI 2006), Shanghai, China, June 21–23 2006; 181–86.

[133] Lagomarsino G, Schweitzer J. E-health in low and middle-income countries: Findings from the Center for Health Market Innovations. Bulletin of the World Health Organization. 2012;90(5):332–40.

[134] Lewis T, Synowiec C, Garai A, Candidate M. Role of mHealth in rural health in India and opportunities for collaboration. Technology. 2011;4:1–5.

[135] Bäck I, Mäkelä K. Mobile phone messaging in health care – where are we now? Journal of Information Technology & Software Engineering. 2012;1(2):1–8.

Chapter 7
Sustainable healthcare system using biomedical engineering

Abstract: As global healthcare systems face rising challenges, sustainability has emerged as the primary concern. Biomedical engineering (BME) is a field of study that integrates ideas from both engineering and biomedical sciences, resulting in an interdisciplinary approach to solving problems in healthcare and technology. This field encompasses a wide range of subjects in which biomedical engineers are engaged, including medicine, regenerative medicine, associated sectors, and the advancement of improved products and services. The field of BME is characterized by its rapid growth and interdisciplinary nature, encompassing multiple fields such as medicine, biology, chemistry, engineering, nanotechnology, and informatics. BME is currently engaged in the development of advanced medical devices, immunizations, disease control products, robotics, and algorithms to improve global human health. The chapter explores the complex field of sustainable healthcare systems, particularly emphasizing the fundamental contribution of BME. It provides a thorough analysis of innovative approaches, technologies, and practices that are transforming the field of healthcare to be more environmentally friendly, economically viable, and socially responsible. The desire of individuals for extended lifetimes has served as an incentive for the rapid advancement of biomedical and healthcare technology. The integration of advanced technologies and artificial intelligence is increasingly being observed in healthcare systems and used in fundamental medical research within the field of BME. The field of BME is of great significance in the endeavor to achieve sustainability in healthcare, as it leads the way in developing advancements that minimize the environmental, economic, and societal consequences associated with healthcare. Biomedical engineers engage in the development of environmentally sustainable medical devices that exhibit improved energy efficiency. However, traditional biomedical and healthcare technologies exhibit several limitations, including limited period, large dimensions, and an increased chance of safety concerns. The primary objective of this work is to examine applications of biomedical engineering within healthcare sustainability.

7.1 Introduction

The growing need for healthcare that is safe, comfortable, sustainable, and of high quality has resulted in a significant rise in the development and utilization of biomedical technologies and healthcare services [1–4]. They monitor the heartbeat, blood pressure, pulse, body temperature, blood flow, and the number of beats per minute [5–14]. The application of engineering principles to biological systems is known as biomedical engi-

https://doi.org/10.1515/9783111436432-007

neering (BME). BME is a subfield within the broader discipline of bioengineering, which focuses on the application of engineering principles and techniques to develop medical devices and design procedures aimed at improving patient healthcare [15, 16].

The field of BME offers a structured platform for the education of future individuals who will assume leadership roles in the development and application of health technology. This field is particularly valuable due to its multidisciplinary nature and emphasis on design principles [17]. Universities in the United States have incorporated a component focused on BME inside their global health programs [17–19]. Additionally, they have integrated elements of global health and development into their BME programs [20]. The concept, commonly known as global health, in industrialized nations is frequently equivalent to local public health in underdeveloped regions. Developing nations exhibit significant socioeconomic gaps, indicating that they are not uniform in society. Biomedical engineers integrate principles and methodologies from the fields of engineering and medicine. Biomedical engineers are responsible for the development of medically relevant technology, such as surgical robots, biosensors, and cost-effective respirators, which have the potential to save lives. The scope of the field of BME encompasses a wide range of contributions that biomedical engineers can provide to society. Engineers conduct an assessment and evaluation of the appropriateness of materials for medical devices, with a specific focus on their biocompatibility and biodegradability. This sector is expected to offer competence in modeling, simulation, and material design [21–23]. The basic objective of BME is to improve healthcare, which includes all therapies. The primary responsibility of a biomedical engineer working in a healthcare facility is to ensure the correct maintenance and security of medical equipment. This includes providing the necessary information to the individuals responsible for managing these technologies [24, 25]. Bioengineers are actively engaged in the advancement of agricultural technology, equipment, and procedures that serve the purpose of sustaining individuals, and promoting modern and sustainable farming practices. The organization offers environmental solutions and aids in advancing advances in industrial production and energy sectors. Bioengineers contribute to improving the quality of life and health through various endeavors such as the creation and evaluation of new technologies, the development of innovative patient care procedures, the investigation of biological processes, and addressing challenges in agriculture and the environment [26–29].

To achieve healthcare sustainability, biomedical engineers must lead the way in developing new methods to lessen healthcare's negative effects on the environment, the economy, and society as a whole. Biomedical engineers engage in the development of environmentally sustainable medical devices that exhibit improved energy efficiency. Additionally, they encourage the use of telemedicine and remote monitoring practices in order to mitigate the carbon emissions connected with patient care. Furthermore, these engineers play a pioneering role in the exploration and implementation of sustainable materials for both devices and packaging purposes. One of the key supported technologies in the field of BME is additive manufacturing. Artificial intelligence (AI), the Internet of things (IoT), virtual reality (VR), holography, and various medical imaging technologies are among the notable advancements in the field. These technologies offer efficient assistance to the healthcare industry. The development of sustainable solutions focused on enhancing health and decreasing global health inequality involves consideration of the cultural and socioeconomic environment within which these interventions are intended to be implemented. There is a significant inequality, sometimes referred to as the "implementation gap," that arises when attempting to apply health advances in developing regions. This gap stems from the fact that the implementation of these innovations is frequently unproven, inadequate, or incomplete [30]. Several educational institutions in the developed world have implemented global health programs to educate students about the health challenges in the poor world and encourage them to work toward finding effective solutions. In this particular scenario, the utilization of biomedical sensors and implantable medical devices is progressively supplanting conventional medical instruments [31–35]. Biomedical sensors are devices used in the field of biomedicine to detect and measure physiological parameters [36–40] or biomarkers in living organisms. The equipment is a combination of engineering and biomedicine [41–44]. These devices are designed to either detect or respond to physiological signals.

The integration of a comprehensive communication platform has been established to facilitate collaboration among researchers specializing in several areas of BME, healthcare, and sustainability. In recent times, the healthcare industry has experienced an extensive reorganization due to advancements in computing, networking technology, big data, and AI. The healthcare industry is undergoing a transformation, characterized by a shift away from a reactive and hospital-centric approach toward a preventive and individualized one. Additionally, there is a transition from a disease-focused perspective to one that prioritizes overall well-being. Healthcare systems, along with fundamental medical research, are increasingly being enhanced and empowered through the integration of BME. Moreover, the utilization of advanced sensors and computer technology in healthcare delivery has the potential to enhance efficiency, improve quality, and reduce costs. The subject matter encompasses various fields, such as the ecological, financial, and societal sustainability of individuals, hence offering an advanced infrastructure for research relevant to this area. The field of study encompassing environmental health sciences and public health is of significant educational significance. Furthermore, biomedical implanted equipment, catheters, prostheses, contact lenses, medical instruments, respiratory machines, and other healthcare tools can also be contaminated with harmful agents [45].

BME plays a crucial role in enhancing the resilience and cost-effectiveness of the healthcare system by imparting knowledge to healthcare professionals and continually promoting research and innovation for sustainability. They promote individualized medicine to reduce unnecessary treatment and advocate for the optimization of healthcare facilities to save energy use. Furthermore, biomedical engineers play an essential part in the integration of renewable energy, improved imaging techniques, and environmental monitoring in healthcare environments. This discipline contributes to the well-being of both patients and the environment. BME combines physical, mathematics, and life sciences with engineering concepts to investigate biology, medicine, and health systems, and applies technology to enhance wellness and sustainability of health [46]. Traditional education provides BME students with a foundational understanding of physiology and anatomy and other aspects of engineering such as mechanical, electrical, and computer engineering. Subsequently, this field offers individuals the opportunity to focus their studies and expertise in various specialized areas, including but not limited to bio instrumentation and biosignal processing, biomechanics and biomaterials, rehabilitation engineering, medical imaging and radiology, medical and health informatics, and clinical engineering, as mentioned in reference [47]. The World Health Organization (WHO) has observed variations in the rate of evolution of BME education and training across different geographical regions, despite the presence of a shared body of knowledge worldwide. Developed countries have been identified as leading the way in developing a curriculum that is more comprehensive and adaptable to the latest industry and research advancements [47]. These tasks encompass the selection, deployment, and management of medical equipment and technology. Additionally, they are involved in the planning of clinical areas for healthcare delivery. Clinical engineers also provide support to other healthcare professionals in ensuring the proper utilization of diagnostics, therapies, and rehabilitation technologies. Furthermore, they contribute to the development of specialized or customized devices or equipment for investigation or treatment purposes [48]. Biomechatronics includes aspects of biology, mechanics, and electricity. Additionally, it encompasses domains within the fields of neurology and robotics. To effectively address health challenges, bioengineers engage in cooperation with other healthcare providers, such as physicians, nurses, surgeons, and technologists [49–51]. As a result, a range of indispensable tools and devices have been created, such as MRI (magnetic resonance imaging) machines, dialysis machines, diagnostic instruments, and ultrasound technology. The field of BME focuses on analyzing the movement and signals of the human body to gain a deeper understanding of the functioning of biological systems and the underlying reasons for the body's behavior [49–51]. The primary objective of this work is to examine notable applications of BME within healthcare sustainability.

7.2 Biomedical engineering

The field of BME has shown significant growth in terms of research, understanding development, educational programs, and its use in the industrial and healthcare sectors – the ability to rapidly adapt to novel information and identify the possibilities for utilizing this knowledge in the field of human health and sustainability development. The biomedical engineering field applies basic principles of engineering. The utilization of machinery, robotics, and innovation has significantly increased the demand for medical engineering in modern healthcare society. In the field of biology, innovative concepts were employed, some of which played a role in lowering mortality rates. The advancement of BME has been greatly influenced by the development of artificial organs, enhanced prostheses, surgical robots, renal dialyzers, and medications [52–54]. As a result of substantial progress in scientific research, advancements in healthcare, and technological innovations, the field of BME has emerged as a highly desirable interdisciplinary subject. The discipline of BME is very interesting since it encompasses the use of technical and scientific methodologies, the examination and provision of healthcare services, as well as the exploration of biological and physiological concerns. BME encompasses the development of products and processes that address the diverse issues encountered in the fields of medicine and health. This is achieved via the integration of biological and medical knowledge with technical principles and techniques [55–57]. The field of BME is experiencing significant growth in terms of research and knowledge expansion, educational programs, and its use in both industrial and healthcare settings. A biomedical engineer applies principles and techniques from the field of engineering to study and address challenges in the fields of biology and medicine, thereby contributing to the overall improvement of healthcare. The area of BME aims to contribute to the well-being of individuals, engage in the dynamic realm of dealing with biological systems, and utilize modern technology to address the intricate challenges in healthcare. The biomedical engineer collaborates with other healthcare professionals, such as physicians, nurses, therapists, and technicians. Biomedical engineers are often tasked with a diverse array of responsibilities, encompassing the design of equipment, devices, and software, the integration of information from other technological domains, the advancement of novel processes, and the undertaking of research initiatives aimed at resolving clinical issues.

The field of BME encompasses a wide range of activities, including the development and design of various biomedical products and systems, organ replacement devices, and diagnostic tools for artificial organs, as well as the provision of support, maintenance, adaptation, repair, and installation services for biomedical equipment.

The utilization of engineering standards within healthcare institutions applies to several components of biological systems. Research consultation is commonly conducted in collaboration with professionals such as chemical engineers, physicians, veterinary professionals, and other healthcare practitioners. Mechanical engineering principles, including forward interactions, reverse kinematics, mechanics of fluids, material characterization tests, thermodynamics, and alternative failure methodologies such as fracture mechanics, are frequently employed in the development of biomedical devices [27–30]. Biomedical engineers work in a wide variety of environments, including the pharmaceutical and biotechnology industries, health facilities, and other healthcare providers (including ambulatory surgical centers, dialysis treatment services, imaging and radiation therapy facilities, and government institutions) [48], and national regulatory agencies. When employed in sustainable healthcare establishments, clinical engineers are responsible for various tasks. The expected growth in demand for biomedical equipment and procedures within the healthcare system is apparent. Biomechanics is employed as a means to expand the limits of persons who have had an amputation or require organ transplantation. Tissue engineering has been shown to provide significant assistance to individuals with severe injuries [58–61]. The integration of engineering principles with biological and physiological understanding facilitates advancements in the field of geriatric care and the unraveling of enigmatic diseases that persistently affect the human population. An increasing number of individuals are consistently seeking biological treatments in anticipation of newer, more convenient, and more effective treatment. Scientists are currently investigating innovative approaches to specifically transport medications to specific areas of the body, to minimize damage to normal cells and tissues [62–64]. The field of BME encompasses a distinct specialization in the area of tissue engineering. The process involves the assembly and restoration of impaired organs or tissues. A combination of scaffold and biologically active compounds is utilized to facilitate the sustenance or regeneration of a compromised tissue or organ. The sector has transformed to accommodate the growing demand for organ transplants. Instead of utilizing animal or preexisting models, researchers have recently built and utilized a 3D-engineered tissue model for diagnostic purposes [65–67]. Tissue engineering integrates the disciplines of molecular biology, materials science, chemistry, and medicine. Tissue has the potential to serve as a therapeutic intervention for individuals afflicted with a range of medical conditions, encompassing severe burns as well as organ dysfunction. The tissues that have been effectively developed encompass liver and muscle tissue, skin, and cartilage. Professionals in the field of bioengineering might be involved in two primary roles: conducting research to develop innovative platforms or providing maintenance expertise for the repair and supervision of biomedical electronic equipment [68–71]. Designing implant materials requires understanding the characteristics and behavior of biomaterials, including live tissue and artificial materials. One of the biggest challenges for biomedical engineers is choosing suitable materials for human body implantation. Various types of implantable materials are commonly used, including metal alloys, ceramics, polymers, and composites [72, 73]. Biomaterials that are nontoxic, noncarcinogenic, chemically inert, stable, and mechanically robust are necessary to effectively withstand repeated physiological stresses. However BME holds considerable promise for advancing healthcare practices in the future. Several studies have demonstrated the effective production of cartilage, skin, liver, and muscle tissue [74–76].

7.3 Biomedical engineering for healthcare advancements

BME has shown significant growth in terms of studies and research development, as well as the development of educational programs and its use in industrial and healthcare organizations.

The development of novel medical goods and a growing number of industries, including small- and medium-sized enterprises (SMEs), are greatly impacted by the new information gained by fundamental BME research (at the gene, molecular, cellular, organ, and system level). SMEs are predicted to introduce novel healthcare products and services to the market. Medicine and healthcare have undergone significant transformations in recent decades, relying heavily on advanced technology for disease prevention, diagnosis, treatment, and patient rehabilitation. Modern biomedical research and healthcare provision are facilitated by multidisciplinary teams in which biomedical engineers contribute to the advancement of knowledge alongside medical professionals. BME is considered one of the fastest-growing industries in the developed world, alongside sustainable and renewable energy sources.

BME is an interdisciplinary field that emerges from the integration of various engineering sciences and principles with the study of biology, medicine, behavior, and health. As such, BME is committed to enhancing human well-being and health sustainability. New devices, materials, processes, and software are developed as a result of BME research, which expands our understanding of the human body, from the molecular and cellular level to the level of organs and the body as a system. Prevention, prediction, diagnosis, treatment, and care for patients, as well as rehabilitation, all benefit from the incorporation of advanced technologies. Many subspecialties exist in BME, just as they do in the medical field. BME is a field that has traditionally encompassed logical sciences aimed at advancing healthcare treatment, specifically in the areas of diagnosis, monitoring, and therapy. The user has provided two references, indicated by the numbers [77, 78]. In addition, the responsibilities of a biomedical engineer encompass the oversight and maintenance of existing medical equipment within hospital settings, while ensuring compliance with pertinent industry regulations and guidelines. The responsibilities encompassed in this domain include the acquisition of equipment, regular testing, proactive maintenance, and providing suggestions for equipment upgrades. This position is commonly referred to as a biomedical equipment technician or clinical engineering.

In recent years, the field of BME has grown as an independent area of study, distinguishing itself from various other branches of engineering. This phenomenon is frequently observed when a nascent discipline progresses from being an interdisciplinary subfield within established domains to attaining recognition as an independent field. A significant portion of the activities within the field of BME involves the undertaking of research and development endeavors, which encompass a wide range of specialized subfields (as indicated subsequently). Prominent applications within the field of BME encompass the advancement of biocompatible prostheses, a diverse array of diagnostic and therapeutic medical devices, spanning from clinical equipment to micro-implants to widely used imaging equipment such as MRIs and EKG/ECGs, the promotion of regenerative tissue growth, and the development of pharmaceutical drugs and therapeutic biologicals.

Numerous advancements and innovative activities have been made within the growing field of BME. Figure 7.1 demonstrates the progression of advancements in BME over the specified time frame. A significant portion of biomedical advancements originated between the 1950s and the present era. Additionally, this field encompasses advancements in pacemaker technology, X-ray diagnostics, computed tomography (CT) scanning, endoscopic procedures, hearing aid support, the application of pulse oximeters, tomographic imaging, and various other innovations. These sequential actions additionally guarantee the promising trajectory of BME in bolstering healthcare and its various fields [79–83]. BME rehabilitation is an emerging area within the field of biotechnology. Rehabilitation engineers aim to enhance the functional abilities and overall healthcare of individuals with physical and cognitive impairments. The individuals engage in the process of designing and producing prosthetic devices, making modifications to residential and occupational environments, as well as to transportation systems. Additionally, they are involved in the advancement of assistive technologies aimed at enhancing seating arrangements, mobility, and communication [84–87]. Rehabilitation engineers are engaged in the development of hardware, computer software, and cognitive assistance tools to provide support for individuals facing cognitive impairments. The field of system physiology encompasses the application of engineering methods, techniques, and tools to comprehensively study and integrate the functioning of living organisms, ranging from bacteria to humans [88–90].

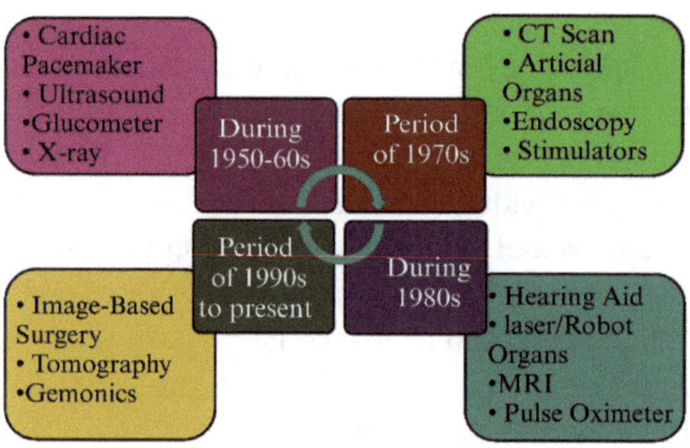

Figure 7.1: Several innovations in biomedical engineering for healthcare.

BME has made significant advancements that contribute to the development in the field of healthcare. Nanotechnology, as an illustrative example, presents the possibility of achieving targeted drug administration and accurate molecular-level diagnostics, thereby reducing adverse effects and improving the effectiveness of treatments. Furthermore, the utilization of 3D printing technology has brought about a significant transformation in the development of personalized prosthetics, implants, and tissue scaffolds, providing patients with individualized remedies that cater to their specific medical requirements. Moreover, the development of AI and machine learning has facilitated the examination of extensive healthcare datasets for predictive diagnostics, enhanced treatment strategies, and more efficient administrative operations, ultimately leading to advancements in patient care and the overall efficiency of healthcare systems. These advanced technologies serve as prime examples of the profound impact that BME can have, leading to the development of healthcare solutions that are more tailored to individual needs, easily accessible, and highly efficient.

Nanorobotics is a field that focuses on the development of small surgical robots that have components at the nanometer scale. These advanced technologies have the potential to revolutionize biotechnology by allowing scientists to manipulate and interact with biological systems at the atomic or molecular level, leading to significant advancements in our ability to effectively combat diseases. Medical nanobots are introduced into the human body through minimally invasive surgical procedures, such as delivery through the circulatory system or other body cavities. Once inside the body, these nanobots perform essential tasks such as investigating for infections, following preprogrammed instructions, or being guided by a human surgeon. Additionally, nanobots can monitor vital signs like temperature and blood pressure, enabling clinicians to diagnose, detect, and track bloodstream bacteria, tissues, and cells. Furthermore, disposals that facilitate the control of neurological prosthetics allow for communication between prostheses, the central nervous system, and exoskeletal devices, which can enhance overall functionality and performance.

As researchers in the field of BME, it has the potential to either enhance or worsen the existing health disparities. This article aims to offer guidelines for biomedical engineers to promote healthcare equity. These guidelines, illustrated in Figure 7.2, are based on three fundamental principles:

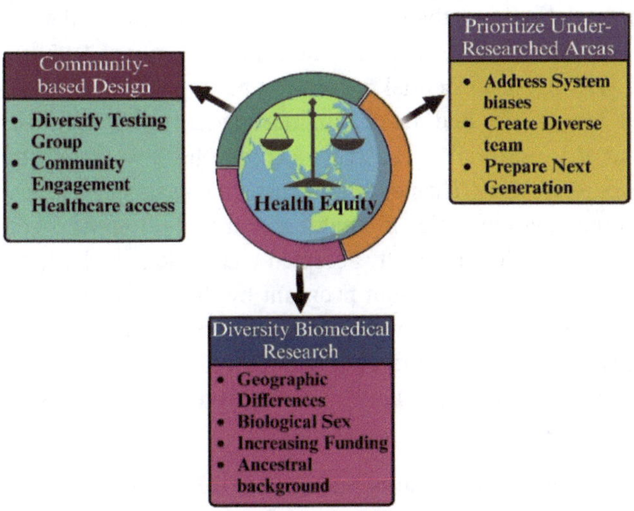

Figure 7.2: An overview of basic biomedical engineering guidelines for improving healthcare equality.

7.3.1 Enhancing diversity and equity within the field of science, technology, and biomedical engineering

– **Rule 1: Acknowledge systemic biases and monitor health inequities:** Systemic biases, such as systemic racism, gender discrimination, disability, age discrimination, and biases related to sexual orientation, gender identity, and socioeconomic position, play a significant role in the emergence of health disparities [91–93]. To address these concerns, it is imperative for us as a society to publicly acknowledge their presence and engage in concerted efforts to eradicate them. The aforementioned disparities extend beyond instances of individual discrimination and permeate social systems that systematically place members of historically marginalized groups at a disadvantage [91, 94, 95]. Throughout history, racism has had a significant influence on the formation of the United States, as seen by the exploitation of territory belonging to indigenous peoples and Mexicans, the forced captivity of Africans, and the deplorable treatment of Chinese immigrants who were employed as workers [96]. Efforts labeled as "color blind" inadequately tackle the deeply ingrained racial disparities that are embedded within the institutional framework of our nation. At the individual level, implicit prejudice contributes to the perpetuation of systemic racism [97–100]. Consequently, it is imperative to demolish these implicit biases through proactive efforts such as reframing negative associations, actively pursuing educational opportunities, and fostering positive relationships and contact. The presence of systemic racism has permeated the academic sphere, resulting in the marginalization of scholars belonging to historically excluded groups [101]. A total of more than 10,000 academic faculty members demonstrated their support by affixing their signatures to a commitment that acknowledges the existence of systematic racism within the academic sphere [102]. To achieve healthcare equity universally, it is essential to establish an improved methodology for identifying healthcare inequalities and delineating the specific issues that need attention. To achieve that goal, it is essential to construct a comprehensive system for assessing equality in healthcare technology. The resulting structure will serve as a foundation for evaluating and monitoring inequities in healthcare across different demographic groups. This system must include universally accepted terminology and criteria [103]. The establishment of an Equity and Inclusion program by the National Cancer Institute (NCI) has been reported. This program encompasses the implementation of systematic monitoring and assessment of equity initiatives [104]. Additionally, an equity council has been formed to provide criteria for evaluating the advancement of NCI's equity endeavors based on quantifiable outcomes.

– **Rule 2: Enhancement of diversity within teams and the establishment of inclusive environments:** The composition of the scientific workforce in the United States is not consistent with the socioeconomic status of society. Specifically, women make up just 28.4% of the scientific workforce, while black and Hispanic scientists represent only 5.4% and 6%, respectively [105]. The promotion of diversity within teams has emerged as a prominent objective in several regions, often accompanied by the notion that variety fosters creativity [106]. In the field of medical services, the inclusion of different teams that reflect the demographic composition of the patient community has been shown to provide supplementary advantages in enhancing patient outcomes [107]. Their unique personality as well as private observations of an investigator often shape their field of interest and competence [108]. Women's participation in biomedical research has led to a surge in studies on the health of women since the late 1970s [109, 110]. Science, as it is, does not do enough to meet the healthcare requirements of several groups [108]. By diversifying the viewpoints and interests represented within research teams, it is possible to enhance the probability of effectively addressing the healthcare requirements of all communities. Efforts have been directed toward enhancing the representation and achievements of engineering teams by implementing intentional recruitment strategies and fostering an inclusive environment that appreciates the contributions of all team members. To diversify the scientific pipeline, institutions have promoted healthcare research as a career through local K-12 community outreach and recruitment at events like the Annual Biomedical Research Conference for Minority Students. Additional initiatives involve partnerships with Minority-Serving Institutions, where students are enrolled in both institutions simultaneously. One example is the Atlanta University Center Consortium Dual-Degree Program, which aims to facilitate the admission of liberal arts students from historically Black colleges and universities into one of the nine traditional engineering institutions [110–112].

– **Rule 3: Prepare present and future professionals to address health disparities:** Significant effort is required to alter the training and education of medical care professionals and researchers. The approach to engineering education should be based on a humanistic perspective, emphasizing the significance and addressing systematic biases and health disparities. It should be done without compromising the scientific study that supports the field. The use of service learning has been employed as a means to familiarize students with health disparities within the community, accompanied by guided reflection [113]. This approach further facilitated the expansion of the campus and the establishment of a strong relationship with neighborhood groups who face health disparities. Moreover, it offered educational opportunities to these populations. The incorporation of education on a range of themes about distinct community needs may be researched and integrated into the conventional curriculum. These topics include public health, health disparities, the historical impact of trauma, women's health, and cultural variations. An instance was previously examined in a case study whereby the incorporation of public health subjects, namely health inequalities and socioeconomic determinants of health, was explored inside a controlled medication delivery course [114]. This intervention facilitated the active involvement of students in inclusive practices and emphasized the need to recognize health inequalities while designing engineering solutions. Furthermore, further educational opportunities may be offered through featuring health inequality researchers at conferences and professional gatherings, organizing journal clubs that emphasize the research and achievements of underrepresented scientists, and promoting literature that provides knowledge on health disparities.

7.3.2 Expanding research in underserved regions

– **Rule 4: Prioritize and enhance research funding for illnesses and communities that are underserved:** There is a tendency within biomedical research to overlook illnesses that have a higher prevalence among historically underprivileged populations, resulting in a delay in the development of medical advancements aimed at improving healthcare for these groups. Common causes of mortality or impairment in such patients include errors in diagnosis and a lack of treatment. Tropical illnesses, such as dengue, lymphatic filariasis, trachoma, and leishmaniasis, affect over 1 billion people worldwide in poor socioeconomic situations but are overlooked in studies outside the United States [115]. Numerous biomedical engineers operating in highly competitive domains have the potential to use their specialized knowledge and skills to address the needs of these underprivileged areas. The study and understanding of illnesses may have a significant impact on improving health outcomes. To enhance research efforts in these domains, it is essential to consider the financial framework as well. Insufficient financing sometimes leads to a lack of investigation into less frequent illnesses. However, this assertion does not hold for every circumstance. Cystic fibrosis, a medical condition predominantly seen in individuals of Caucasian descent, gets thrice more financial support per patient compared to sickle cell disease, a condition often linked with those of African descent [116]. Hence, the amount of variety among researchers will impact the degree of exertion in addressing health inequalities. The National Institutes of Health (NIH) focuses significant emphasis on reducing health inequalities. However, there is a notable disparity in funding rates between underrepresented investigators, particularly black scientists, and their white counterparts of comparable academic standing. Studies have shown that underrepresented investigators are around 55% less likely to get financing compared to white scientists [117, 118]. The decrease in funding rate significantly impacts the rates of tenure and promotion, eventually influencing the diversity within the field [119]. Several potential strategies have been suggested to mitigate this disparity. These include modifying the grant evaluation procedure to guarantee thorough deliberation of all applications by the review panel, as well as taking into account the diversity of the research team when assigning scores to the "Investigator" part. Private sector and charitable funding organizations provide other possible sources of money that are sometimes committed to the promotion of global health and the reduction of health inequities [120].

– **Rule 5: Investigate sex-based health determinants:** Women have a variety of health inequities, including greater rates of illness incidence and death, as well as increased susceptibility to adverse consequences from therapeutic interventions [121].

These discrepancies may emerge due to a range of variables, including inequities rooted in gender and sex-related distinctions in both health and healthcare. The concept of gender is a socially constructed phenomenon that has significant implications for individuals' perceptions, relationships, power dynamics, and access to resources. It plays a role in perpetuating inequities via several sociocultural variables, including biases in diagnosis and treatment, psychological stresses, exposure to violence, and levels of physical activity [122]. Sex is a fundamental biological characteristic that is determined by genetic factors, sex steroid hormones, and the presence of reproductive organs. It has significant relevance for biomedical researchers engaged in the development of healthcare technology [123]. Recent studies have uncovered disparities, depending on sex, in several aspects, including immune system reactions [124], makeup of the microbiome [125], manifestation of different illnesses and conditions [126], administration of therapeutic doses and metabolism [127–129], and distribution and efficacy of nanotechnology [130, 131]. Sex-based differences have been postulated to arise from disparities in the X and Y chromosomes. Specifically, the X chromosome is responsible for encoding more than 1,100 genes, while the Y chromosome encodes only around 100 genes. Additionally, the phenomenon of X inactivation occurs in female cells, wherein one X chromosome is inactivated in each cell, resulting in a mosaic pattern that is absent in males. Furthermore, variations in levels of sex steroid hormones and disparities in reproductive organs contribute to these differences [124, 132]. Furthermore, variations in pharmacokinetics and pharmacodynamics have been ascribed to disparities in adipose tissue and body fluid composition, levels of steroidal sex hormones, genetic phenotype, the activity of hepatic drug-metabolizing enzymes, and additional dissimilarities in the process of absorption, distribution, metabolism, and elimination, resulting from variations in bodily functions [127–129]. The aforementioned disparities may give rise to significant variations in performance, depending on sex, necessitating attention and consideration from engineers. For instance, the presence of these factors may have an impact on the effectiveness of nanotechnologies [130, 131] or the level of acceptability of biomedical devices, when implanted in the human body [133]. Sexual dimorphism in skeletal structure, bone strength, and the characteristics of muscles, tendons, and ligaments contribute significantly to the consideration of sex as a crucial variable in the design of implants and biomaterials [134]. However, a significant portion of research studies fail to include sex as a scientific component in their analysis, save in cases about reproduction [135].

Biomedical research often adopts a universal strategy that has mostly been influenced by male factors, resulting in a disproportionate vulnerability for women. To foster inclusive designs that yield benefits for all individuals, it is imperative to enhance our comprehension of these disparities. To facilitate researchers' comprehension of these distinctions, funding organizations have initiated the provision of financial resources for this particular area of study. The NIH has implemented procedures aimed at achieving gender balance in cell and animal experiments, as well as establishing projects focused on addressing women's health concerns [136]. The Food and Medication Administration

(FDA) strongly advises studies to consider the impact of sex and gender variations on medication metabolism [127].

– **Rule 6: Investigate the biological factors that determined health in the past:** Race is a societal construct that has given rise to many injustices and inequalities, both inside and outside the field of healthcare. While the categorization of populations, based on race, initially assisted in recognizing health inequalities, it is important to note that this framework is more closely associated with the social factors that influence health rather than physiological variables [137]. The conceptualization of race is mostly influenced by an individual's physical appearance and how they are socially perceived, rather than their genetic lineage. An illustrative instance is individuals with a combination of African and European ancestry, such as Barack Obama, who are often categorized as black. This serves as evidence that one's perceived racial identity does not always align with their ancestral heritage [138]. Race groupings are comprised of separate populations that have diverse cultural and biological distinctions based on their ancestral origins. For instance, individuals from China and India are often categorized as "Asian" within Western societies [139]. Moreover, the use of race as a determinant of health is detrimental, since it fails to include the genetic mixing and corresponding allele frequencies that exist in the contemporary global population. Additionally, only examining a patient through the lens of their racial background overlooks important factors. The aforementioned scenario pertains to a kid of African descent who had persistent respiratory issues, which went undetected as cystic fibrosis for an extended period. It was only upon reviewing the child's X-rays without knowledge of their racial background that the diagnosis was made [140]. Several clinical decision-making systems also include race correction variables that are deemed inappropriate, resulting in worse health outcomes for patients belonging to racial minority groups [141].

7.3.3 Consider various populations while designing research

– **Rule 7: Create designs that facilitate adoption by a diverse community:** When developing a novel treatment or medical equipment, we need to include the community at every stage, beginning with the conception of the first idea and continuing through to the implementation phase. The acceptance of treatments and enhancement of patient compliance may be achieved via the adaptation of technology to accommodate the specific demands and lifestyles of patients. Treatment regimens that are intricate and involve regular visits to healthcare providers for infusions or laboratory tests, adherence to strict medication schedules, self-monitoring, and maintenance of concentrated medical equipment, can result in decreased patient adherence and pose significant challenges to a patient's ability to work and uphold a desired lifestyle [142]. It is essential to develop treatment strategies that are grounded in realism, considering factors such as the administration method, frequency, and level of intricacy involved in the therapy process. The advent of community-based participatory research has posed a significant challenge to the conventional approach of scientific inquiry, as it promotes a deliberate and structured endeavor to foster connections between biomedical researchers and the community. The implementation of the Belmont Report in 1979 enhanced the ethical standards of scientific research by emphasizing the need for informed permission. However, levels of skepticism towards the US Healthcare System and Clinical Research exhibit significant variation, ranging from 20% to 80%, depending on factors such as age, ethnic background, and socioeconomic standing [143–145]. To address the prevailing lack of faith in biomedical research and identify potential cultural obstacles to early adoption, it may be beneficial to include the community as active collaborators rather than as passive subjects. This approach may facilitate the rectification of public distrust and enable the identification of cultural barriers at an earlier stage in the design process.

– **Rule 8: Consider healthcare access while designing devices for biomedical applications:** In the field of BME, it is essential for practitioners to include considerations of healthcare access through the first stages of design. Individuals of a lower socioeconomic position or residing in rural regions within the United States have barriers to accessing healthcare technology, mostly attributable to financial constraints or limited accessibility to medical establishments. The financial burden associated with treatment or medical equipment has historically been a significant factor in the unequal distribution of healthcare outcomes across different socioeconomic groups. The partial mitigating of this issue is attributed to the presence of health insurance, which is, however, limited due to the absence of universal healthcare and disparities in reimbursement. In addition to financial factors, there are concerns about access, as a result of travel limitations and the availability of services. The availability of transportation and sick leave may significantly limit the range of treatments and services that a patient may access. In 2014, the American College of Gynecologists and Obstetricians published research indicating that a majority of rural women, specifically less than 50%, were outside a 30-min driving distance from the closest healthcare facility providing prenatal treatment. The distances to healthcare services tend to increase for remote communities that have a significant proportion of black or indigenous residents [146]. Furthermore, technologies that need refrigeration, specialized apparatus, or substantial electrical use are not readily available. The availability of cold storage for COVID-19 vaccines was significantly constrained, resulting in varying levels of access to and administration of the vaccinations within various communities. Hence, engineers must take into account the necessary storage conditions for their therapeutics and technologies during the design stage.

To address the challenges associated with cold chain manufacturing, various initiatives are underway to develop a more insulated carrier box or design vaccine formulations that exhibit stability at elevated temperatures for extended periods, thereby effectively resolving this problem [147, 148]. The use of point-of-care technology has emerged as a crucial domain in the effort to mitigate healthcare access inequities. In the United States, there is a wide array of smartphone apps and peripherals available that are specifically designed for monitoring, tracking, and enhancing an individual's health. The advent of smartphones and internet connectivity in rural areas of India has established the nation as a rapidly expanding technological market. Interestingly, it is worth noting that about 20% of the worldwide burden of illness may be attributed to the 39 million individuals living in poverty in India [149]. Smartphones are becoming prevalent in a nation where 70% of the population dwells in rural regions [150]. The Richards-Kortum lab has developed screening and diagnostic tools for underdeveloped nations, pioneering their development [151]. It is recommended to prioritize the development of straightforward solutions, while still acknowledging the value of conventional approaches. Additionally, it is important to consider long-term implications while addressing short-term issues, including students' involvement in the process of economic design and tailor designs to specific contextual factors [152].

7.4 Research aspect in biomedical engineering for healthcare

BME has contributed much to healthcare, but there are still many areas that may use further study to make them more effective and efficient. Several topics are covered in Figure 7.3, including digital approaches in healthcare, significant data-enabled healthcare processes, the use of robots for precise operations in healthcare, the employment of smart and invisible sensing systems, and more. These subsequent innovations in BME will provide an advanced and effective healthcare system [153–157]. BME is crucial in the development of diagnostic medical equipment [158–161]. VR tool may be used to make 3D views of inside organs, robotics surgeons may be used to assist doctors, and other systems may be used to deliver drugs or build tissues. Bioengineering, an interesting area, bridges the gap between engineering and medicine. Bioengineers employ a wide range of approaches to help patients with a wide range of ailments, including the development of novel digital tools, software platforms, equipment, and other technologies. It helps in the research and development of healthcare technology [162–166].

Healthcare developments are driven by several important research areas in the dynamic field of BME. Medical treatments have the potential to be more sustainable and successful if scientists can develop biocompatible materials for use in implantable devices, biological engineering, and drug delivery systems. Bioinformatics and data analytics are becoming increasingly important in healthcare, allowing for the mining of large healthcare databases to support predictive analytics and precision medicine. Medical robots and wearable health technology are also advancing, allowing for noninvasive and remote patient monitoring. Research into the moral implications of technologies like gene editing, AI, and data privacy is also expanding, guaranteeing that the discipline of BME will continue to develop in a morally responsible manner. Collectively, these varied research efforts are propelling innovation that improves healthcare delivery on many fronts, including individualization, effectiveness, and accessibility. As the

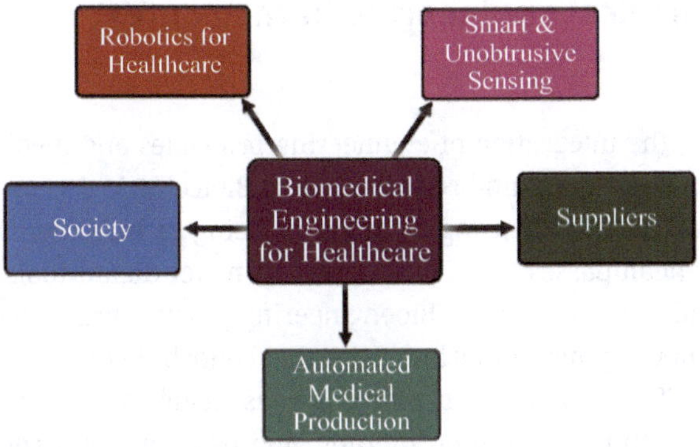

Figure 7.3: Several innovations in biomedical engineering for healthcare.

field of 3D printing continues to advance, there has been significant progress in the production of microscopic bioreactors that include distinct flow pathways and personalized growth chambers. Biomedical researchers want to integrate key technological advancements, such as reduced size, material progress, personalized medicine, and additive manufacturing, into their design methodologies. These technologies have the potential to introduce a diverse set of innovative design options that are not attainable through traditional manufacturing methods [167–171]. Some bioengineers engage in the development of innovative equipment and tools that aid in medical treatment, while others focus on improving healthcare delivery systems. Bioengineers seek to get a comprehensive understanding of the functioning of the human body and the mechanisms behind biological systems. In pursuit of this knowledge, they also investigate the signals generated by vital organs, like the heart and brain. Numerous individuals have defective tissues, calling for substitution by artificial organs, limbs, and valves. Biotechnologists are engaged in the field of rehabilitation by improving the functionality of therapeutic devices [172–177]. Furthermore, biomedical engineers are actively involved in enhancing the efficacy of therapeutic devices, including prosthetic limbs and medical implants, through the optimization of their design and functionality. Their contributions also encompass the field of rehabilitation. For example, researchers in this field focus on the creation of sophisticated rehabilitation equipment and technologies, such as exoskeletons and neuroproteins, which assist persons with mobility impairment in the process of restoring their functional abilities and achieving independence. Biomedical engineers are expanding advances in the field of healthcare by combining technology and biology, thereby not only improving diagnoses and therapies but also augmenting the quality of life and rehabilitation for people afflicted with diverse medical disorders.

7.5 The application of biomedical engineering in the context of healthcare

The field of BME is characterized by the integration of engineering principles and medical considerations, resulting in a dynamic and multidisciplinary area. Biotechnologists operate within the wider field of life sciences, focusing on the well-being of individuals living in various conditions. This encompasses the development of novel digital tools, software platforms, instruments, and other apparatus. Bioengineering practice primarily involves the conceptualization and development of technologies that contribute to the facilitation of healthcare procedures [178–183]. Biological technology has yielded several recent advancements, such as genetic testing, medication therapy, and tissue growth. The rapid progress in medical innovation has given rise to a variety of emerging challenges. Numerous aspects necessitate investigation and monitoring within this rapidly advancing area, including areas such as finance and ethics. Biotechnology plays a substantial role in advancing stem cell research by significantly contributing to the cultivation of stem cells in laboratory settings or in vitro. The potential benefits of stem cell implantation are for patients afflicted with a medical condition or ailment, which could be addressed through this intervention, leading to the restoration of their overall well-being and the provision of an alternative way of life. The ability of stem cells to undergo proliferation and differentiation into different cell types provides valuable insights for biotechnologists in the development of specific cells with unique characteristics [184–193].

The field of BME involves the application of principles and problem-solving approaches to the areas of biology and medicine. The increasing prevalence of implantable medical devices, including pacemakers and artificial hips, in conjunction with advancements in futuristic technologies like stem cell engineering and 3D printing of biological organs, has become evident across various aspects of healthcare, ranging from diagnostic and analytical procedures to therapeutic interventions and restorative measures [194–199]. Tissue and stem cell engineers endeavor to develop human organs, facilitate transplantation procedures, and contribute to the improvement of the global quality of life for millions of individuals. Professionals in the field of medical equipment are currently working on the creation of innovative implanted and external instruments. These instruments include pacemakers, coronary stents, orthopedic implants, prosthetics, dental goods, and ambulatory equipment [200–206]. Clinical engineers are actively engaged in ensuring the safety and dependability of medical equipment intended for use in healthcare facilities. The field of BME encompasses a wide array of specialized areas. The field of BME often combines problem-solving skills and technical expertise with focused research in the medical, health, and related domains. The process of hybridization has led to the emergence of novel advancements in the field of BME, presenting numerous prospects within the healthcare sector [207–211]. A method of instruction that holds potential for application in BME education involves the increased utilization of case-based, problem-based, and project-based learning methodologies. This approach entails structuring students' learning experiences around the resolution of genuine problems that arise within the area of BME. Table 7.1 presents a comprehensive overview of the recognized uses of BME within healthcare sustainability.

Table 7.1: Applications of biomedical engineering in healthcare.

S. no.	Application	Explanation
1.	Develop remedies for healthcare issues	Biomedical engineers may use their special knowledge to solve global health issues. This improves patient care and reduces healthcare costs via competence in physiology, biology, healthcare IT, mechanics, and engineering. Bioengineered products, such as wheelchairs and radioscopes, seamlessly integrate into daily life. Bioengineers innovate to address difficulties at the intersection of biology, medicine, and engineering. Some bioengineers develop new medical equipment and technology, while others improve healthcare procedures. Bioengineers study heart and brain impulses to comprehend bodily function and biological systems. For defective tissues, many make artificial organs and limbs. By improving therapeutic equipment designs, biotechnologists contribute to rehabilitation.
2.	Development and management of machinery	In the biomedical electronics business, bioengineers build and maintain devices for medical centers and hospitals. Biomedical electronics has greatly benefited the medical industry with the invention and introduction of equipment, including intensive unit monitoring systems, CT imaging systems, dialysis machines, and surgical lasers. Biomedicine covers almost all clinical testing and treatment devices. Pharmacists measure and administer medicines, while lab personnel examine equipment thoroughly. The managers ensure smooth facility operations. With thousands of staff at health institutions, biomedical technicians' jobs are often simple to understand. Engineers, largely focused on manufacturing, may appear unimportant in healthcare; however, biomedical professionals play a vital role as team members.

Table 7.1 (continued)

S. no.	Application	Explanation
3.	Improve patient care	Bioengineered MRI and dialysis devices are common medical equipment. Innovation in engineering aligns with technology advancements, enhancing patient care and outcomes. Careers in biotechnology include a wide variety of sectors with innovative applications. Biomechanical experts study and create products for sports and athletics, focusing on the human body and its functions. The field of biomechanics has significantly enhanced healthcare for athletes and other physically active persons via study. Sports equipment, including training gear and footwear, are typically influenced by biomechanical studies. Professional athletes hire biomechanics to analyze their movements and devise methods. Biomedical engineers design medical treatments, systems, and products, such as 3D imaging machines, replacement knee biomaterials, and accelerated rehabilitation for physical limitations. Biomedical engineering is crucial due to the growing life expectancy, aging population, and healthcare needs.
4.	Artificial organ development	Biomedical engineers handle diverse tasks, including developing artificial organs, diagnostic tools, medical technology support, clinician training, and studying biological structures technology in animals and humans, including endocrine systems. 3D bioprinting combines calls, growth factors, and biomaterials to create biomedical components that mimic natural tissue qualities. The layer-by-layer method stores bioinks and creates tissue architectures for medicinal and tissue treatments. There is huge potential for 3D bioprinting. Doctors will soon be able to manufacture replacement skin cells for burn victims. Surgeons may print new organs in one day, removing the need for risky organ transplantation at scale.
5.	Healthcare sustainability	By developing eco-friendly medical equipment, developing new materials, and improving healthcare procedures, biomedical engineering advances healthcare sustainability. Engineers are reducing healthcare facilities' environmental impact by developing energy-efficient medical equipment with recyclable components. Biocompatible and biodegradable medical implants and gadgets enhance patient care and reduce environmental impact. Biomedical engineers optimize healthcare operations via automation and data-driven solutions to improve efficiency, resource consumption, and sustainability.

Table 7.1 (continued)

S. no.	Application	Explanation
6.	Healthcare administration	Biomedical engineers must be well-trained since it affects the management of healthcare facilities globally. Most biomedical engineers gain from saving lives and improving the quality of life for millions globally via current medicines. Healthcare facilities and treatment centers need competent biomedical engineers to effectively use this technology. These devices directly impact severe coronavirus survival. Miscalibrations in equipment may harm patients and even cause death. The worldwide health crisis highlights the critical need for biomedical engineers. Engineers use reverse engineering methods to analyze and comprehend the composition of practical components. Their strategy for developing devices without biological waste is optimized. Speeding up manufacturing is biomedical engineers' biggest challenge.
7.	Deliver life-changing devices	Biomedical engineering is crucial for creating life-changing devices in the rapidly evolving technological environment. Bioengineering is multidisciplinary biomedical engineering. Using biology, medicine, and engineering principles and methods, it aims to enhance human health. Medicinal students learn and use biotechnology for ethical design, development, and translation of medicinal innovations. Through engineering team connections, students get insight into developing and translating medical innovations for clinical use. It assists in the development of pharmaceutical and diagnostic products, employing live cells and their components. These things aid in treating and avoiding diseases. Medical biotechnology has advanced, benefiting millions via vaccinations, DNA mapping, and agricultural benefits.
8.	Technological advancement	Medical biotechnology has the potential to enhance cancer research and agriculture, benefiting many persons. Scientists may alter DNA and genetic engineering functions. Many applications include genetic fault repair, illness therapy, prevention, agricultural enhancement, etc. Genome change has several ethical issues. Biomedical engineers have developed innovative medical breakthroughs such as prosthetic arms and legs, artificial livers, hearts, bionic contacts, and the camera pill, which records internal processes with a color camera, battery, light source, and transmitter. Researching physiological systems may lead to innovative treatments for diseases like cancer. Bioengineers have contributed to the development of innovative therapies for chronic health concerns, such as laser surgery.

Table 7.1 (continued)

S. no.	Application	Explanation
9.	Research in medicine	Safety requirements guide biomedical engineering's diagnostic tool development. Engineering and medical research capabilities are used to develop specialized equipment for artificial organs and limb replacements. Biomedical engineers learn about medical device technologies, human anatomy, and biology. Specialized software is developed to analyze patient bodies. They create innovative medical devices and drugs. Biomedical engineers may easily enter biomedical engineering research and clinical research, including device testing. They may create new testing tools or technologies themselves. Job growth is driven by several sources, such as advanced cardiac imaging technologies that allow physicians to pinpoint the exact location and function of heart tissue. Additionally, medical technological advancements enable practitioners to provide superior therapies with less patient risk.
10.	Clinical engineering	Clinical engineering applies technology to hospital treatment. The healthcare team includes the clinical engineer, physicians, nurses, and other hospital staff. Clinical engineers are responsible for building and managing computer records, as well as acquiring and using new medical technology. Biomedical engineers play a crucial role in healthcare facilities, with a diverse variety of applications. Biomedical engineers keep hospitals, clinics, and care facilities running and helping patients. This involves investigating biological processes, inventing medical devices, managing facility networks, and educating on treatment approaches. Biomedical engineering is crucial for healthcare delivery and facility operations worldwide, requiring competent professionals with practical training.

7.6 An outline of healthcare procedures for biomedical engineering education

The field of BME education involves surpassing conventional BME activities to enhance abilities and professional growth. The study provides a detailed analysis of the definition of applicable instruction within health services and medical processes. Based on the studies of Kolker and Story [212], common pressing concerns in hospital areas refer to the adequacy of current infrastructure in delivering healthcare services, the availability of skilled medical personnel to meet operational and service performance goals, the optimization of personnel schedules to ensure safe and efficient treatment of patients, the impact of patient flow on waiting periods and productivity, the effective allocation and utilization of resources, and the accurate prediction of patient demand and transaction volumes in both short and long terms. These various factors influence the quality, timeliness, financial implications, safety, and even long-term viability of delivering services within healthcare establishments [213].

These problems, which affect healthcare administration in both the public and commercial sectors and are not limited to developing nations, are a universal challenge. Competitive pressure in the healthcare sector [214] has led to issues in nations, including the United States, the United Kingdom, Sweden, and Brazil, all of which center on cost reductions or efficiency improvements. In other cases though, shortages of personnel, facilities, and supplies are to blame for excessive wait times and inadequate medical treatment [215–217].

The study on the problems now facing various hospital services (such as admissions and discharge, triage, phlebotomy, sterilization and launderette, ambulance, X-ray, and MRI) is replete with instances of such situations. Over the past 20 years, there have been various initiatives aimed at addressing the increasing challenges faced by healthcare services on a global scale. These challenges have highlighted the importance of applying the principles of industrial engineering and operations management, as the underlying issues primarily stem from processes and operations. For example, methodologies such as lean healthcare and healthcare management engineering, which have their origins in the manufacturing sector, have been employed to combat concerns related to quality, time, cost, and safety. These methodologies utilize various tools such as discrete simulations, process mapping, seven wastes, visual aids, continuous flow, error-proof devices, quick setups, and pull-Kanban systems, among others [218–220]. Consequently, there is a chance to go in the particular direction to provide educational opportunities for engineering students, particularly BME scholars, to broaden their potential contributions to the healthcare industry. The study suggests identifying relevant learning opportunities in healthcare and hospital operations, emphasizing active and experiential learning. It maps these processes and activities according to their degree of situational complexity and patient care criticality.

Complexity in a real-world healthcare environment or process is measured by the number of relevant research factors, kinds of activities, allotted resources, and their linkages.

The complexity requires effective problem-solving, communication, and coordination among players to improve instances and coordinate activities. Complexity-low situations often call for procedures with predictable or preset forms of behavior. Conversely, high-level complexity refers to interactions that establish complex causal relationships with multidimensional consequences throughout time. The criticality of circumstances relates to patient care, including medical staff engagement and the influence of activities on health conditions. Direct health services, delivering medical treatments, ethical concerns, and data protection rules all fall under the category of "very important." Medium-level criticality refers to conditions that indirectly impact patient health, healthcare treatments, ethics, and data protection rules. Low-criticality instances do not affect patient health, medical treatments, ethics, or data protection rules. Table 7.2 displays activities and processes categorized by complexity and criticality. Billing, a high-complexity, low-criticality process, involves integrating information and decision-making from various healthcare facilities but has minimal impact on patient health. The structure serves as a tool for categorizing relevant research settings, which may be used to conceptualize credible learning experiences but are limited by educational, medical, operational, legal, and ethical constraints. Furthermore, the framework might help in the identification of appropriate study conditions for BME education objectives that need validation to comply with healthcare requirements and applicable regulations within hospitals and healthcare providers. As a result, it is critical to consider this constraint while choosing any probable research environment.

Table 7.2: Healthcare process mapping based on situational complexity and patient criticality.

Complexity	Criticality		
	High	**Medium**	**Low**
Low	Outpatient appointment	Food sector	Adequate cleanliness
Medium	Ambulatory surgeries and sterilization	Hospital room cleaning and laundry	Relations with the public
High	Emergencies and unexpected situations	Diagnostic medical imaging	Medical invoices

7.7 Conclusion

Biologic engineering is a specialized discipline that combines engineering and healthcare principles. These aid medical sciences in streamlining national healthcare services. With improved knowledge and population growth, healthcare expenditures are expected to climb. The WHO recognizes health as a fundamental human right, and healthcare researchers are responsible for ensuring its protection for everyone. This requires acknowledging structural prejudices and health inequities in society and working to remove them. In conclusion, the discipline of BME plays a crucial role in driving the progress of healthcare sustainability across all domains. Biomedical engineers are actively engaged in addressing significant healthcare issues, ranging from the development of breakthrough medical technologies to the enhancement of healthcare procedures and the mitigation of ecological consequences. The chapter analyzed the role BME plays in the development of sustainable medical technologies. Engineers can substantially reduce the carbon emissions associated with healthcare systems by developing energy-efficient medical devices, including low-power diagnostic instruments and ecologically sustainable materials. Furthermore, by emphasizing the recyclability and durability of medical equipment, it is possible to reduce the amount of waste produced by the healthcare sector.

Through the development of advanced monitoring systems and data analytics tools, researchers and healthcare professionals can facilitate the early diagnosis of diseases and provide individualized treatment. This eventually results in improved patient outcomes and a decrease in the consumption of healthcare resources. The significance of BME in enhancing healthcare procedures are numerous. Engineers are enhancing healthcare delivery, minimizing mistakes, and enhancing efficiency by using automation, robotics, and AI technologies. These improvements have the additional advantage of improving patient outcomes and promoting the adoption of environmentally conscious and sustainable practices within healthcare institutions. Interdisciplinary cooperation is key to BME-based healthcare sustainability. The need for promoting collaboration among engineers, healthcare professionals, legislators, and environmental specialists is essential in advancing the development of comprehensive solutions that take into account the well-being of patients as well as their influence on the environment. Collectively, these many stakeholders possess the capacity to promote innovation, execute sustainable practices, and influence the trajectory of healthcare in the long-term. In summary, BME serves as a potent driver in the progression of healthcare sustainability. Biomedical engineers are actively contributing to the advancement of a sustainable and environmentally aware healthcare system via the development of eco-friendly technology, enhancement of patient care, optimization of healthcare processes, and promotion of cooperation. As the discipline continues to advance, it can fundamentally transform healthcare methodologies and address the environmental obstacles faced by the sector, eventually yielding advantages for both individuals and the environment.

References

[1] Chung HU, Kim BH, Lee JY, et al. Binodal, wireless epidermal electronic systems with in-sensor analytics for neonatal intensive care. Science. 2019;363(6430):eaau0780.

[2] Rajewsky N, Almouzni G, Gorski SA, et al. Lifetime and improving European healthcare through cell-based interceptive medicine. Nature. 2020;587(7834):377–86.

[3] Haque A, Milstein A, Fei-Fei L. Illuminating the dark spaces of healthcare with ambient intelligence. Nature. 2020;585(7824):193–202.

[4] Smethurst DP, Williams HC. Power laws. Are hospital waiting lists self-regulating? Nature. 2001;410(6829):652–53.

[5] Nayeem MOG, Lee S, Jin H, et al. All-nanofiber-based, ultrasensitive, gas-permeable mechanoacoustic sensors for continuous long-term heart monitoring. Proceedings of National Academy of Science USA. 2020;117(13):7063–70.

[6] Yao MT, Cheng YX, Zhou ZY, Liu M. Recent progress on the fabrication and applications of flexible ferroelectric devices. Journal of Materials Chemistry C. 2020;8(1):14–27.

[7] Li G, Lee DW. An advanced selective liquid–metal plating technique for stretchable biosensor applications. Lab Chip. 2017;17(20):3415–21.

[8] Tai YL, Yang ZG. Flexible pressure sensing film based on ultra-sensitive SWCNT/PDMS spheres for monitoring human pulse signals. Journal of Materials Chemical B. 2015;3(27):5436–41.

[9] Wei P, Yang X, Cao Z, et al. Flexible and stretchable electronic skin with high durability and shock resistance via embedded 3D printing technology for human activity monitoring and personal healthcare. Advanced Materials Technologies. 2019;4(9):1900315.
[10] Mahmud MS, Fang H, Wang HG. An integrated wearable sensor for unobtrusive continuous measurement of the autonomic nervous system. IEEE Internet of Things. 2019;6(1):1104–13.
[11] Zhang Y, Webb RC, Luo H, et al. Theoretical and experimental studies of epidermal heat flux sensors for measurements of core body temperature. Advanced Healthcare Materials. 2016;5(1):119–27.
[12] Guo J, Zhou B, Yang C, Dai Q, Kong L. Stretchable and temperature-sensitive polymer optical fibers for wearable health monitoring. Advanced Functional Materials. 2019;29(33):1902898.
[13] Cheong JH, Ng SS, Liu X, et al. An inductively powered implantable blood flow sensor microsystem for vascular grafts. IEEE Transactions on Biomedical Engineering. 2012;59(9):2466–75.
[14] Huang C, Gu Y, Chen J, et al. A wearable fiberless optical sensor for continuous monitoring of cerebral blood flow in mice. IEEE Journal of Selected Topics in Quantum Electronics. 2019;25(1):1–8.
[15] Wang W, Pang J, Su J, Li F, Li Q, Wang X, et al. Applications of nanogenerators for biomedical engineering and healthcare systems. InfoMat. 2022;4(2):e12262.
[16] Ghomi ER, Khosravi F, Neisiany RE, Singh S, Ramakrishna S. Future of additive manufacturing in healthcare, Curr. Current Opinion in Biomedical Engineering. 2021;17:100255.
[17] Oden M, Mirabal Y, Epstein M, Richards-Kortum R. Engaging undergraduates to solve global health challenges: A new approach based on bioengineering design. Annals of Biomedical Engineering. 2010;38(9):3031–41.
[18] Palamountain KM, Stewart KA, Krauss A, Kelso D, Diermeier D. University leadership for innovation in global health and HIV/AIDS diagnostics. Global Public Health. 2010;5(2):189–96.
[19] Vermund SH, Sahasrabuddhe VV, Khedkar S, Jia YJ, Etherington C, Vergara A. Building global health through a center-without-walls: The Vanderbilt Institute for global health. Academic Medicine. 2008;83(2):154–64.
[20] Guilbeau EJ, Pizziconi VB. Increasing student awareness of ethical, social, legal, and economic implications of technology. Journal of Engineering Education. 1998;87(1):35–45.
[21] White JA, Gaver DP, Butera RJ, Choi B, Dunlop MJ, Grande-Allen KJ, et al. Core competencies for undergraduates in bioengineering and biomedical engineering: Findings, consequences, and recommendations. Annals of Biomedical Engineering. 2020;48(3):905–12.
[22] Coulby G, Clear A, Jones O, Young F, Stuart S, Godfrey A. Towards remote healthcare monitoring using accessible IoT technology: State-of-the-art, insights and experimental design. BioMedical Engineering OnLine. 2020;19(1):80.
[23] Chen RJ, Lu MY, Chen TY, Williamson DFK, Mahmood F. Synthetic data in machine learning for medicine and healthcare. Nature Biomedical Engineering. 2021;5(6):493–97.
[24] Wang B, Wang L, Lin SJ, Wu D, Huang BY, Zhang YT, et al. A body sensor networks development platform for pervasive healthcare. In: 3rd International Conference on Bioinformatics And Biomedical Engineering. IEEE Publications; 2009. p. 1–4.
[25] Andreu-Perez J, Leff DR, Ip HM, Yang GZ. From wearable sensors to smart implants–toward pervasive and personalized healthcare. IEEE Transactions on Bio-medical Engineering. 2015;62(12):2750–62.
[26] Al Asif MR, Roy S, Abdullah A, Raihan M, Akter R, Hossain MZ. Role and impact of biomedical engineering discipline for developing country perspective. International Journal of Innovative Research in Computer Science and Technology;6(4):87–90.
[27] Goyal R, Khosla C, Sood M, Singh K, Punam K, Goyal K. A review on artificial intelligence-driven biomedical engineering implants in healthcare. In: 2nd International Conference on Advance Computing and Innovative Technologies in Engineering (ICACITE). IEEE Publications; 2022. p. 1705–08.

[28] Sun M, Han K, Hu R, Liu D, Fu W, Liu W. Advances in micro/nanoporous membranes for biomedical engineering. Advanced Healthcare Materials. 2021;10(7):e2001545.

[29] Kumar SR, Gayathri N, Muthuramalingam S, Balamurugan B, Ramesh C, Nallakaruppan MK. Medical Big Data Mining and Processing in e-Healthcare. In: Internet of Things in Biomedical Engineering (pp. 323–39). Academic Press; 2019.

[30] Madon T, Hofman KJ, Kupfer L, Glass RI. Implementation science. Science. 2007 Dec 14;318(5857):1728–29.

[31] Zeggini E, Gloyn AL, Barton AC, Wain LV. Translational genomics and precision medicine: Moving from the lab to the clinic. Science. 2019;365(6460):1409–13.

[32] Hinchet R, Yoon HJ, Ryu H, et al. Transcutaneous ultrasound energy harvesting using capacitive triboelectric technology. Science. 2019;365(6452):491–94.

[33] Cook-Deegan R. Public health. Boosting health services research. Science. 2011;333(6048):1384–85.

[34] Kang SK, Murphy RK, Hwang SW, et al. Bioresorbable silicon electronic sensors for the brain. Nature. 2016;530(7588):71–76.

[35] El-Ali J, Sorger PK, Jensen KF. Cells on chips. Nature. 2006;442:7101):403–411.

[36] Rong G, Corrie SR, Clark HA. In vivo biosensing: Progress and perspectives. ACS Sensors. 2017;2(3):327–38.

[37] Song Y, Min J, Gao W. Wearable and implantable electronics: Moving toward precision therapy. ACS Nano. 2019;13(11):12280–86.

[38] Curtis AB, Karki R, Hattoum A, Sharma UC. Arrhythmias in patients >/=80 years of age: Pathophysiology, management, and outcomes. Journal of the American College of Cardiology. 2018;71(18):2041–57.

[39] Liu S, Zhang XD, Gu X, Ming D. Photodetectors based on two-dimensional materials for biomedical application. Biosensors and Bioelectronics. 2019;143:111617.

[40] Rebelo R, Barbosa AI, Caballero D, et al. 3D biosensors in advanced medical diagnostics of high mortality diseases. Biosensors and Bioelectronics. 2019;130:20–39.

[41] Luo J, Gao W, Wang ZL. The triboelectric nanogenerator is an innovative technology for intelligent sports. Advances in Materials. 2021;33(17):2004178.

[42] Zhao K, Ouyang B, Bowen CR, Wang ZL, Yang Y. One structure-based multi-effects coupled nanogenerators for flexible and self-powered multi-functional coupled sensor systems. Nano Energy. 2020;71:104632.

[43] Zhang R, Hummelgård M, Örtegren J, et al. Human body constituted triboelectric nanogenerators as energy harvesters, code transmitters, and motion sensors. ACS Applied Energy Materials. 2018;1(6):2955–60.

[44] Gao L, Hu D, Qi M, et al. A double-helix-structured triboelectric nanogenerator enhanced with positive charge traps for self-powered temperature sensing and smart-home control systems. Nanoscale. 2018;10(42):19781–90.

[45] Donlan RM. Biofilms and device-associated infections. Emerging Infectious Diseases. 2001;7:277–81.

[46] Nasir M, Seta J, Meyer EG. Introducing high school students to biomedical engineering through summer camps. In: 2014 ASEE Annual Conference & Exposition 2014 Jun 15 (pp. 24–809).

[47] Voigt HF, Hosea F, Nkuma-Udah KI, Pallikarakis N, Kang PL, Secca M, Zequera M. Education and Training. In Human Resources for Medical Devices, the Role of Biomedical Engineers (pp. 39–52). Geneva, Switzerland: World Health Organization; 2017.

[48] Malkin RA. Design of health care technologies for the developing world. Annual Review of Biomedical Engineering. 2007 Aug 15;9:567–87.

[49] Tiwana I, Redmond SJ, Lovell NH. A review of tactile sensing technologies with applications in biomedical engineering, Sensor Actuator. Physiology. 2012;179:17–31.

[50] Li Y, Zhang X, Ying B. On textile biomedical engineering. Science China Technological Sciences. 2019;62(6):945–57.

[51] Nisha SS, Meeral MN. Applications of Deep Learning in Biomedical Engineering. In: Handbook of Deep Learning in Biomedical Engineering (pp. 245–70). Academic Press; 2021.
[52] Sahada A, Tekindor AN, Abbadi MB, Malluhi MA, Huri PY. Role of biomedical engineering during COVID-19 pandemic. Journal of Applied and Natural Science. 2020;3(1):1–16.
[53] Akay M. Biomedical engineering for global healthcare. In: 8th IEEE International Conference on BioInformatics And Bio-Engineering. IEEE Publications; 2008. p. 1.
[54] Lin CH, Yang PK, Lin YC, Fu PK. On machine learning models for heart disease diagnosis. In: 2nd Eurasia Conference on Biomedical Engineering, Healthcare and Sustainability (ECBIOS). IEEE Publications; 2020. p. 158–61.
[55] Ahluwalia A, Maria CD, Diaz Lantada A, Madete J, Makobore PN, Ravizza A, et al. Biomedical engineering project-based learning: Euro-African design school focused on medical devices. International Journal of English and Education. 2018;34(5):1709–22.
[56] Haleem A, Javaid M, Khan RH, Suman R. 3D printing applications in bone tissue engineering. Journal of Clinical Orthopaedics and Trauma. 2020;11(Suppl 1):S118–24.
[57] Mishu MM. A patient-oriented framework using big data & c-means clustering for biomedical engineering applications. In: International Conference on Robotics, Electrical and Signal Processing Techniques (ICREST). IEEE Publications; 2019. p. 113–15.
[58] Yu KH, Beam AL, Kohane IS. Artificial intelligence in healthcare. Nature Biomedical Engineering. 2018;2(10):719–31.
[59] De Maria C, Mazzei D, Ahluwalia A. Open source biomedical engineering for sustainability in African healthcare: Combining academic excellence with innovation. In: Proceedings of the ICDS; 2014. p. 23–27.
[60] Francis A. Biological evaluation of preceramic organosilicon polymers for various healthcare and biomedical engineering applications: A review. Journal of Biomedical Materials Research B, Applied Biomaterials. 2021;109(5):744–64.
[61] Fermin JL, Tan MJT. The need for the establishment of biomedical engineering as an academic and professional discipline in the Philippines – A quantitative argument. IEEE Access. 2020;9:3097–111.
[62] Subasi A. Biomedical Signal Analysis and its Usage in Healthcare. In: Biomedical Engineering and Its Applications in Healthcare (pp. 423–52). Singapore: Springer; 2019.
[63] Boulos MNK, Wheeler S, Tavares C, Jones R. How smartphones are changing the face of mobile and participatory healthcare: An overview, with example from eCAALYX. BioMedical Engineering OnLine. 2011;10(1):24.
[64] Pecchia L, Pallikarakis N, Magjarevic R, Iadanza E. Health technology assessment and biomedical engineering: Global trends, gaps and opportunities. Medical Engineering & Physics. 2019;72:19–26.
[65] Sienko KH, Kaufmann EE, Musaazi ME, Sabet Sarvestani A, Obed S. Obstetrics-based clinical immersion of a Multinational Team of Biomedical Engineering Students in Ghana. Gynecology Journal. 2014;127(2):218–20.
[66] Abu-Faraj ZO. Bioengineering/biomedical engineering education and career development: Literature review, definitions, and constructive recommendations. International Journal of English and Education. 2008;24(5):990.
[67] Shirmohammadi S, Barbe K, Grimaldi D, Rapuano S, Grassini S. Instrumentation and measurement in medical, biomedical, and healthcare systems. IEEE Instrumentation & Measurement Magazine. 2016;19(5):6–12.
[68] Ortiz-Posadas MR, Ortiz-Pedroza MR, Martínez-Licona FM, Garcíagonzalez MT, Gaitan-Gonzalez MJ. Development of projects by clinical biomedical engineering students. In: Proceedings of the 25th annual international conference of the IEEE Engineering in Medicine and Biology Society. Vol. 4. IEEE Publications; 2003, September. p. 3529–32.
[69] Yang X, Chung E, Johnston I, Ren G, Cheong YK. Exploitation of antimicrobial nanoparticles and their applications in biomedical engineering. Applied Sciences. 2021;11(10):4520.

[70] Namoglu N, Ülgen Y. Network Security vulnerabilities and personal privacy issues in healthcare information systems: A case study in a private hospital. In: 18th National Biomedical Engineering Meeting. IEEE Publications; 2014. p. 1–3.
[71] Amin MR, Zuhairi MF, Saadat MN. Transparent data dealing: Hyperledger fabric-based biomedical engineering supply chain. In: 15th International Conference on Ubiquitous Information Management And Communication (IMCOM). IEEE Publications; 2021. p. 1–5.
[72] Hassan R, Hasan S, Hasan MJ, Jamader MR, Eisenberg D, Pias T. Human attention recognition with machine learning from brain-EEG signals. In: 2nd Eurasia Conference on Biomedical Engineering, Healthcare and Sustainability (ECBIOS). IEEE Publications; 2020. p. 16–19.
[73] Friebe MH, Fritzsche H, Morbach O, Heryan K. The PLH-purpose launch-pad health-meta-methodology to explore problems and evaluate solutions for biomedical engineering impact creation. In: 44th Annual International Conference of the IEEE Engineering in Medicine & Biology Society (EMBC). IEEE Publications; 2022. p. 3299–302.
[74] Zadpoor AA, Malda J. Additive manufacturing of biomaterials, tissues, and organs. Annals of Biomedical Engineering. 2017;45(1):1–11.
[75] Van Alste JA. A new educational program on biomedical engineering. In: Proceedings of the 22nd annual international conference of the IEEE Engineering in Medicine and Biology Society. Vol. 4. IEEE Publications; 2000, July. p. 2952–54.
[76] Deperlioglu O. Intelligent Techniques Inspired by Nature and Used in Biomedical Engineering. In: Biotechnology: Concepts, Methodologies, Tools, and Applications (pp. 51–77). IGI Global; 2019.
[77] Enderle JD, Bronzino JD. Introduction to Biomedical Engineering (p. 16). Academic Press; 2012.
[78] Geng W, Wang L, Jiang N, Cao J, Xiao YX, Wei H, Yetisen AK, Yang XY, Su BL. Single cells in nanoshells for the functionalization of living cells. Nanoscale. 2018;10(7):3112–29.
[79] Wickramasinghe N, Choudhary S, Geisler E. Bionanotechnology: Its applications and relevance to healthcare. International Journal of Biomedical Engineering & Technology. 2007;1(1):41–58.
[80] Acevedo FJ, Fuentes JE, Enderle JD. Design of computerized maintenance management system for the Chilean Naval Hospital biomedical engineering department. In: Conference Proceeding IEEE Eng Med Biol Soc 27th Annual Conference;2006, January:174–77.
[81] Hashimoto S, Ohsuga M, Yoshiura M, Tsutsui H, Akazawa K, Mochizuki S, et al. Parallel curriculum of biomedical engineering subjects with rotational experimental project for interdisciplinary study field. In: Proceedings of the 11th World Multiconference on Systemic Cybernetics and Informatics. Vol. 4; 2007. p. 39–44.
[82] Bielefeldt AR, Canney NE, Swan C, Polmear M, Knight D. Microethics and macro ethics education of biomedical engineering students in the United States. Ethics in Biology Engineering and Medicine. 2016;7(1–2).
[83] Sun L, Guo J, Chen H, Zhang D, Shang L, Zhang B, et al. Tailoring materials with specific wettability in biomedical engineering. Advanced Science. 2021;8(19):e2100126.
[84] Singh A, Ferry D, Balasubramanian S. Efficacy of clinical simulation-based training in biomedical engineering education. Journal of Biomechanical Engineering. 2019;141(12):121011.
[85] Hu C, Shi Q, Liu L, Wejinya U, Hasegawa Y, Shen Y. Robotics in biomedical and healthcare engineering. Journal of Healthcare Engineering. 2017:1610372.
[86] Elfani M, Putra NK. Biomedical engineering and its potential for employment in Indonesia. International Journal of Technology. 2013;4(1):34–44.
[87] Lhotska L, Hozman J. Biomedical engineering education in the Czech Republic. In: Mediterranean Conference on Medical and Biological Engineering and Computing. Vol. XIV. Cham: Springer; 2016. p. 1323–28.
[88] Bhatia D, Mishra A, Prasad AK. Blockchain Technology for Biomedical Engineering Applications. In: Applications of Blockchain and Big IoT Systems (pp. 245–66). Apple Academic Press; 2022.

[89] Chavali M, Palanisamy P, Nikolova MP, Wu RJ, Tadiboyina R, Rao PP. Inorganic composites in biomedical engineering. Materials for Biomedical Engineering. 2019:47–80.
[90] Setiawan AW, Wicaksono AA. Implementation of health technology assessment and management course in an undergraduate program in biomedical engineering ITB. In: Annual Int Conf IEEE Eng Med Biol Soc 44th Annual International Conference Of the IEEE Engineering in Medicine & Biology Society (EMBC). 2022;2022:3281–84.
[91] Feagin J, Bennefield Z. Systemic racism and U.S. health care. Social Science & Medicine. 2014;103:7–14.
[92] Chrisler JC, Barney A, Palatino B. Ageism can be hazardous to women's health: Ageism, sexism, and stereotypes of older women in the healthcare system. Journal of Social Issues. 2016;72:86–104.
[93] Rapp KS, Volpe VV, Hale TL, Quartararo DF. State–level sexism and gender disparities in health care access and quality in the United States. Journal of Health and Social Behavior. 2022;63:2–18.
[94] Bailey ZD, Krieger N, Age´nor M, Graves J, Linos N, Bassett MT. Structural racism and health inequities in the USA: Evidence and interventions. Lancet. 2017;389:1453–63.
[95] Homan P. Structural sexism and health in the united states: A new perspective. Health Inequality and the Gender System. 2019;84:486–516.
[96] Clair M, Denis JS. Sociology of racism. The International Encyclopedia of the Social and Behavioral Sciences. 2015;19(2015):857–63.
[97] Payne BK, Hannay JW. Implicit bias reflects systemic racism. Trends in Cognitive Sciences. 2021;25:927–36.
[98] Banaji MR, Fiske ST, Massey DS. Systemic racism: Individuals and interactions, institutions and society. Cognitive Research: Principles and Implications. 2021;6:1–21.
[99] Williams DR, Mohammed SA. Discrimination and racial health disparities: Evidence and needed research. Journal of Behavioral Medicine. 2009;32:20–47.
[100] Brewer LPC, Carson KA, Williams DR, Allen A, Jones CP, Cooper LA. Association of race consciousness with the patient–physician relationship, medication adherence, and blood pressure in urban primary care patients. American Journal of Hypertension. 2013;26:1346–52.
[101] Johnson K, Jackson M, Vaughan D. We exist We are your peers. Nature Reviews Materials. 2020;5:783–84.
[102] Barber PH, Hayes TB, Johnson TL, Ma´rquez-magaña L. Systemic racism in higher education. Science. 2020;369:1440–41.
[103] Braveman P. Health disparities and health equity: Concepts and measurement. Annual Review of Public Health. 2006;27:167–94.
[104] Garton EM, Cira MK, Loehrer PJ, Eldridge L, Frank A, Prakash L, Chang S, Salloum RG, Ciolino H, He M, Gopal S. Global oncology research and training at US National Cancer Institute-designated cancer centres: Results of the 2021 Global Oncology Survey. Lancet Oncology. 2023 Oct 1;24(10): e407–14.
[105] Lucena JC. Making women and minorities in science and engineering for national purposes in the United States. Journal of Women and Minorities in Science and Engineering. 2000;6(1).
[106] Gibbs KD, Han A, Lun J. Demographic diversity in teams: The challenges, benefits, and management strategies. Strategies for Team Science Success. 2019:197–205.
[107] Gomez LE, Bernet P. Diversity improves performance and outcomes. Journal of the National Medical Association. 2019;111:383–92.
[108] Kozlowski D, Lariviere V, Sugimoto CR, Monroe-White T. Intersectional inequalities in science. Proceedings of National Academy of Science USA. 2022:119.
[109] Ah-King M, Barron AB, Herberstein ME. Genital evolution: Why are females still understudied? PLOS in Biology. 2014:12.

[110] Koning R, Samila S, Ferguson J-P. Who do we invent for? Patents by women focus more on women's health, but few women get to invent – working paper – faculty & research – Harvard Business School. Science (80-). 2021;372:1345–48.
[111] Maton KI, Pollard SA, McDougall TV, Hrabowski FA. Meyerhoff scholars program: A strengths-based, institution-wide approach to increasing diversity in science, technology, engineering, and mathematics. Mount Sinai Journal of Medicine. 2012;79:610–23.
[112] Maton KI, Beason TS, Godsay S, Mariano MR, Bailey TSC, Sun S, et al. Outcomes and processes in the Meyerhoff scholars program: STEM PhD completion, sense of community, perceived program benefit, science identity, and research self-efficacy. CBE-Life Sciences Education. 2016;15:2.
[113] Sabo S, De Zapien J, Teufel-Shone N, Rosales C, Bergsma L, Taren D. Service learning: A vehicle for building health equity and eliminating health disparities. American Journal of Public Health. 2015:105:S38–S43.
[114] Sánchez-López AL, Franco-Villareal H, Perfecto-Avalos Y. Design and conduct of clinical research: Raising awareness of ethical aspects in biopharmaceutics for students of Biotechnology Engineering. In: 2021 IEEE Global Engineering Education Conference (EDUCON) 2021 Apr 21 (pp. 388–92).
[115] Mukherjee S. The United States food and drug administration (FDA) regulatory response to combat neglected tropical diseases (NTDs): A review. PLOS Neglected Tropical Diseases. 2023 Jan 12;17(1): e0011010.
[116] Farooq F, Mogayzel PJ, Lanzkron S, Haywood C, Strouse JJ. Comparison of US federal and foundation funding of research for sickle cell disease and cystic fibrosis and factors associated with research productivity. JAMA Network. 2020:3:e201737.
[117] Ginther DK, Schaffer WT, Schnell J, Masimore B, Liu F, Haak LL, et al. Race, ethnicity, and NIH research awards. Science (80-). 2011;333:1015–19.
[118] Stevens KR, Masters KS, Imoukhuede PI, Haynes KA, Setton LA, Cosgriff-Hernandez E, et al. Fund Black scientists. Cell. 2021;184:561–65.
[119] Cropsey KL, Masho SW, Shiang R, Sikka V, Kornstein SG, Hampton CL. Why do faculty leave? Reasons for attrition of women and minority faculty from a medical school: Four-year results. Journal of Women's Health Care. 2008;17:1111–18.
[120] Viergever RF, Hendriks TCC. The 10 largest public and philanthropic funders of health research in the world: What they fund and how they distribute their funds. Health Policy and Systems Research. 2016;14:1–15.
[121] Watson S, Caster O, Rochon PA, Den Ruijter H. Reported adverse drug reactions in women and men: Aggregated evidence from globally collected individual case reports during half a century. EClinicalMedicine. 2019:17.
[122] Gahagan J, Gray K, Whynacht A. Sex and gender matter in health research: Addressing health inequities in health research reporting. International Journal for Equity in Health. 2015;14:1–4.
[123] Ritz SA, Antle DM, Coˆte´ J, Deroy K, Fraleigh N, Messing K, et al. First steps for integrating sex and gender considerations into basic experimental biomedical research. FASEB Journal. 2014;28:4–13.
[124] Klein SL, Flanagan KL. Sex differences in immune responses. Nature Reviews Immunology. 2016 Oct;16(10):626–38.
[125] Cabal A, Wassenaar TM, Ussery DW. Gender differences in the gut microbiome and how these affect cardiovascular diseases. Gender Difference in Pathology Management Heart Disease. 2018:89–100.
[126] Walker CJ, Schroeder ME, Aguado BA, Anseth KS, Leinwand LA. Matters of the heart: Cellular sex differences. Journal of Molecular & Cellular Cardiology. 2021;160:42–55.
[127] Strobino DM, Grason H, Minkovitz C. Charting a course for the future of women's health in the United States: Concepts, findings and recommendations. Social Science & Medicine. 2002 Mar 1;54(5):839–48.

[128] Owens NJ, Hume AL. Pharmacotherapy in women: Do clinically important gender-related issues exist? Ri-Med. 1994;77:412–16.
[129] Waxman DJ, Holloway MG. Sex differences in the expression of hepatic drug-metabolizing enzymes. Molecular Pharmacology. 2009;76:215–28.
[130] Hajipour MJ, Aghaverdi H, Serpooshan V, Vali H, Sheibani S, Mahmoudi M. Sex as an important factor in nanomedicine. Nature Communications. 2021:12.
[131] Vulpis E, Giulimondi F, Digiacomo L, Zingoni A, Safavi-Sohi R, Sharifi S, et al. The possible role of sex as an important factor in development and administration of lipid nanomedicine-based COVID-19 vaccine. Molecular Pharmaceutics. 2021;18:2448–53.
[132] Angum F, Khan T, Kaler J, Siddiqui L, Hussain A. The prevalence of autoimmune disorders in women: A narrative review. Cureus. 2020:12.
[133] Hsich EM. Sex differences in advanced heart failure therapies. Circulation. 2019;139:1080–93.
[134] Hansen M. Female hormones: Do they influence muscle and tendon protein metabolism? The Proceedings of the Nutrition Society. 2018;77:32–41.
[135] Madla CM, Gavins FKH, Merchant HA, Orlu M, Murdan S, Basit AW. Let's talk about sex: Differences in drug therapy in males and females. Advanced Drug Delivery Reviews. 2021;175:113804.
[136] Clayton JA, Collins FS. Policy: NIH to balance sex in cell and animal studies. Nature. 2014;509:282–83.
[137] Ford CL, Airhihenbuwa CO. Critical race theory, race equity, and public health: Toward antiracism praxis. American Journal of Public Health. 2010:100.
[138] Alim HS, Reyes A, Kroskrity PV. Coloniality of Mixed Race and Mixed Language. In: The Oxford Handbook of Language and Race (pp. 533). New York: Oxford University Press; 2020.
[139] Popejoy AB. Too many scientists still say Caucasian. Nature. 2021;596:463.
[140] Washington H. Medical apartheid: The dark history of medical experimentation on Black Americans from colonial times to the present. Anchor. 2008.
[141] Sellers SL, Cunningham BA, Bonham VL. Physician knowledge of human genetic variation, beliefs about race and genetics, and use of race in clinical decision-making. Journal of Racial and Ethnic Health Disparities. 2019 Feb 15;6:110–16.
[142] Trakoli A. Treatment burden and ability to work. Breathe. 2021;17:1–8.
[143] Smirnoff M, Wilets I, Ragin DF, Adams R, Holohan J, Rhodes R, et al. A paradigm for understanding trust and mistrust in medical research: The community voices study. AJOB Empirical Bioethics. 2018;9:39.
[144] Durant RW, Legedza AT, Marcantonio ER, Freeman MB, Landon BE. Different types of distrust in clinical research among Whites and African Americans. Journal of the National Medical Association. 2011;103:123.
[145] Armstrong K, Rose A, Peters N, Long JA, McMurphy S, Shea JA. Distrust of the health care system and self-reported health in the United States. Journal of General Internal Medicine. 2006;21:292.
[146] Eberth JM, Hung P, Benavidez GA, Probst JC, Zahnd WE, McNatt MK, et al. The problem of the color line: Spatial access to hospital services for minoritized racial and ethnic groups. Health Affairs. 2022;41:237–46.
[147] Leung V, Mapletoft J, Zhang A, Lee A, Vahedi F, Chew M, et al. Thermal stabilization of viral vaccines in low-cost sugar films. Scientific Reports. 2019;9:1–11.
[148] Devrani S, Tiwari R, Khan N, Sankar K, Patil S, Sridhar K. Enhancing the insulation capability of a vaccine carrier box: An engineering approach. Journal of Energy Storage. 2021;36:102182.
[149] Balarajan Y, Selvaraj S, Subramanian S. Health care and equity in India. Lancet. 2011;377:505.
[150] DeSouza SI, Rashmi MR, Vasanthi AP, Joseph SM, Rodrigues R. Mobile phones: The next step towards healthcare delivery in rural India? PLoS One. 2014;9:e104895.

[151] Brown J, Machen H, Kawaza K, Mwanza Z, Iniguez S, Lang H, et al. A high-value, low-cost bubble continuous positive airway pressure system for low-resource settings: Technical assessment and initial case reports. PLoS One. 2013;8:e53622.
[152] Niemeier D, Gombachika H, Richards-Kortum R. How to transform the practice of engineering to meet global health needs. Science (80-). 2014;345:1287–90.
[153] Noblitt J. Introduction to biomedical engineering technology. Biomedical Instrumentation & Technology. 2008;42(6):446.
[154] Sharma A, Sharma P, Roy S. Elastin-inspired supramolecular hydrogels: A multifaceted extracellular matrix protein in biomedical engineering. Soft Matter. 2021;17(12):3266–90.
[155] Zvarova J, Hanzlicek P, Nagy M, Preckova P, Zvara K, Seidl L, et al. Biomedical informatics research for Individualised lifelong shared healthcare. Biocybernetics and Biomedical Engineering. 2009;29(2):31–41.
[156] Denend L, Xu S, Yock P, Venook R. Biomedical technology innovation education and its effect on graduate student careers over 17 years. Biomedical Engineering Education. 2021 Jul;1:291–300.
[157] Nkuma-Udah KI, Agoha EE, Ejeta K, Ndubuka GI. Biomedical Engineering in Nigeria: A Developmental Overview. In: World Congress on Medical Physics and Biomedical Engineering (pp. 1643–48). Toronto, Canada; 2015 June 7–12.
[158] Nigeria: A Developmental Overview. Cham: World Congress on Medical Physics and Biomedical Engineering (pp. 1643–48). Toronto, Canada: Springer; 2015.
[159] Zaman MH. Biomedical engineering education in resource-limited settings: Creating an innovation eco-system. In: 7th International Conference on Appropriate Healthcare Technologies for Developing Countries. Institution of Engineering and Technology; 2012, 75.
[160] Goldberg JR, Malassigné P. Lessons learned from a 10-year collaboration between biomedical engineering and industrial design students in capstone design projects. International Journal of Engineering Education. 2017;33(5):1513–20.
[161] Douglas TS. Biomedical engineering education in developing countries: Research synthesis. In: Annual International Conference of the IEEE Engineering in Medicine and Biology Society. 2011:3628–30.
[162] Settapat S, Achalakul T, Ohkura M. Web-based 3D medical image visualisation framework for biomedical engineering education. Computer Applications in Engineering Education. 2014;22(2):216–26.
[163] Nocera TM, Ortiz-Rosario A, Shermadou A, Delaine DA. How do biomedical engineering graduates differ from other engineers? Bridging the gap between BME and industry: A case study. In: 2018 ASEE Annual Conference & Exposition 2018 Jun 23.
[164] Javaid M, Haleem A. Exploring smart material applications for COVID-19 pandemic using 4D printing technology. Journal of Industrial Integration and Management. 2020;04:481–94.
[165] Mishra S. Social and ethical concerns of biomedical engineering research and practice. Biomedical Engineering and Information Systems. 2011:54–80.
[166] Zhong Q, Ding H, Gao B, He Z, Gu Z. Advances of microfluidics in biomedical engineering. Advanced Materials Technologies. 2019;4(6):1800663.
[167] Filgueiras TP, Bertemes-Filho P. Intelligent agents in biomedical engineering: A systematic review. International Journal of Biosensors & Bioelectronics. 2020;6(5):123–28.
[168] Akujuobi CM. Nanotechnology Applications in Biomedical Engineering. In: Handbook of Research on Diverse Applications of Nanotechnology in Biomedicine, Chemistry, and Engineering (pp. 50–63). IGI Global; 2015.
[169] Magjarevic R, Diaz ML. Biomedical engineering education – Status and perspectives. In: 2014 36th Annual International Conference of the IEEE Engineering in Medicine and Biology Society 2014 Aug 26 (pp. 5149–52).

[170] Ahluwalia A, Atwine D, De Maria C, Ibingira C, Kipkorir E, Kiros F, et al. Open biomedical engineering education in Africa. In: Annu Int Conf IEEE Eng Med Biol Soc. 2015, 3687–90.
[171] Estrela VV, Franca RP, Monteiro AC, Khelassi A, Iano Y, Razmjooy N, et al. Sdr-based high-definition video transmission for biomedical engineering. Medicine in Novel Technology. 2020;4(3):584–85.
[172] Montesinos L, Santos-Diaz A, Salinas-Navarro DE, Cendejas-Zaragoza L. Experiential learning in biomedical engineering education using wearable devices: A case study in a biomedical signals and systems analysis course. Education Sciences. 2022;12(9):598.
[173] Rao RR, Jensen H, Schulz T, McElfish P. Novel clinical needs finding course brings biomedical engineering students together with regional medical campus students, residents, and faculty to solve real-world problems: Clinical needs finding and observations. Journal of Residences Campuses. 2018;1(4).
[174] Krishnan S. Project-based learning with international collaboration for training biomedical engineers. In: Annual International Conference of the IEEE Engineering in Medicine and Biology Society. 2011; 2011:6518–21.
[175] Fernandes F, Barbalho I, Barros D, Valentim R, Teixeira C, Henriques J, et al. Biomedical signals and machine learning in amyotrophic lateral sclerosis: A systematic review. BioMedical Engineering OnLine. 2021;20(1):1–22.
[176] Boskovic D, Badnjevic A. Opportunities and challenges in biomedical engineering education: Focus on Bosnia and Herzegovina. In: 4th Mediterranean Conference on Embedded Computing (MECO). IEEE Publications; 2015. p. 407–10.
[177] Mattsson J, Ostlund B, Williamsson A, Eriksson A. Interprofessional € learning for enhanced patient safety: Biomedical engineering students and nursing students in joint learning activities. Journal of Interprofessional Education & Practice. 2019;9(1).
[178] Ghista DN, Kabinejadian F. Coronary artery bypass grafting hemodynamics and anastomosis design: A biomedical engineering review. BioMedical Engineering OnLine. 2013;12(1):129.
[179] Lee JW, Ostrowski A, Daly SR, Huang-Saad A, Seifert CM. Idea generation in biomedical engineering courses using Design Heuristics. European Journal of Engineering Education. 2019;44(3):360–78.
[180] Armentano RL, Chatterjee P. MOOC on biomedical engineering for Latin American students–unleashing the potential of virtual learning. In: International Conference on Computational Intelligence and Communication Networks (CICN). IEEE Publications; 2015. p. 405–10.
[181] Purnomo MH, Kristian Y, Setyati E, Delfana Rosiani UD, Setiawan EI. Limitless possibilities of pervasive biomedical engineering: Directing the implementation of affective computing on automatic health monitoring system. In: 8th International Conference on Information Technology And Electrical Engineering (ICITEE). IEEE Publications; 2016. p. 1–4.
[182] Vazquez-Estrada P, Sanchez-Sosa G, Tlacuilo-Parra Y, Díaz-Domínguez R, Garcia-Gonzalez A. Creation and analysis of technological intelligence reports as an educative tool in biomedical engineering. In: 43rd Annual International Conference Of the IEEE Engineering in Medicine & Biology Society (EMBC). 2021;2021:7639–42.
[183] Abu-Faraj ZO. A recommended model of an undergraduate biomedical engineering curriculum for the MEDA region. In: Proceedings of the international medical informatics and biomedical engineering symposium, 2006, March. p. 14–20.
[184] Gerdsri N, Kongthon A, Puengrusme S. Profiling the research landscape in emerging areas using bibliometrics and text mining: A case study of biomedical engineering (BME) in Thailand. International Journal of Innovative Technology & Adaptive Management. 2017;14(2):1740011.
[185] Carleton PF, Schachter S, Lash TB, Chen J, Merchak T, Brand O, et al. Point-of-care technology research network: An evolving model for collaborative translational research in biomedical engineering. Current Opinion in Biomedical Engineering. 2019;11:145–48.

[186] Bamigboye AA, Bello KA. Biomedical engineering in Nigeria: The genesis, present and the future. Clinical Reviews and Opinions. 2021;10(1):1–4.
[187] Jones H, Nazarpour K. Gaining NHS ethical approval from the perspective of a biomedical engineering team. British Journal of Healthcare Management. 2018;24(2):71–76.
[188] Singh A, Ferry D, Mills S. Improving biomedical engineering education through continuity in adaptive, experiential, and interdisciplinary learning environments. Journal of Biomechanical Engineering. 2018;140(8):810091–98.
[189] Shi J, Lam KY, Gu M, Li M, Chung SL. Towards energy-efficient secure communications using biometric key distribution in wireless biomedical healthcare networks. In: 2nd International Conference on Biomedical Engineering And Informatics. IEEE Publications; 2009. p. 1–5.
[190] Nkuma-Udah KI, Mazi EA. Developing biomedical engineering in Africa: A case for Nigeria. In: World Congress on Medical Physics and Biomedical Engineering. Berlin, Heidelberg: Springer; 2006. p. 3828–31.
[191] Hong JH, Kim NJ, Cha EJ, Lee TS. Development brief of a body area network for ubiquitous healthcare: An introduction to ubiquitous biomedical systems development center. Journal of Biomedical Engineering. 2005;26(5):331–35.
[192] Badnjevic A, Gurbeta L. Development and perspectives of biomedical engineering in South East European countries. In: 39th International Convention on Information and Communication Technology, Electronics And Microelectronics (MIPRO). IEEE Publications; 2016. p. 457–60.
[193] Nagel JH. IUPESM: The international umbrella organisation for biomedical engineering and medical physics. Biomedical Imaging and Intervention Journal. 2007;3(3).
[194] Iakovidis I, Le Dour O, Karp P. Biomedical engineering and ehealth min Europe outcomes and challenges of past and current EU research programs. IEEE Engineering in Medicine and Biology Magazine. 2007;26(3):26–28.
[195] Lujerdean C, Baci GM, Cucu AA, Dezmirean DS. The contribution of silk fibroin in biomedical engineering. Insects. 2022;13(3):286.
[196] Hashimoto S, Ohsuga M, Yoshiura M, Tsutsui H, Akazawa K, Fujisato T, Mochizuki S, Kobayashi H, Nakaizumi F, Kawai T, Uto S. Parallel curriculum between application and fundamental subjects with rotational experimental project for multidisciplinary study field of biomedical engineering. In: Proc. 12th World Multi-conference on Systemics Cybernetics and Informatics 2008 (Vol. 2, pp. 98–103).
[197] Gatchell DW, Linsenmeier RA, Murphy RL, Coker AO, Osuntoki AA. Developing innovative interdisciplinary biomedical engineering programs in Nigeria: Lessons learned. In: ASEE International Forum, 2016 June; 2016.
[198] Escayola J, Martínez I, Trigo J, García J, Martínez-Espronceda M, Led S, et al. Recent innovative advances in biomedical engineering: Standard-based design for ubiquitous p-health. In: Fourth international multi-conference on computing in the global information technology. IEEE Publications; 2009. p. 109–17.
[199] Makobore PN, Madete J, Leibovits A. The UBORA project: Euro-African open biomedical engineering eplatform for innovation through Education. In: 13th International CDIO Conference. F, A. Díaz Lantada, M. Mridha. Calgary, Canada 2017, June.
[200] Saddow SE. Recent advances in SiC biomedical devices: Healthcare applications. Silicon Carbide Technology for Advanced Human Healthcare Applications. 2022:1–48.
[201] Tamura T, Maeno S, Hattori T, Kimura Y, Kimura Y, Yoshida M, et al. Assessment of participant compliance with a Web-based home healthcare system for promoting specific health checkups. Biocybernetics and Biomedical Engineering. 2014;34(1):63–69.
[202] Waples LM, Ropella KM. University-industry partnerships in Biomedical engineering. IEEE Engineering in Medicine and Biology Magazine. 2003;22(4):118–21.
[203] Aruna M, Gunasilan U, Naeem S. Efficient equipment management for the biomedical engineering department in the hospital. International Review of Management Mark. 2018;8(3):69.

[204] Houssein A, Lefor AK, Veloso A, Yang Z, Ye JC, Zeugolis DI, et al. BMC Biomedical Engineering: A home for all biomedical engineering research. BMC Biomedical Engineering. 2019;1(1):1.
[205] Istanbullu A, Güler I. Multimedia-based medical instrumentation course in biomedical engineering. Journal of Medical Systems [Internet]. 2004;28(5):447–54.
[206] Gomez AW, Nyassi E, Mullally S. Towards a biomedical engineering quality system: Writing documentation to support clinical & laboratory accreditation at MRC unit in the Gambia. In: 7th International Conference on Appropriate Healthcare Technologies for Developing Countries. Institution of Engineering and Technology; 2012, 1–5.
[207] Li C, Un KF, Mak PI, Chen Y, Munoz-Ferreras JM, Yang Z, et al. Overview of recent development on wireless sensing circuits and systems for healthcare and biomedical applications. IEEE Journal on Emerging and Selected Topics in Circuits and Systems. 2018;8(2):165–77.
[208] Hayaliev R, Nurkhodjaev S, Nazarova N, Rizayev J, Rahimberdiyev R, Timokhina T, et al. Interdisciplinary approach of biomedical engineering in the development of technical devices for medical research. Journal of Biomimetics, Biomaterials and Biomedical Engineering. 2021;53:85–92.
[209] Harris TR, Bransford JD, Brophy SP. Roles for learning sciences and learning technologies in biomedical engineering education: A review of recent advances. Annual Review of Biomedical Engineering. 2002;4(1):29–48.
[210] Cinelli I. Space medicine requirements open to innovation in biomedical engineering. In: 40th Annual International Conference Of the IEEE Engineering in Medicine And Biology Society (EMBC). 2018:973–76.
[211] Darwish A, Sayed GI, Hassanien AE. The Impact of Implantable Sensors in Biomedical Technology on the Future of Healthcare Systems. Intelligent Pervasive Computing Systems for Smarter Healthcare. 2019:67–89.
[212] Denton T. Frontiers of medical decision-making in the modern age of data analytics. IISE Transactions. 2023 Jan 2;55(1):94–105.
[213] Kimsey DB. Lean methodology in health care. AORN Journal. 2010 Jul 1;92(1):53–60.
[214] Engineer LD, Winters BD, Weston CM, Zhang A, Sharma R, Bass E, Jones D, Rosen A, Yoon FB, Borzecki A, Dy SM. Hospital characteristics and the agency for healthcare research and quality inpatient quality indicators: A systematic review. The Journal for Healthcare Quality (JHQ). 2016 Sep 1;38(5):304–13.
[215] Miners C, Hundert M, Lash R. New structures for challenges in healthcare management. Healthcare Management Forum. 2015;28(3):114–17.
[216] Nair SS. Operational excellence: The healthcare management imperative. Management. 2021;9 (2):19–30.
[217] Peimbert-García RE, Gutiérrez-Mendoza LM, García-Reyes H. Applying lean healthcare to improve the discharge process in a Mexican Academic Medical Center. Sustainability. 2021;13(19):10911.
[218] Tapping D. Value stream management for lean healthcare: Four steps to planning, mapping, implementing, and controlling improvements in all types of healthcare environments; MCS media, incorporated; ISBN 978-1-4619-0426-7; 2014.
[219] Womack JP, Jones DT. Lean thinking – Banish waste and create wealth in your corporation. Journal of the Operational Research Society. 1997 Nov 1;48(11):1148.
[220] Fine B, Golden B, Hannam R, Morra D. Leading lean: A Canadian healthcare Leader's guide. Hydroxychloroquine. 2009;12(3):32–41.

Chapter 8
Developing sustainable healthcare in waste management: a comprehensive approach

Abstract: The healthcare sector, although essential for the population's well-being, plays a substantial role in environmental deterioration due to the production of both hazardous and nonhazardous waste materials. The purpose of the analysis is to establish a complete framework for the management of hospital waste that is sustainable, integrating considerations of ecology, economics, and public health. The approach involves reducing waste at the source, employing effective segregation techniques, incorporating innovative treatment technology, and utilizing resources obtained from waste. To analyze the environmental, social, and economic implications of such an integrated model, life cycle evaluations and cost-benefit analyses were performed. The study shows that advanced waste treatment methods and careful use of resources may significantly reduce carbon emissions and environmental harm. The sustainable model provided fiscal responsibility by showing how savings and income can be generated through long-term investments in environmentally responsible hospital waste management. The chapter focuses on the critical function of healthcare organizations as catalysts for operational sustainability. Healthcare providers have a responsibility to protect the public's health as well as to be efficient managers of the environment and good managers of tax money. By implementing this strategy, hospital waste management may become a catalyst for sustainability, contributing to worldwide efforts to reduce the healthcare sector's environmental impact. The present study aims to investigate the primary determinants of hospital waste sustainability through the use of a comprehensive methodology. Additionally, it seeks to propose suggestions for enhancing the waste management system. The chapter contributed to the understanding of garbage management strategies by investigating the various manners in which policy, organizational, and personal dynamics affect system effectiveness.

8.1 Introduction

Healthcare services have an essential role in the management, prevention, and treatment of illnesses, making them a significant aspect of sustainability. The investments of both human and financial resources in the healthcare sector of any given nation are important, indicating the economic aspect of sustainability. Therefore, hospitals and medical facilities use considerable quantities of materials and unavoidably generate waste, which has an impact on the sustainability of the environment [1–2]. The proper management of medical waste not only yields significant environmental advantages but also leads to economic savings [3–4]. Diseases are becoming more wide-

https://doi.org/10.1515/9783111436432-008

spread as a result of global warming, chemical pollution, and the overexploitation of natural resources. The increasing issues in the field of ecological health have had an extensive effect on healthcare systems. Interestingly, the healthcare sector itself contributes to the development of several ecological issues while simultaneously dealing with the resultant impacts. The healthcare industry, via the utilization of goods and technology, as well as the utilization of resources, generation of waste, and everyday construction and operations, significantly contributes to pollution on a global scale. Consequently, it has an impact on public health [5]. The healthcare sector has a significant impact on the environment through the generation of about 2.4 million tons of material waste each year [6]. Hospitals may produce hazardous waste as a byproduct of the medical treatments they provide. While a significant proportion of the waste generated by hospitals is categorized as general waste, offering similarities with municipal solid garbage and thus being suitable for disposal in municipal landfills, a small fraction of infectious waste requires appropriate management to reduce potential risks to public health. Healthcare waste is waste product that is produced by various healthcare institutions, including clinics, nursing facilities, hospitals, pathological laboratories, and healthcare research institutes [7]. A study indicate [8] that, the waste includes body parts, chemical substances, cytotoxic dressings, contaminated syringes, bleeding cotton, and radioactive substances. Healthcare waste becomes a harmful substance for both people and animals when infectious and dangerous substances are present [9].

Globally, there has been a substantial rise in the production of healthcare waste (HCW), mostly caused by the increase in population and the subsequent expansion of healthcare institutions [10]. The destruction and management of waste at hospitals and healthcare institutions are essential to prevent potential health concerns for individuals, animals, plants, and the environment [11]. Healthcare waste creates the highest risk of diseases and injury compared to other types [12]. Studies show that 75–90% of HCW is harmless compared to household waste. Wastes from administrative and management tasks at these institutions are mostly hazardous (10–25%), as reported [13]. The production waste of medical and clinical institutes, known as specialized waste or clinical garbage, has been neglected, particularly in developing nations [14]. Due to the negative financial and ecological effects of waste materials, early discussions about the future sustainability of the healthcare sector focused on the importance of implementing waste reduction and recycling programs. The Sustainable Development Goals of the United Nations (UNSDGs) are designed to achieve this by maintaining a balance between social, economic, and ecological advancement and enhancing system efficiency by addressing the requirements of both the present and the future [15, 16]. Inappropriate disposal of contaminated syringes and scalpels has been identified as a significant contributing factor to the rise in HIV infections among those engaged in rag-picking activities [17]. According to the World Health Organization [7], there were 33,800 new cases of HIV in 2010, around 1.7 million cases of hepatitis B, and 315,000 cases of hepatitis C. Many health-related environmental challenges arise from waste generation and dumping methods. Hospitals worldwide struggle to provide sustainable environments; underdeveloped nations face additional challenges. One of the most significant challenges is that hospitals frequently lack the facilities and equipment required to properly dispose of medical waste. The concept of adopting environmentally sustainable practices, sometimes referred to as "going green," encompasses several strategies such as waste elimination as well as the saving of energy and resources [18]. The growing understanding of environmental concerns and the implementation of strict regulations, together with the present need to decrease public expenditures, has amplified the significance of environmental issues within the health sector. Hence, the establishment and execution of a sustainable hospital waste disposal system need a comprehensive strategy that involves several stakeholders, such as policymakers, governmental and nongovernmental entities, as well as the community [19–21]. To establish an effective hospital waste management system, it is important to consider several elements, such as public health, economic conditions, social dynamics, regulatory frameworks, and hospital administration. It is necessary to comprehensively assess the interrelationships among these components to ensure the successful implementation of the system. The majority of prior studies have only examined individual components, avoiding exploring the potential interconnections between these elements. The objective of the chapter is to start the establishment of a system dynamics framework for hospital waste handling, taking into account the comprehensive integration of all relevant components. The present case study represents hospital garbage management practices in a developing nation. Hospital waste management in poor countries receives fewer resources than other waste types including solid waste from municipalities and e-waste [22, 23].

8.2 Healthcare waste

Healthcare waste, also known as medical waste or clinical waste refers to any waste generated by healthcare facilities, such as hospitals, clinics, laboratories, dental offices, and pharmaceutical companies, during the diagnosis, treatment, or immunization of humans or animals or research activities. Healthcare waste is considered a different kind of waste due to its potentially hazardous nature and the need for careful handling and disposal. Healthcare services improve people's quality of life by preventing and treating illness and helping them live longer. However, healthcare procedures produce considerable amounts of medical waste; 20% of this waste material is infectious or otherwise potentially harmful [24]. The World Health Organization (WHO) [25] established regulations regarding hospital waste disposal in its report "Safe Handling of hazardous waste via medical treatment activities." The WHO described HCW as "waste generated by healthcare organizations, hospitals, labs, and biomedical research centers, as well as waste from smaller or scattered sources" in their recommendations. Multiple terminologies are now in use to characterize the garbage produced by hospitals and clinics [26]. Due to the recognition that clinical waste involves all waste generated within healthcare institutions, a prevalent issue of inadequate waste separation occurs in the majority of healthcare institutions.

8.2.1 Infectious waste

It involves waste that has been infected with blood, body fluids, or other substances that have the potential to cause infection and the capacity to spread infectious illnesses to both people and animals. Examples include used syringes, cultures, and swabs. Improper handling and management of this particular form of garbage might give rise to substantial health hazards.

8.2.2 Hazardous waste

Hazardous waste is any waste product that is poisonous, reactive, combustible, or infectious and poses a serious risk to human health, the environment, or both. Chemicals, solvents, insecticides, heavy metals, radioactive compounds, and industrial wastes are examples. Industrial, healthcare, laboratory, and domestic procedures can produce hazardous waste. Hazardous waste generated and disposed of properly prevents pollution, environmental damage, and health problems, and rigorous restrictions prevent risks to people and the surrounding environment [25, 27].

8.2.3 Nonhazardous waste

It is often known as general or solid waste, and does not pose an immediate risk to human beings or the environment. This category contains nontoxic industrial, commercial, and healthcare trash and home waste products such as paper, cardboard, plastics, glass, and organic debris. Nonhazardous waste does not exceed regulation criteria for flammable, reactive, corrosive, or poisonous compounds. Proper disposal and management of nonhazardous waste are still important to prevent litter, reduce landfill usage, and promote recycling and waste reduction practices, contributing to overall environmental sustainability [25, 27].

Only 10–25% of waste generated by the healthcare industry is clinical, as reported by the WHO [26]. Clinical waste is “all waste material collected by medical facilities,” and most hospitals do not do a good job of sorting their waste materials. Most health workers are unaware of the Clinical Waste Management Code of Practice [28], which contributes to the issue. New findings in medical centers and clinic consulting areas indicate that most healthcare institutions exclusively store red bags. This implies that nonclinical items, including the packaging for medications, gloves, and syringes, end up in the red bags while they belong in the black bags designated for general garbage. Due to this misunderstanding, hospitals and other medical institutions produce more "red-bag" garbage, which contains a combination of clinical and nonclinical materials. Some hospitals require all patients to use red bags for their medical waste disposal. According to Hossain et al. [29], HCW consists of "any waste generated by healthcare providers and related businesses," including clinics, hospitals, and pharmacies. The term "medical waste" is used here to describe materials discarded during the care of patients (Table 8.1). Infectious medical waste is separated into categories such as organization, country, and date as given in Table 8.1.

Table 8.1: Classifications of healthcare wastes and related information.

Institution/enactment year	Classifications	Information on contaminated waste
The World Health Organization published "Safe Handling of Wastes from Medical Activities" in 1992 [25] Concept: Healthcare waste	– Hazardous medical waste – Sharp waste – Infectious materials – Pathological items – Pharmaceutical waste – Cytotoxins – Chemicals	– Pathogen cultures or strains – Surgical or dissection waste from an infectious patient – Waste from a pathogenic patient in an isolation ward – Waste from an infectious patient during surgery – Waste from animals which contaminated with viruses or transmissible diseases
United States, 1990 Environmental Protection Act [30] Concept: Biomedical discards	– Isolation trash – Cultures, stores – Human plasma and blood products – Pathological waste – Contaminated dangerous objects – Infectious animal carcasses, body parts	– The Centers for Disease Control and Prevention (CDC) defines quarantine waste material – Infection-related media and strains – Blood – Pathological debris (human tissue, etc.) Waste that might cause harm – Experimental waste from animals

Table 8.1 (continued)

Institution/enactment year	Classifications	Information on contaminated waste
1992 European Waste Catalogue [31] Concept: Healthcare waste	– Separable anatomical waste (biological waste) – Disposal of chemical, hazardous waste – Pharmaceutical waste disposal, including cytotoxins – Hazardous waste (injection needles, scalpels, sharply damaged materials, etc.) – Radiation waste (radioactive waste directive)	– Childbirth waste, as well as infectious diseases detection, treatment, and prevention – Waste generated during the diagnosis, treatment, and prevention of animal illnesses
Environmental Protection Act of 1990 in the United Kingdom [32] Concept: Waste from healthcare	– Harmful medical waste – Waste materials within the healthcare sector that have the potential to induce illness	– Microorganisms or poisons that possess the capability to induce illness in humans or other living organisms – Items that have been infected with biologically active drugs – Sharp objects, human fluids, and materials that have been polluted with hazardous chemicals

8.3 Environmental impacts of hospital waste

Hospitals play a vital role in healthcare, but they also have significant environmental impacts. The generation of garbage and the use of energy by hospitals have significant implications for both the environmental and physical well-being of those working in the affected region. Hospital owners have the responsibility for not only providing medical care inside the hospital premises but also for monitoring the management of the hospital's physical infrastructure, ensuring its environmental sustainability, and promoting the health and well-being of both patients and staff. The organization's stakeholders have assigned them the responsibility of safeguarding and preserving the environmentally sustainable hospital [18]. Approximately 75–85% of medical wastes exhibit similarities to typical municipal garbage and cause little risks when subjected to incineration. The next most significant category is pathogenic wastes, comprising approximately 5–25% of the overall waste composition. In general, infectious garbage may be classified into two main categories: infectious and sharp waste, which constitutes around 1% of the entire garbage, and highly infectious waste, anatomical waste, and pathological waste, which also accounts for approximately 1% of the total waste. Biochemical and radioactive contaminants, including medical trash, laboratory chemicals, cleansers, heavy metals like mercury from fractured thermometers, and pesticides with diverse health and environmental impacts, constitute approximately 3% of the whole medical waste [33].

Additionally, hospitals use a lot of water and electricity, which strains local resources and increases greenhouse gas emissions. In addition, the establishment and management of healthcare facilities frequently result in the clearance of forests and disruption of natural habitats. Hospitals are using energy-efficient construction, waste management programs, and eco-friendly purchasing to reduce their environmental impact and develop a healthier planet. Figure 8.1 illustrates the correlation between pollution and increased demands for healthcare facilities, thereby resulting in a significant increase in contamination [33].

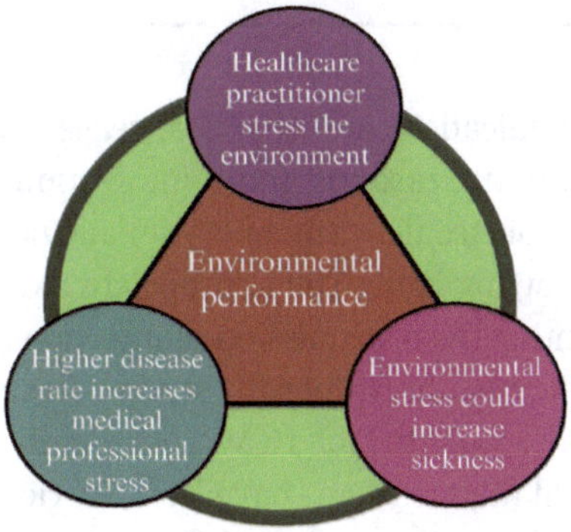

Figure 8.1: The correlation between environmental degradation, elevated prevalence of illnesses, and the environmental effects of medical therapeutic services.

The removal of hospital waste from the medical waste category warrants further consideration. According to a study [34], the discharge from sanitary facilities that contain drug-resistant bacteria and a wide range of chemicals poses a greater risk compared to home sewage. Healthcare administrators need to know which types of waste are being produced and what kinds of risks they pose to people and the environment [18]. The many categories of hospital waste are shown in Table 8.2.

8.4 Composition of healthcare waste (HCW)

The materials produced in healthcare facilities, such as hospitals, clinics, and laboratories, are included in the composition of HCW. HCW commonly comprises infectious waste, encompassing materials that have been contaminated with blood, bodily fluids, or potentially infectious agents. Examples of such waste include used syringes, soiled dressings, and cultures. Hazardous waste constitutes a significant classification that includes chemicals, pharmaceuticals, and various items that present potential dangers to

Table 8.2: Different kinds of hospital waste products [35].

1) Solid waste	Waste is generally referred to as trash, garbage, or municipal solid waste. Such materials are recycled or disposed of in a sanitary landfill.
2) Biohazardous waste material	Waste that has the potential to transmit infectious illnesses. These waste materials are widely recognized as infectious waste or red bag trash. They encompass sharps, cultures, human blood, and human products, among others.
3) Harmful chemical substances	Waste exhibiting potentially hazardous properties. The classification and management of these wastes are carried out by the Resource Conservation and Recovery Act, as regulated by the United States Environmental Protection Agency (US EPA).
4) Medical waste	Every waste material produced in healthcare facilities, including hospitals, clinics, physician offices, dentist offices, blood banks, veterinary hospitals/ clinics, and medical research centers.

human health or the environment, such as expired medications and chemical reagents. HCW encompasses a variety of materials, including nonhazardous and noninfectious substances such as paper, packaging, and food waste. Electronic waste (e-waste), including discarded medical devices and equipment, is a growing issue. The appropriate handling and disposal of HCW plays a vital role in preventing the transmission of infections, reducing environmental harm, and protecting public health. Numerous studies have been conducted on various aspects of estimating the quantity of HCWs on a daily, weekly, monthly, and annual basis. In addition, several investigations have been carried out to determine the composition of HCW from a specific healthcare institution or hospital. The rate of HCW production in India was estimated by Patil and Shekdar [36] to be between 0.5 and 2 kg per bed per day. Total annual waste product was predicted to amount to 0.33 million tons, with 7–10% coming from plastics, 3–5% from glass, and 30–35% from infectious HCW. In 2003, Altin et al. provided details on the HCW staffing in four Turkish hospitals. They determined that 92% of HCW could be incinerated and 8% could not. They calculated a waste production rate of 1.25–2.6 kg per bed per day [37]. To forecast the birthrate and demographic makeup of HCW throughout Iran's regions, Reza Sabour et al. [38] developed a regression model based on hospital and bed counts. The content and rate of production of HCW in Hamadan, Iran, were estimated in research by Nabizadeh et al. [39]. They determined that dental waste comprised primarily of general garbage (71.15), infectious waste (21.40%), chemical and pharmaceutical waste (7.26%), and hazardous waste (0.18%). Zhang et al. [40] researched healthcare institutions in Gansu Province, China, to determine their HCW generation levels and composition. They discovered that the amounts of HCW that were generated in primary, secondary, and tertiary healthcare facilities were, respectively, 0.61, 0.59, and 0.79 kg per bed per day, with 0.61 being the highest and 0.79 being the lowest. Maamari et al. [41] researched the production rates and nature of HCW in Lebanon and gathered data from

57 different hospitals. The outcomes of the investigation indicated that small private medical facilities with a bed capacity of 100 or less and big private hospitals with a bed capacity of 200 or more produced infectious HCW at rates of 1.14 and 2.45 kg per bed per day, respectively. Using data from primary healthcare organizations in Sao Paulo (Brazil), Moreira and Gunther [42] evaluated the HCW management strategy based on HCW composition and generation. In 2010, Graikos et al. [43] investigated the production rate and composition of HCW at the Social Insurance Institute of Xanthi, Greece. The Greek pharmaceutical and chemical waste composition and production rate were estimated in research by Voudrias et al. [44].

8.5 Management of hospital waste

A hospital is a facility that offers a variety of healthcare services to the local population. The day-to-day activities of the organization can include curative, rehabilitative, preventative, and health promotion education. During its operations, the hospital has the potential to produce waste materials originating from both medical and nonmedical sources. The presence of infectious elements in hospital waste has the potential to cause a risk for patients, nearby communities, hospital staff (particularly garbage employees), visitors, and the surrounding environment. The waste produced by hospitals varies not just between nations but even within the same nation depending on the type of facility, the percentage of reusable products that are utilized, and the percentage of patients that are treated on an outpatient basis. Along the same lines as the trends of municipal solid waste creation, a report that was issued by the WHO in 1999 demonstrates that the amount of hospital trash produced is exactly related to the degree of wealth [45]. To ensure the safe management of HCW, there are strict regulations and protocols in place that govern its handling, transportation, treatment, and disposal. High-income nations are likely to have a wider range of values for countries with similar income levels than low-income ones do [45]. The production of hospital garbage in different countries is shown in Table 8.3.

Table 8.3: The amount of waste produced by hospitals in different countries.

Country	Generation rate (kg per bed per day)	GNP per capita (US$)	Reference
Tanzania	0.84	320	[46]
India	1.60	620	[36]
Iran	1.25	2,320	[47]
Thailand	1.75	2,490	[48]
Japan	2.05	37,050	[49]

Hospital waste management is a big and important system that tries to solve the many problems that come with HCW materials. To prevent cross-contamination, it starts with the important step of waste segregation, in which different waste types, such as infectious, hazardous, pharmaceutical, and general garbage, are segregated at the source. Waste is collected in color-coded containers and stored in specific facility locations after segregation. Strict protocols are followed during transportation to ensure the safety of the garbage, and different types of waste require different treatments, such as autoclaving and cremation for infectious waste and chemical treatments for hazardous waste. Municipal landfills are the normal destination for general trash. To maintain community and environmental safety, the disposal procedure is strictly governed. Staffs at hospitals receive training on how to properly manage trash to protect both the public and the environment. One major difference between hospital waste and other types of garbage is that hospital waste often contains dangerous waste that could spread disease. Infectious illnesses can only spread if specific circumstances are met at the same time. The pathogenic microorganism, sufficient viral transmission, dosage, the route of entry, and the susceptibility of the host are all necessary components. Infectious waste from hospital operations is defined differently across nations [50]. The garbage is temporarily deposited at the source before being collected and processed at a treatment facility. In general, the waste segregation process in hospitals makes use of different containers (Figure 8.2): a red bag for infectious waste, a yellow bag or container for anatomical waste, a black bag for general garbage (noninfectious waste), a white container for sharp waste (needles, broken glass) and a blue container for sharp waste that is recyclable.

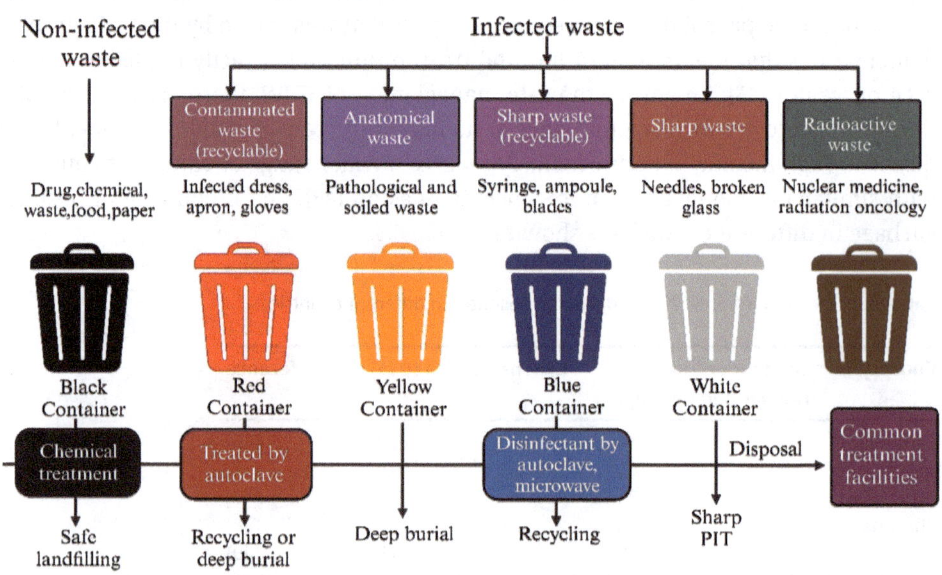

Figure 8.2: Color coding system for hospital waste disposal and management.

The national guideline for segregating hospital waste is not being effectively applied across the country, according to research done in Tanzania [51]. When the storage containers are three-quarters filled, they should be withdrawn from the point of generation. When the waste management facility is situated off-site, as is the case with most hospitals, a storage area on the grounds is necessary to provide the time gap between garbage accumulation and treatment operating hours. There are numerous methods for treating infectious refuse, such as incineration, chemical disinfection, moist thermal treatment, and microwave irradiation. One of the most widely used methods of waste disposal worldwide is incineration. The World Health Organization said that incineration technology, though expensive, might be utilized to handle pathogenic, anatomical, sharps, pharmaceuticals, cytotoxic, chemical substances, and low-level radioactive contaminants. The management of all different kinds of trash has become a significant issue that is relevant to every aspect of society. The handling of hospital trash is not an exception [45]. There has recently been increased demand for waste managers to adopt a sustainable approach to hospital waste handling and to incorporate policies aimed at achieving a sustainable society. It is a very difficult and complex work to plan a hospital waste removal system as it constantly deals with the changing character of waste and variability. In addition, it is vital to take into consideration health hazards, as well as economic, technological, and environmental factors. In order to examine the connection between different factors in hospital waste management, the present systemic dynamics method is provided.

8.6 Sustainable medical waste management

Based on data provided by the WHO [52], it has been determined that around 15% of the total medical waste produced can be classified as hazardous. In high-income nations, the daily generation of harmful HCW per bed in hospitals amounts to 0.5 kg, but in low-income countries, the figure stands at 0.2 kg. The production of medical waste has increased during the COVID-19 epidemic. The amount of medical waste produced by COVID-19-related medical institutions is 3.4 kg per person and around 2.5 kg per hospital bed each day globally, according to the Environment Program of the United Nations [54]. China produced an estimated daily quantity of 469 tons of medical waste during the period of the epidemic [53]. According to sources [54] and [55], Japan, India, Indonesia, and South Korea produced daily waste quantities of 876, 608, 290, and 476 tons, respectively. The World Health Organization (WHO) has released guidelines that outline the safe and sustainable management of HCW as a requirement in public health concerns, as well as the process to implement all connected measures to provide the required financial resources [56].

The World Health Assembly has also urged governments throughout the world to adopt more measures to deal with medical waste [57]. The United Nations Special Rapporteur on the Commission on Human Rights has invited both persons and countries to engage in the development of a comprehensive international legislative framework aimed at safeguarding human health and the environment against the detrimental consequences arising from inadequate management and disposal practices of hazardous medical wastes [58]. However, HCW management should not continue to suffer from inadequate funding and implementation. Regrettably, the substantial risks posed by the combining toxic and infectious nature of medical waste to both the environment and public health have not received considerable attention. According to a recent literature study, it was determined that over 50% of the global population is susceptible to the health consequences associated with HCW [59]. According to Hassan et al. [60], the issue of medical waste arises due to a combination of factors, including a lack of understanding and motivation among healthcare workers, as well as inconsistent rules and legislation regarding the appropriate handling of medical waste. Hossain et al. [29] addressed the significance of staff misbehavior and inadequate medical waste disposal practices in hospitals, as they might lead to elevated health hazards and environmental contamination due to the infectious nature of such material.

Consequently, healthcare establishments need an operational plan to educate those engaged in the production of medical waste on how to effectively address this growing problem. While earlier studies in medical waste management mostly concentrated on the handling of hazardous waste, there has been a recent change in attention towards operational solutions for the management of all categories of medical waste. The underlying cause for this phenomenon may be attributed to the fact that ensuring the proper management and elimination of medical waste is a crucial measure in mitigating possible risks, such as the spread of diseases or injuries, as well as the contamination of the natural environment [61]. However, analysis on the transmission of blood-borne viruses as well as respiratory and other illnesses caused by improper disposal of medical waste has not been completed [25], The possible risks to human health as well as the ecological problem are quite significant [62]. So, managing medical trash is now seen as an important part of providing high-quality medical services [63]. This transformation is a consequence of studies that have illustrated the detrimental impact of environmental contaminants produced during waste management on both the ecosystem and human well-being. It was estimated that approximately eight million tons of plastic garbage associated with the COVID-19 epidemic had been produced all over the world, with more than 25,000 tons being released into the sea. The potential consequences of this phenomenon may lead to harmful and enduring impacts on the marine ecosystem [53]. A uniform categorization technique was developed by Windfeld and Brooks [64] to teach medical professionals how to efficiently handle medical waste. According to Thakur et al. [65] the six elements of medical waste management procedures are as follows: "experience, relationship, environmental variables, technology and qualifications, financial considerations, and firm's capabilities." Healthcare organizations need to establish comprehensive medical waste management strategies that encompass the routine collection, processing, segregation, and storage of HCW. Additionally, these institutions need to implement consistent monitoring and training initiatives [29, 62, 66, 67].

To establish a sustainable healthcare ecosystem by effectively minimizing and managing medical waste, it is imperative to cultivate a corporate environment that promotes the active involvement of all stakeholders engaged in the production of medical waste [68]. The successful implementation of this initiative necessitates the active participation and collaboration of all relevant parties, including the many professional groups and organizations within the healthcare facility, as well as the participation of patients, their family members, subcontractors, and the wider community [69]. It is recommended that healthcare organizations establish a comprehensive strategy for the management of medical waste [65, 66]. Hence, it is important that the duty for medical waste reduction not be only assigned to a single department. Rather, it is recommended that all members of the hospital actively engage in these practices as part of their routine duties. For example, it is mandatory for the department responsible for medical waste disposal to adhere to appropriate separation protocols to avoid the inclusion of general trash in the medical waste stream. Healthcare institutions should limit infectious trash and one-time-use product emissions. These broad practices can help healthcare organizations reduce HCW and operating expenses, creating a sustainable service environment.

Technology for managing waste should be accessible and compatible with waste characteristics, operations, and upkeep of legal requirements. Waste reduction may be accomplished by the use of the subsequent strategies:

I. Whenever possible, it is advisable to purchase products produced from recycled materials such as glass and metals that possess the capacity to undergo disinfection and subsequent reuse.
II. The proposed approach includes implementing measures to sterilize reusable goods, enhancing sterilization methods, ensuring quality assurance, implementing control mechanisms, and validating cleaning and disinfection processes for patient care. Additionally, the aim is to decrease reliance on pre-sterilized disposable items.
III. Establish and implement rules and procedures for the effective management of produced garbage.
IV. Develop and implement efficient regulations in collaboration with the authorized producer of plastics to facilitate recycling processes.
V. Employ the steam sterilization technique, which is recommended above chemical disinfection, to mitigate the production of harmful chemical waste [36].

8.7 Comprehensive strategies for healthcare waste disposal, treatment, and management

A well-planned strategy can minimize medical waste without compromising healthcare worker quality. Kwikiriza et al. [70] highlighted the need for clinical staffs' comprehensive understanding of their crucial role in ensuring efficient medical waste management. This is due to their direct involvement in trash categorization at the site of creation. Additionally, it was stated that nonclinical personnel often possess little understanding and expertise regarding the management of medical waste, including treatment, segregation, and proper disposal methods. In order to effectively reduce the production of hazardous waste in their routine practices, healthcare providers must possess precise data on the quantity of medical waste created by their facilities. Reducing the amount of garbage that needs to be treated is a clear way to decrease waste management costs and make the company run more smoothly. Efforts aimed at identifying and eliminating unnecessary sources of waste creation can have a beneficial impact on the effectiveness of building a sustainable healthcare environment. Hence, enhancing the efficacy of medical waste management may be achieved by implementing accurate waste categorization and sorting practices at the point of material utilization. When a systematic and efficient control and management protocol is implemented, HCW is expected to have no negative impact on both human beings and the surrounding ecosystem. The handling of medical waste is a complicated task, and the effectiveness of this significant sector depends heavily on modifying staff behaviors. Within this particular environment, it is imperative to decrease the generation of waste materials and establish a systematic categorization process.

Effective waste management reduces disposal costs, environmental risks, and energy consumption by recycling nonmedical trash while minimizing the need for replacement products [71]. However, the combination of common and infectious contaminants under inappropriate waste management practices can result in a large quantity of garbage that poses a significant risk of infection. However, it should be noted that the generation of hospital waste has the potential to pose health concerns. Unfortunately, many poor nations lack a safe and dependable infrastructure to effectively manage this waste [72]. The Organization for Economic Co-operation and Development (OECD) developed the sustainable waste treatment system, which encourages effective resource management over the whole lifespan of a resource by building on previously established policies that are directed toward waste management [73]. The G7 Toyama Environment Ministers' Meeting in 2016 implemented a strategy on resource efficiency to advance the sustainable development agenda via the promotion of effective resource utilization [73]. In their study, Kim et al. [74] proposed many strategies for the implementation of a resource-recycling economy. These strategies include the prevention of waste formation, the reuse of waste materials, the promotion of waste recycling, energy recovery, and proper disposal methods.

In order to find in-hospital activities that have the potential to minimize the development of medical waste, it is important to first analyze the sequential processes involved in waste processing. Safe disposal of hospital waste involves four steps: segregation, collection, storage, containment, transportation, and disposal [72]. The procedure of separating waste in hospitals has distinct characteristics. Temporary storage occurs at the factory site before garbage is collected and processed [72]. Hospitals must dispose of general wastes with infectious material if separation of sources and recycling efforts are not in place due to unexpected disposal expenses. The implementation of a well-defined strategy will enable hospitals to effectively manage their infectious waste, thus mitigating the additional expenses [75]. Typically, a three-colored container system is commonly employed by numerous hospitals worldwide for waste segregation. This system involves the use of yellow or occasionally red bags for infectious waste, black bags for general waste, and designated containers, referred to as safety boxes, for the disposal of sharps. Needles and other sharp objects are safely discarded in these containers and disposed of following specific instructions [72]. Sharps are one of the most dangerous types of medical waste. This is true for doctors, nurses, midwives, other healthcare workers, recyclers, and the community as a whole. Injury from sharp needles can be avoided during disposal or recovery, which is something that should be done to the greatest extent acceptable [75]. A portion of possible dangerous plastic-based clinical wastes can be stored or repurposed instead of burnt, reducing greenhouse gas emissions and toxic pollutants including dioxin and furan [76]. However, incineration has been widely considered environmentally incompatible in recent years. According to environmentalists, garbage incinerators just alter waste forms, leaving hazards unchanged. In general, incinerators burn garbage and release dangerous gases and ash into the air [75].

Wastes may be properly disinfected, buried, and disposed of using a variety of non-incineration procedures. Autoclaves are most commonly associated with sterilizing instruments for use in medicine and surgery, but they are also frequently employed as an antiseptic alternative to flame. Autoclaves are inexpensive and widely accepted by healthcare systems because of these factors [4, 10]. However, it appears that non-incineration technology can reduce pollution emissions while also being cost-effective, dependable, and able to minimize secondary contamination. Most non-incineration methods leave behind nonhazardous solid residues [75]. An autoclave is a pressurized and heated container. During this procedure, a vacuum is created using water steam, the microorganisms within the packages are killed with heat, and the surplus steam is released. Materials-handling equipment, disinfection equipment, and environmental control equipment are the three essential components of a microwave disinfection system. Pathogenic microorganisms in the waste were killed by the steam generated from the wastewater, thanks to the microwave energy [75]. Positive alternatives to incineration include the use of an autoclave or microwave. However, pathogenic, irradiation, and chemotherapy wastes are often incompatible with microwaving and autoclaving [77].

The choice of landfill disposal options for the management of HCW in developing nations is dependent upon the availability of appropriate disposal facilities. The facilities provided by open veterinary clinics are designed to ensure regulated and healthy environments for the animals being treated. The methods employed for waste disposal, such as landfill and encapsulation, exhibit a variety of approaches [75].

8.8 Criteria for waste management

Hospitals have a direct impact on the health of both their patients and their surroundings. The present environmental concerns regarding healthcare are closely linked to the nature of waste production and the strategies employed for its disposal [78]. Greening refers to actions or behaviors that can have a positive impact on environmental results. Lausten has presented concrete strategies for making the perioperative environment greener. The three primary categories that he proposes are known as the "three Rs," which are reduction, recycling, and reuse [79]. To decrease waste and promote greener practices in hospitals, consider using environmentally preferred purchasing. Sustainable consumption is described as "the selection of goods and services that minimize adverse effects on the natural and human-made environments" [79]. Various solutions have been recommended for minimizing waste generation. These include reducing or eliminating the use of disposable items, such as overshoes, except in chemotropic environments. Another suggestion is to repurpose empty items of cleaning products and antimicrobial materials as containers for the disposal of needles and syringes. Additionally, it is advised to avoid unnecessary packaging and take measures to prevent the expiration of laboratory chemicals, among other strategies [33].

Hospitals can effectively handle equipment maintenance by employing strategies such as utilizing materials with smaller packaging, opting for reusable items over disposable ones, and improving training programs on waste separation and classification [79]. Moreover, the absence of recycling initiatives contributes to a rise in the quantity of waste materials that are either burned or deposited in landfills [80, 81]. Sterilization is a technique employed to enable the repeated utilization of surgical tools. Sterilization procedures and instruments are employed to effectively decrease the risk of contracting possible infectious diseases, such as HIV or hepatitis, enabling users to securely reuse these devices. Recycling represents an additional approach to promoting environmental sustainability. Facility managers are tasked with making decisions on the selection of goods to be recycled, determining the appropriate storage location for these items before pick up, identifying the individual responsible for their removal, and allocating the necessary funds for their disposal. Recyclable items should be gathered from various sources, including stores, kitchens, laundries, pharmacies, and workshops, and thereafter transported to a central storage facility for transportation [33].

8.9 Discussion

To the extent of our current understanding, this investigation represents the initial effort to present a thorough examination of the sustainability aspects of the management of hospital waste. The utilization of a multi-method approach, encompassing both primary and secondary data, revealed an absence of comprehensive study on the topic matter. The study discovered that in the nine national and regional policy papers examined, HCW was mentioned. However, it was observed that these documents had limited coverage since only one document discussed four out of the five parts of the environmental sustainability framework [82]. In general, the publications reveal significant oversight in acknowledging the distinct characteristics of hospital waste management sustainability, irrespective of its association with policies, organizations, or persons [83–90]. The lack of sustainability in the national HCW system may mostly be attributed to issues with policy, objectives, procedures, organizational framework, and environmental monitoring [91]. Globally, the management of HCW is governed by either local regulations or dedicated policies [50]. Ali et al. and others report that hospital waste policies are relatively new in poor countries, leading to absence of waste practices [10, 21, 92]. Furthermore, the national policy objectives must be formulated in a manner that takes into account the principles of sustainability and encompasses a holistic approach. The approach should ensure that the mission and vision statements of healthcare organizations are aligned with the values deeply ingrained at the organizational level. Inconsistencies were found between the clarity of hazardous and medical waste (HMW) policy at the macro level and operational process at the meso level. Despite the requirement outlined in the policy for the separate collection and control of HMW by the hazardous waste legislation of the GCC (Articles 4 and 5), a lack of awareness and adherence to the policy and its processes was identified through document analysis and observations [82, 85–89]. Nevertheless, it is important to emphasize that HCW misconduct is not exclusive but has also been seen in several other developed and developing nations [10, 20, 21, 92, 93].

In developing countries like Brazil, Scavard and coworkers developed a medical waste disposal strategy customized to the country's culture and economic circumstances, emphasizing the social dimension. For this reason, it was necessary to launch awareness campaigns and give the locals more control over their environment. The population's standard of living went up as a result of this activity, which also increased the potential for long-term sustainability of environmental management [94]. Implementing a hospital waste strategy should engage all staff levels, including healthcare employees, to improve environmental sustainability in healthcare industries. Training should also be customized to the needs of individual departments and their workers [95, 96], as the nature of HCW differs across different environments. One of the best ways to get employees to adopt a greener attitude and lifestyle is to offer them environmental management training. However, it is not always the case that a purely passive training method would result in lasting changes in behavior [97].

8.10 Conclusion

Efficient hospital waste handling is a strategic focus for healthcare providers owing to rising concerns about communicable and infectious illnesses, climate change, and medicine resistance. Packaging materials for medical gadgets are recyclable. Environmental protection needs eco-friendly management of medical waste, mostly cremated. Healthcare establishments should also classify and separate general hospitals and medical trash. The investigation analyses the sustainability of the existing hospital waste management system from many aspects. The primary crucial points were determined by a study of the material gathered from the analysis of documents, conversations with key informants, a session with a focus group, and observations. In summary, the chapter presents a complete theoretical structure for the management of hospital waste with a focus on sustainability. It effectively addresses the complex issues associated with environmental degradation, economic responsibility, and public health concerns. It shows the significance of waste reduction at its origin, efficient segregation, inventive treatment methods, and the usage of resources. The life cycle evaluations and cost-benefit assessments conducted in the present study provide evidence of the considerable potential for reducing carbon emissions and environmental damage. Additionally, these evaluations represent the financial prudence of making long-term investments in waste management practices that prioritize environmental responsibility and the importance of healthcare organizations in driving operational sustainability, highlighting their duty to safeguard public health and effectively oversee environmental resources. The implementation of this particular method within the context of hospital waste management has the potential to make a significant contribution to worldwide endeavors aimed at reducing the environmental effects of the healthcare sector. Ultimately, it can facilitate the development of a more sustainable healthcare system.

References

[1] Jameton A, McGuire C. Toward sustainable healthcare services: Principles, challenges, and a process. International Journal of Sustainability in Higher Education. 2002;3:113–27.

[2] Ambali AR, Bakar AN, Merican FM. Environmental policy in Malaysia: Biomedical waste, strategies and issues. Journal of Administrative Science. 2013;10(1):1–7.

[3] Dace E, Stibe A, Timma L. A holistic approach to managing environmental quality by using the Kano model and social cognitive theory. Corporate Social Responsibility and Environmental Management. 2020;27(2):430–43.

[4] Duane B, Ramasubbu D, Harford S, Steinbach I, Swan J, Croasdale K, Stancliffe R. Environmental sustainability and waste within the dental practice. British Dental Journal. 2019;226(8):611–18.

[5] Farzianpour F, Hosseini SH, Hosseini S. Global Change and Human Health. In: 2nd International Congress on Energy Efficiency and Energy Related Materials Liberty Hotels Lykia, Oludeniz, 16–19, 2014 (365).

[6] Barbara Sattler RN, Kathryn Hall RN. Healthy choices: Transforming our hospitals into environmentally healthy and safe places. Online Journal of Issues in Nursing. 2007;12:2.

[7] Caniato M, Tudor T, Vaccari M. International governance structures for health-care waste management: A systematic review of scientific literature. Journal of Environmental Management. 2015 Apr 15;153:93–107.

[8] Kenny C, Priyadarshini A. Review of current healthcare waste management methods and their effect on global health. Healthcare. 2021 Mar 5;9(3):284.

[9] Chauhan A, Singh A. A hybrid multi-criteria decision-making method approach for selecting a sustainable location of healthcare waste disposal facility. Journal of Cleaner Production. 2016;139:1001–10.

[10] Ali M, Wang W, Chaudhry N, Geng Y. Hospital waste management in developing countries: A mini-review. Waste Management Research. 2017;35:581–92.

[11] Droste A. Lean thinking, banish waste and create wealth in your corporation. Journal of the Operational Research Society. 2007;48(11):105–06.

[12] Cheng YW, Sung FC, Yang Y, Lo YH, Chung YT, Li KC. Medical waste production at hospitals and associated factors. Waste Management. 2009;29(1):440–44.

[13] Abah SO, Ohimain EI. Healthcare waste management in Nigeria: A case study. Journal of Public Health and Epidemiology. 2011 Mar 31;3(3):99–110.

[14] Shojaei S, Tehrani H. Textbook of Preventive Andocial Medicine. Samt Public, Persian. 2004, 3.

[15] Fonseca LM, Domingues JP, Dima AM. Mapping the sustainable development goals relationships. Sustainability. 2020;12:3359.

[16] Bernstein S. The United Nations and the governance of sustainable development goals. Governing through Goals: Sustainable Development Goals as Governance Innovation. 2017 May;5:213–40.

[17] Ananth AP, Prashanthini V, Visvanathan C. Healthcare waste management in Asia. Waste Management (New York, N.Y.). 2010;30(1):154–61.

[18] Harris N, Pisa L, Talioaga S, Vezeau T. Hospitals going green: A holistic view of the issue and the critical role of the nurse leader. Holistic Nursing Practice. 2009;23:101–11.

[19] Alharbi NS, Qattan MY, Alhaji JH. Towards sustainable food services in hospitals: Expanding the concept of 'plate waste' to 'tray waste'. Sustainability. 2020;12:6872.

[20] Caniato M, Tudor T, Vaccari M. Understanding the perceptions, roles, and interactions of stakeholder networks managing health-care waste: A case study of the Gaza Strip. Waste Management. 2015;35:255–64.

[21] Caniato M, Tudor TL, Vaccari M. Assessment of health-care waste management in a humanitarian crisis: A case study of the Gaza Strip. Waste Management. 2016;58:386–96.

[22] Coker A, Sangodoyin A, Sridhar M, Booth C, Olomolaiye P, Hammond F. Medical waste management in Ibadan, Nigeria: Obstacles and prospects. Waste Management (New York, N.Y.). 2009;29(2):804–11.

[23] Adedigba MA, Nwhator SO, Afon A, Abegunde AA, Bamise CT. Assessment of dental waste management in a Nigerian tertiary hospital. Waste Management & Research: The Journal of the International Solid Wastes and Public Cleansing Association, ISWA. 2010;28(9):769–77.

[24] Muhwezi L, Kaweesa P, Kiberu F, Eyoku EL. Healthcare waste management in Uganda: A case study of soroti regional referral hospital. International Journal of Waste Management and Technology. 2014;2(2):1–2.

[25] Lee CC, Huffman GL. Medical waste management/incineration. Journal of Hazardous Materials. 1996 Jun 1;48(1):1–30.

[26] Al-Khatib IA, Sato C. Solid healthcare waste management status at healthcare centers in the West Bank–Palestinian Territory. Waste Management. 2009 Aug 1;29(8):2398–403.

[27] Lattanzio S, Stefanizzi P, D'ambrosio M, Cuscianna E, Riformato G, Migliore G, Tafuri S, Bianchi FP. Waste management and the perspective of a green hospital – A systematic narrative review. International Journal of Environmental Research and Public Health. 2022 Nov 28;19(23):15812.

[28] Bendjoudi Z, Taleb F, Abdelmalek F, Addou A. Healthcare waste management in Algeria and Mostaganem department. Waste Management. 2009 Apr 1;29(4):1383–87.

[29] Hossain M, Santhanam A, Norulaini N, Omar A. Clinical solid waste management practices and its impact on human health and environment: A review. Waste Management. 2011;31:754–66.

[30] Ozymy J, Jarrell ML. Predicting the US Environmental Protection Agency's criminal enforcement outcomes, 1983–2019. Environmental Politics. 2021 Nov 10;30(7):1237–58.

[31] Lee SM, Lee D. Effective medical waste management for sustainable green healthcare. International Journal of Environmental Research and Public Health. 2022 Nov 10;19(22):14820.
[32] Taberner J, Burns S, Taubman A. European Union emissions trading scheme. Australian Environment Review. 2005 Jun 1;20(6):3–5.
[33] Sandrick K. Hospitals go green. Trustees discover that being gentle to the environment can save money. Trustee: The Journal for Hospital Governing Boards. 2009 May 1;62(5):12–16.
[34] Stringer R. Medical waste and human rights (Special Rapporteur). Health Care Without Harm. 2011 6(19).
[35] Strashok C, Dale A, Herbert Y, Foon R Greening Canadian Hospitals. Community research connections discussion paper series. 2010:7.
[36] Patil AD, Shekdar AV. Health-care waste management in India. Journal of Environmental Management. 2001;63(2):211–20.
[37] Altin S, Altin A, Elevli B, Cerit O. Determination of hospital waste composition and disposal methods: A case study. Polish Journal of Environmental Studies. 2003;12(2):251–55.
[38] Reza Sabour M, Mohamedifard A, Kamalan H. A mathematical model to predict the composition and generation of hospital wastes in Iran. Waste Management. 2007;27(4):584–87.
[39] Nabizadeh R, Koolivand A, Jafari AJ, Yunesian M, Omrani G. Composition and production rate of dental solid waste and associated management practices in Hamadan, Iran. Waste Management & Research: The Journal of the International Solid Wastes and Public Cleansing Association, ISWA. 2012;30(6):619–24.
[40] Zhang H-J, Zhang Y-H, Wang Y, Yang Y-H, Zhang J, Wang Y-L, et al. Investigation of medical waste management in Gansu Province, China. Waste Management & Research: The Journal of the International Solid Wastes and Public Cleansing Association, ISWA. 2013;31(6), 655–59.
[41] Maamari O, Brandam C, Lteif R, Salameh D. Health care waste generation rates and patterns: The case of Lebanon. Waste Management (New York, N.Y.). 2015;43:550–54.
[42] Moreira AMM, Gunther WMR. Assessment of medical waste management at a primary health-care center in São Paulo, Brazil. Waste Management. 2013;33(1):162–67.
[43] Graikos A, Voudrias E, Papazachariou A, Iosifidis N, Kalpakidou M. Composition and production rate of medical waste from a small producer in Greece. Waste Management (New York, N.Y.). 2010;30(8–9):1683–89.
[44] Voudrias E, Goudakou L, Kermenidou M, Softa A. Composition and production rate of pharmaceutical and chemical waste from Xanthi General Hospital in Greece. Waste Management (New York, N.Y.). 2012;32(7):1442–52.
[45] Orloff K, Falk H. An international perspective on hazardous waste practices. International Journal of Hygiene and Environmental Health. 2003 Jan 1;206(4–5):291–302.
[46] Mato RR, Kassenga GR. A study on problems of management of medical solid wastes in Dar es Salaam and their remedial measures. Resources, Conservation and Recycling. 1997;21(1):1–6.
[47] Askarian M, Vakili M, Kabir G. Results of a hospital waste survey in private hospitals in Fars province, Iran. Waste Management. 2004;24:347–52.
[48] Adsavakulchai S. Study on waste from hospital and clinics in Phitsanulok. Online Journal of Health Allied Sciences. 2002;3:3–11.
[49] Tanaka M, Kaneko N, Takahara N, Shekdar AV. Sustainable healthcare waste management in Japan. In: Presented paper in ISWA World Congress, Rome, Italy, 2004 (pp. 17–21).
[50] Muhlich M, Scherer M, Daschner FD. Comparison of infectious waste management in European hospitals. Journal of Hospital Infection. 2003;55:260–68.
[51] Mato RR, Kaseva ME. Critical review of industrial and medical waste practices in Dar es Salaam City. Resources, Conservation and Recycling. 1999;25(3–4):271–87.
[52] Chowdhury T, Chowdhury H, Rahman MS, Hossain N, Ahmed A, Sait SM. Estimation of the healthcare waste generation during the COVID-19 pandemic in Bangladesh. Science of the Total Environment. 2022 Mar 10;811:152295.

[53] Peng Y, Wu P, Schartup AT, Zhang Y. Plastic waste release caused by COVID-19 and its fate in the global ocean. Proceedings of the National Academy of Sciences. 2021, 118(47):2111530118.

[54] Tsukiji M, Gamaralalage PJ, Pratomo IS, Onogawa K, Alverson K, Honda S, Ternald D, Dilley M, Fujioka J, Condrorini D. Waste management during the COVID-19 pandemic from response to recovery. In: United Nations Environment Programme, International Environmental Technology Centre (IETC) IGES Center Collaborating with UNDP on Environmental Technologies (CCET), 2020 (pp. 1–60).

[55] Rhee SW, Min DK. Waste Management Related to COVID-19 in the Noncontact Era: Case Study of the Republic of Korea. Health Care Waste Management and COVID 19 Pandemic: Policy, Implementation Status and Vaccine Management. 2022 Nov;13:43–70.

[56] Khan BA, Cheng L, Khan AA, Ahmed H. Healthcare waste management in Asian developing countries: A mini review. Waste Management & Research. 2019 Sep;37(9):863–75.

[57] Muduli K, Barve A. Barriers to green practices in healthcare waste sector: An Indian perspective. International Journal of Environmental Science and Development. 2012 Aug 1;3(4):393.

[58] Esteves AM, Franks D, Vanclay F. Social impact assessment: The state of the art. Impact Assessment and Project Appraisal. 2012 Mar 1;30(1):34–42.

[59] Harhay MO, Halpern SD, Harhay JS, Olliaro PL. Health care waste management: A neglected and growing public health problem worldwide. Tropical Medicine and International Health. 2009;14:1414–17.

[60] Hassan M, Ahmed S, Rahman K, Biswas T. Pattern of medical waste management: Existing scenario in Dhaka City, Bangladesh. BMC Public Health. 2008;8:36.

[61] Blenkharn JI. Standards of clinical waste management in UK Hospitals. Journal of Hospital Infection. 2006;62:300–03.

[62] Kumar R, Somrongthong R, Ahmed J. Impact of waste management training intervention on knowledge, attitude and practices of teaching hospital workers in Pakistan. Pakistan Journal of Medical Sciences. 2016;32:705–10.

[63] Borowy I. Medical waste: The dark side of healthcare. História, Ciências, Saúde-Manguinhos. 2020;27:231–51.

[64] Windfeld E, Brooks M. Medical waste management: A review. Journal of Environmental Management. 2015;163:98–108.

[65] Thakur V, Mangla S, Tiwari B. Managing healthcare waste for sustainable environmental development: A hybrid decision approach. Business Strategy and the Environment. 2021;20:357–73.

[66] Johannessen LM, Dijkman M, Bartone C, Hanrahan DC, Boyer MG, Chandra C. Health Care Waste Management Guidance Note. Washington, DC, USA: The World Bank; 2000.

[67] Seifert C, Guenther E. Prevention is better than cure-environmental management measures in hospitals. Corporate Social Responsibility and Environmental Management. 2019;26:781–90.

[68] Lee S, Lee D. Developing green healthcare activities in the total quality management framework. International Journal of Environmental Research and Public Health. 2022;19:6504.

[69] McGain F, Naylor C. Environmental sustainability in hospitals: A systematic review and research agenda. Journal of Health Services Research & Policy. 2014;19:245–52.

[70] Kwikiriza S, Stewart A, Mutahunga B, Dobson A, Wilkinson E. A whole systems approach to hospital waste management in rural Uganda. Frontiers in Public Health. 2019;7:136.

[71] Azmal M, Kalhor R, Dehcheshmeh NF, Goharinezhad S, Heidari ZA, Farzianpour F. Going toward green hospital by sustainable healthcare waste management: Segregation, treatment and safe disposal. Health. 2014 Nov 7;6(19):2632.

[72] Chaerul M, Tanaka M, Shekdar AV. A system dynamics approach for hospital waste management. Waste Management. 2008;28:442–49.

[73] Hughes N, Ekins P. The role of policy in unlocking the potential of resource efficiency investments. In: Flachenecker F, Rentschler J (eds.), Investing in Resource Efficiency: The Economics and Politics of Financing the Resource Transition (pp. 247–81). Springer International Publishing; 2018.

[74] Townend WK. Safe management of wastes from healthcare activities. Bulletin of the World Health Organization. 2001;79:171.

[75] Diaz LF, Savage GM, Eggerth LL. Alternatives for the treatment and disposal of healthcare wastes in developing countries. Waste Management. 2005;25:626–37.

[76] Gorchev HG, Ozolins G. WHO guidelines for drinking-water quality. WHO chronicle. 1984;38(3):36–45.

[77] Lee BK, Ellenbecker MJ, Moure-Ersaso R. Alternatives for treatment and disposal cost reduction of regulated medical wastes. Waste Management. 2004;24:143–51.

[78] Kaiser B, Eagan PD, Shaner H. Solutions to healthcare waste: Life-cycle thinking and "green" purchasing. Environmental Health Perspectives. 2001;109:205–07.

[79] Laustsen G. Reduce-recycle-reuse: Guidelines for promoting perioperative waste management. AORN Journal. 2007;85:717–22, 24, 26–28.

[80] Armstrong F. Do no harm?. Health Care and the Environment. Australian Nursing Journal. 2005;12:18–21.

[81] Farzianpour F, Hosseini S, Mirsepasi A, Honary H, Hosseini S, Hosseini S. Internal evaluation of the Department of Environmental Health Science and Engineering. American Journal of Applied Sciences. 2013;10:58–63.

[82] Bdour AN, Tarawneh Z, Al-Momani T, El-Mashaleh M. Analysis of hospital staff exposure risks and awareness about poor medical waste management-A case study of the Tabuk regional healthcare system-Saudi Arabia. Journal of Communicable Diseases. 2015;47:1–13.

[83] Alzahrani D Environmental legislation and the management of medical waste. In: People and the Planet 2013 Conference Proceedings; Global Cities Research Institute, RMIT University, Melbourne, Australia, 2013 (pp. 1–8).

[84] Alharbi NS, Alhaji JH, Qattan MY. Toward sustainable environmental management of healthcare waste: A holistic perspective. Sustainability. 2021 May 9;13(9):5280.

[85] Al-hadlaq A, Ali Z, Balachandran W. Bio-medical waste handling and management in Riyadh, Saudi Arabia. International Journal of Chemical and Environmental Engineering. 2013;4:343–52.

[86] Aljabre SH, Hoffmann F, Almorzog BS, Mikiling L, Alabdulatif M, Al-Quorain AA. Hospital generated waste: An assessment of the awareness of hospital staff. Journal of Family and Community Medicine. 2002;9:47–50.

[87] Habeeb TH, Ahmad S. Handling healthcare waste management and gender differences in the Madinah Primary Healthcare Centers, Kingdom of Saudi Arabia. Alaysian Journal of Society and Space. 2015;6:47–55.

[88] Omer E, Alsubaie A. Medical waste management survey in primary healthcare centers, Saudi Arabia. Majmaah Journal of Health Sciences. 2017;7:7–22.

[89] Aljabre SH. Hospital generated waste: A plan for its proper management. Journal of Family and Community Medicine. 2002;9:61–65.

[90] Alagha O, AlOmari A, Jarrah N. Medical waste management and production rate in the Eastern Province of the Kingdom of Saudi Arabia. Euro-Mediterranean Journal for Environmental Integration. 2018;3:35.

[91] Trumpp C, Endrikat J, Zopf C, Guenther ED. Conceptualization, and measurement of corporate environmental performance: A critical examination of a multidimensional construct. Journal of Business Ethics. 2015;126:185–204.

[92] Ciplak N, Kaskun S. Healthcare waste management practice in the West Black Sea Region, Turkey: A comparative analysis with the developed and developing countries. Journal of the Air & Waste Management Association. 2015;65:1387–94.

[93] Hangulu L, Akintola O. Health care waste management in community-based care: Experiences of community health workers in low resource communities in South Africa. BMC Public Health. 2017;17:448.

[94] Scavarda A, Daú GL, Scavarda LF, Korzenowski AL. A proposed healthcare supply chain management framework in the emerging economies with the sustainable lenses: The theory, the practice, and the policy. Resources, Conservation and Recycling. 2019;141:418–30.

[95] Tudor T, Noonan C, Jenkin L. Healthcare waste management: A case study from the National Health Service in Cornwall, United Kingdom. Waste Management. 2005;25:606–15.

[96] Gao Q, Shi Y, Mo D, Nie J, Yang M, Rozelle S, Sylvia S. Medical waste management in three areas of rural China. PLoS ONE. 2018;13:e0200889.

[97] McKenzie-Mohr D. Fostering sustainable behavior through community-based social marketing. American Psychologist. 2000 May;55(5):531.

Chapter 9
Measurement of social sustainability in healthcare supply chain management

Abstract: Sustainability has become an essential component of healthcare supply chain management, requiring the development of accurate measurement frameworks. The concept of ecological and social sustainability is increasingly recognized as a significant factor that poses a potential risk to overall sustainability, as human actions play a large role in its development. The chapter provides a critical analysis of the many components of sustainability within healthcare supply chains, including environmental, social, and economic aspects. Furthermore, the study delves into the significance of technology, data analytics, and industry standards in improving the measurement of sustainability. The chapter offers a thorough examination of the assessment of sustainability within healthcare supply chain management, including a complete analysis of its importance and continuous development. Social sustainability focuses on the aspects of sustainability that pertain to the well-being and development of human beings. The chapter discusses a growing practice of incorporating social practices into supply chain management. The significance of social sustainability has increased as a compelling concern, prompting supply chain managers to evaluate the societal effects of their operations on communities. It is essential in healthcare supply chains like health care because human beings are involved at every level. The chapter presents a variety of sustainability indicators that aim to assess the sustainability of supply chains and assist stakeholders in making strategic decisions. The primary objective of the study is to construct a model for measuring the performance of a healthcare supply chain sustainably. It will take into account both intangible attributes and sustainability factors, to ensure the happiness of customers and stakeholders. This chapter aims to examine social sustainability within the hospital and healthcare industry, specifically from the viewpoint of stakeholders. The sustainability index is represented as a multidimensional vector and is organized hierarchically, with different dimensions of sustainability being further broken down into sets of sub-indicators and metrics. While the current study is conceptual and lacks empirical evidence, its objective is to propose the use of this index as a means to enhance the evaluation and health coverage of public health supply chains. The promotion of sustainable practices within the healthcare sector involves a dynamic and engaged collaboration between hospitals and their many stakeholders. The study focuses on the healthcare supply chain, specifically the coordination of numerous healthcare providers to offer therapy to patients.

https://doi.org/10.1515/9783111436432-009

9.1 Introduction

The Sustainable Development Goals, established by the United Nations to establish objectives and metrics for the UN 2030 Agenda, acquired worldwide recognition [1]. However, a lack of adequate indicators as well as a strategy for monitoring sustainability was a key criticism [2]. Presently, the concept of sustainable development includes the examination of established ideas while also exploring novel approaches to assess the interconnectivity of economic and social progress, with due consideration for environmental factors [3]. The significance of sustainability concerns has increased in recent years. According to Bird and Rowlands [4], the Millennium Development Goals placed a responsibility on all nations to address the negative impacts experienced by economies and populations. It has become important for all individuals, including company owners, government entities, and significant stakeholders, to prioritize the enhancement of all aspects. Therefore, multiple notions of sustainability based on three separate viewpoints of economic, environmental, and social points of view have emerged as a result of the analyses by Wilcox et al. [5] and Dimond and Webb [6]. However, considerable efforts to apply the economic and environmental components in sustainability assessment have been abandoned, while the social factor has largely been neglected [7, 8]. Governments worldwide are faced with the potent task of ensuring the long-term sustainability of their economies, ecosystems, and, most significantly, societies [9]. If governments fail to act on such a significant problem, the difference between the services they provide and the needs of the community will continue to widen daily. The need to incorporate sustainability into all activities and organizations has transformed it from an independent option to an indispensable requirement.

The introduction of the concept of "sustainable development" has prompted companies to assume a leading role [10]. Stakeholder theory suggests that organizations aiming for social sustainability should convince stakeholders to contribute to social goals. In supply chains, including the healthcare industry, where people's lives are at stake, this is even more essential. Therefore, the aim of social progress has emerged as a fundamental goal and an apparent requirement within the healthcare sector. The major objective of healthcare services is the restoration of patients' health. However, it is equally essential to prioritize the provision of accessibility as well as educational opportunities for preventive behavior and wellness, as emphasized by Isherwood [11]. Services provided along the supply chain need to be coordinated amongst the numerous players that engage with customers and the community to keep them running [12]. The human and social aspects are integral to every phase of the healthcare process, emphasizing the need to ensure that all necessary components are provided to the relevant individuals. Specifically, stakeholders should be encouraged to promote the social sustainability of the supply chain. Sustainability can be described as the deliberate and transparent incorporation and attainment of social, environmental, and economic objectives in the systematic coordination of crucial interorganizational business procedures to enhance the long-term financial performance of both an organization and its supply chains [13]. According to Laguna [14], genuine sustainability involves the relationship between economic, environmental, and social pillars. However, these pillars are not afforded equal levels of attention [15]. Because of the significant focus placed on the economic and environmental pillars of sustainability, there has been a substantial reduction in the amount of research conducted on social sustainability [16, 17]. The concept of social sustainability relates to the human dimension of sustainability, as discussed by Huq et al. [18]. It deals with concerns about the quality of life and encourages decision-makers to think about the impact their choices may have on others [19]. When making these kinds of choices, it should be taken into account that everyone has the chance to live a full life in terms of their mental, social, spiritual, and physical health [20]. Social sustainability, in concept, involves meeting basic human requirements and monitoring the impact of our actions on stakeholders [20]. The involvement of stakeholders in the healthcare business is crucial for the support and achievement of social sustainability objectives. Stakeholders play a significant role in providing valuable insights and guidance on the effective integration of sustainable practices [21]. However, it is important to recognize that social sustainability has been given a lower priority by both academics and practitioners [22]. This suggests the need for a greater focus on a more human-centered approach [23]. The issue of social sustainability has emerged as a significant focal point for supply chain management, mostly driven by the expectations of stakeholders who promote ethical company conduct within the broader social environment [24]. According to stakeholder theory, businesses should not ignore the impact or influence of different stakeholders. The implications in the healthcare business are particularly significant due to the involvement of the human element at every level of healthcare supply chains [25, 26]. As a result, it is crucial to examine social sustainability within a healthcare supply chain.

The management of these supply chains is complicated because of differences in the economies, regulations, and standards, which have an impact on the overall performance, particularly in developing nations [27].

Therefore, it is essential to attain sustainability in supply chains to assist societies in overcoming the challenges caused by stressed health systems and enhancing their performance. The concept of sustainability emerged as individuals began to recognize their influence on the environment. People were unaware of the consequences until they emerged as global warming, climate change, and other challenges to human existence. Various interpretations of sustainability have arisen, with particular focus on practical applications and the identification of significant features associated with a sustainable society [28]. The concept of sustainability has expanded to encompass the well-being of individuals, communities, and the greater environment, as well as environmental degradation. To accomplish sustainability objectives, collaboration between the supply chain members is necessary [29]. The lack of effective coordination among many stakeholders would have a detrimental impact on the overall competitiveness of the supply chain. Developing nations face an added obstacle when financial sustainability is prioritized above other characteristics, resulting in the neglect of those issues [30]. Most studies focus on financial aims, while some explore how companies might develop links between environmental management and social elements to improve economic efficiency [31]. Social sustainability is a specific resource for evaluating the standard of living and possesses the potential to facilitate informed decision-making. The current effort aims to comprehensively and concisely illustrate the application of stakeholder theory in evaluating the social sustainability of a healthcare system. There has been a substantial increase in the recognition of social sustainability over the past decade, as a result of a substantial shift in the perspective observed in many public and private circumstances. The analysis subsequently examines the factors that drive, hinder, and facilitate the effective integration and execution of social sustainability within service supply chains. The study addresses the need for a methodical approach to social sustainability, as suggested by Mueller et al. [32]. They provide these by focusing on the viewpoints of stakeholders to construct a framework that effectively incorporates social sustainability into supply chains. To better understand the relationship between stakeholders and their roles in implementing social sustainability, the study identifies four main stakeholder groups: suppliers, employees, patients/community, and owners/government, and addresses the perspective of each group accordingly. The utilization of the supplier-input-process-output-customer (SIPOC) technique has been employed to establish connections among various stakeholders within the supply chain. This technique involves the compilation of a comprehensive list of healthcare suppliers, the inputs provided by these suppliers (such as employees), and the inputs utilized by the primary healthcare processes. Additionally, the outputs generated by these processes (including products and services) and the recipients of these outputs (namely, patients and the community) are also identified and documented.

9.2 Social sustainability development

The integration of sustainability has emerged as an important factor within the field of modern businesses. The target of sustainability is based on the principle of triple bottom line, which involves the need to achieve a harmonious equilibrium between the goals associated with financial gains, environmental preservation, and societal well-being for corporations as they expand and engage in global market competition [18, 33, 34]. Although the prevailing recommendation is for firms to achieve equilibrium across these three interdependent areas [18], a significant number of corporations have primarily concentrated on economic and environmental concerns, while generally neglecting social considerations [35–37]. The achievement of actual sustainable development occurs when an institution actively and thoroughly integrates its objectives within a strategic vision and long-term mission, including the environment, society, and the economy [18].

The concept of social sustainability relates to the human aspects of sustainability. According to Labuschagne et al. [38], the idea is combined with the consequences that the corporation has on the social systems in which it works as well as its interaction with a variety of stakeholders. Companies are facing mounting demand from stakeholders to integrate social sustainability and accountability systems throughout their supply chains [39]. Meeting stakeholder expectations is crucial for long-term corporate sustainability, and managing interactions with them can relate the organizational approach to social and ethical concerns [40]. This has made people more aware of how they treat stakeholders, which shows that the good and bad effects on people and groups are being managed well [41]. Consequently, there has been greater emphasis on social development within supply chain practices. This shift in focus has been driven by both political agendas and business objectives, with particular interest in the social pillar of sustainability. Consequently, there has been a change in attention toward social development within supply chain practices. In recent times, there has been a growing emphasis on the social aspect of sustainability.

The change is driven by both political targets and company objectives that aim to promote the growth and well-being of communities [42, 43]. To ensure social sustainability, it is essential to meet a comprehensive range of human needs while simultaneously preserving the environment and its capacity for regeneration. This involves the consideration of principles such as upholding social justice, preserving human dignity, and promoting active participation within society [44]. However, social sustainability is often considered the most ambiguous idea among the three pillars [45–47]. The concept of social sustainability has received far less attention than environmental sustainability [47]. This indicates an absence in the present research concerning how social sustainability might be included in modern decision-making procedures [37]. Furthermore, a significant portion of current studies in the field of logistics and supply chain management has focused on individual aspects of social responsibility, such as the environment, safety, and human rights, without adequately considering the potential interconnectedness between these factors and other dimensions of social responsibility [48]. There has been growing focus on the social aspect of sustainability, prompted by legal priorities and corporate goals centered around community development. The drive has led to several essays on the issue from different perspectives. Communities and commercial organizations are displaying a growing attitude toward adopting social sustainability as a means of connecting themselves with global changes [49].

9.3 Supply chain sustainability via social perspective

Social sustainability affects stakeholders, both inside and outside the company's supply chain [50]. Although the social aspect of supply chains is well-known, it is often given less attention than it deserves [50]. A supply chain is commonly perceived as a network of essential connections that interconnects organizations (stakeholders) and establishes relationship between the inputs and outputs of each company [51]. Focusing on supply chain practices can promote social sustainability acceptance and development by emphasizing the need to integrate social sustainability principles within supply chains [52]. Companies operating within supply chains are under increasing pressure from stakeholders to adopt more open policies and report on their efforts to improve social and environmental sustainability [53, 54]. Despite these challenges, many businesses' social activities do not correspond with their overall business plan [55]. Furthermore, it is commonly observed that these efforts are primarily focused on specific aspects of social sustainability, neglecting to adequately consider the interconnectedness between these aspects [24]. Numerous sustainability studies focus on social sustainability in developed countries [24, 56–58]. Moreover, there is limited academic study on the subject of social sustainability practices and their antecedents within the framework of supply chains.

Social challenges in supply chains are due to having three levels of stakeholders (who), focusing on the growing set of social concerns in the supply chain that an organization can affect (which issues), and including management skills that respond to these issues by reducing risk or increasing customer value (how) [59]. Equity, safety, well-being and health, philanthropy, integrity, and rights for people are all essential components of socially sustainable supply chains [60]. The main influences of social sustainability in the supply chain are competitive pressure, consumer demand, financial liquidity, and social concern and sustainability understanding [61, 62]. The importance of social sustainability extends beyond its direct impact on other sustainability activities. It should be recognized that any social injustices occurring within a particular level of a supply chain might have negative effects on enterprises operating throughout the whole supply chain [22]. Social sustainability may undoubtedly improve supply chain performance, as discovered by Mani et al. [63]. There is increasing demand for supply chain managers to adopt sustainable practices across all departments. The increased focus on supply chains represents an essential step toward the wider acceptance and advancement of sustainability [55]. According to a study [64], supply chain management involves integrating company procedures from consumers to suppliers to produce goods, services, and data that provide value for consumers and stakeholders. This indicates that supply networks are comprised of a large number of enterprises and that the continuity of the chain is contingent upon the continuity of each of these businesses. To ensure sustainability in supply chain management, it is necessary to include a greater variety of factors and expand the analysis to encompass the whole supply chain, rather than only focusing on particular groups related to a certain industry. According to Tate et al. [54], a majority of corporations are currently obligated to provide reports on specific dimensions of corporate social sustainability. Although many organizations prioritize social concerns in supply chains, their efforts often slip outside the triple bottom line due to disconnects between social programs and corporate approaches [55]. Stakeholders are placing increasing pressure on managers of supply chains to adopt social criteria.

To create socially ethical supply chains, organizations have to analyze their influence on stakeholders. For example, an employee will give a corporation their skills and knowledge in exchange for monetary compensation. The organization can have a positive influence on its employees and their families via the provision of quality education, childcare, and healthcare services. Further, the organization engages in financial transactions with suppliers in return for products and services. However, it can also influence the ethical principles of suppliers by establishing criteria for social responsibility and opting for suppliers that adhere to the company's employment, health, and safety protocols [65]. As a result, providers would fight to meet the standards set by the company. However, the firm is subject to the influence of the community, which introduces rules and regulations that must be complied with. In addition, the corporation distributes dividends to its shareholders, and the owners demand financial transparency and responsibility from the organization. However, effectively addressing the concerns and priorities of all stakeholders involved in the business is a significant problem in supply chain management [66]. Furthermore, it is important to note that there is an inadequate amount of research that has examined the practices and factors influencing social sustainability in supply chains [8, 16, 17]. This emphasizes the need for further investigation into this topic, especially within the service supply chains [67].

9.4 Social sustainability in the healthcare supply chain

In recent years, there has been significant growth in the healthcare supply chain due to a greater amount of increasing impact on patients, and the need for more efficient and effective delivery of health services [68, 69]. While most healthcare companies have acknowledged the need to implement supply chain management strategies, it is insufficient to just adopt an industrial supply chain strategy. The presence of different parties and the complex nature of diverse technologies contribute to the distinct characteristics of the supply chain [69]. In general, a supply chain encompasses three distinct travels, namely the physical flow, the informational flow, and the financial flow. The primary determinant of supply chain design is often the physical movement of items, together with the corresponding limitations and possibilities. The structure of the healthcare supply chain is defined by the crucial involvement of financial and informational flows [70]. The healthcare industry has transformed in the past decade due to increased competition, patient influence, and the need for efficient and effective service delivery [68, 69].

The healthcare industry is showing growth due to the influence of demographic and economic factors [71]. The existing sustainability model of healthcare is being challenged by the emergence of new situations involving environmentally produced risks and a changing disease environment [72]. Because of the increased emphasis on health, safety, and quality, questions have been raised about the practical application of sustainability ideas in healthcare services and whether or not such concepts can be compatible with healthcare findings [73]. The concept of social sustainability in the healthcare sector has developed as an integral component of maintaining healthy communities. According to Oslen [12], the sustainability of a healthcare service is dependent upon its operation within an organizational structure that is adequately supported by resources and activities focused on addressing both individual and public health requirements. Furthermore, the human factor must play a crucial role in all aspects of the healthcare system [74]. Social sustainability encompasses the human aspect of sustainability, and the promotion of sustainable practices involves ongoing engagement between healthcare institutions and their many stakeholders. This indicates that stakeholders, including workers and suppliers, play an important role in supporting social sustainability objectives by promoting sustainability and offering knowledge on the efficient incorporation of sustainable healthcare methods [22].

Figure 9.1 illustrates the conceptual framework utilized for the assessment of social sustainability within healthcare supply chains. However, studies on the healthcare industry are dispersed from a supply chain administration perspective. While most healthcare professionals have accepted the need to implement supply chain management strategies, it is not sustainable for the healthcare industry to comply with an industrial supply chain model. The different characteristics of the industry contribute to this occurrence, including the complex nature of the technology employed and the presence of several stakeholders [69]. In general, a supply chain encompasses three different movements, namely the movement of physical products, the transmission of information, and the transfer of financial resources. The primary factors influencing supply chain design are often the needs, restrictions, and possibilities related to the physical movement of products. The healthcare industry exhibits distinct characteristics due to the significant influence of financial and information flows on supply chain design decisions [70].

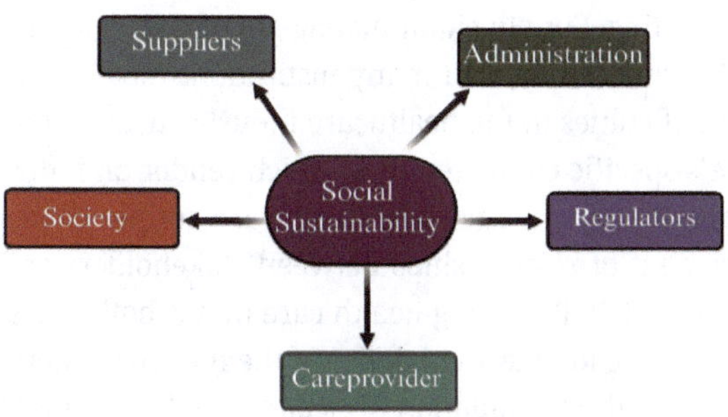

Figure 9.1: Stakeholders in the healthcare supply chain for measuring social sustainability.

Healthcare chain procedures include the participation of several stakeholders. Hence, the utilization of supply chain management strategies within a healthcare context is associated with organizational elements, including the establishment of connections, allocation of authority and duties, and structuring of interaction activities [69]. The field of healthcare encompasses the management of supply chains, which consist of both internal and external components. The internal chain involves various entities such as patient care units, hospital storage facilities, and patients themselves. On the other hand, the external chain involves vendors, manufacturers, and distributors. These concepts have been discussed in previous studies [75, 76]. It encompasses the integration of business activities and operations to provide a seamless and uninterrupted flow of goods and services within the healthcare sector [75, 77]. The concept of social sustainability in healthcare has emerged as an essential element of a well-functioning social infrastructure. Organizational systems that are maintained with appropriate resources and activities to address individual and public health demands are crucial to the long-term viability of health services. In addition, the human aspect is present throughout the whole of healthcare delivery [26]. Hospitals must actively engage with stakeholders to adopt social sustainability strategies. By pushing the transformation toward sustainability and offering insights into the effective integration of sustainable healthcare practices, stakeholders like workers and suppliers play a vital role in supporting social sustainability goals [22].

The healthcare industry has addressed patient health priorities and quality problems [78]. According to Al Hammadi and Hussain [78, 79], healthcare services may be considered sustainable when they are managed by an organizational structure that is supported by an adequate number of resources and activities to fulfill both individual and public health requirements. Supply chains in healthcare provide a seamless movement of resources, information, and services. The healthcare sector is different because of how complicated the technologies are, how many different people are involved in the supply chain, and how the flow of money and information is important to how the supply chain is designed [70]. Therefore, organizational factors such as relationship development, task assignment, and interface process management are relevant to the implementation of supply chain methods in the healthcare setting. Supply chain management strategies are becoming more important in the healthcare industry, and many institutions have begun to use them. Supply chain management difficulties in the healthcare industry need a multidisciplinary approach due to the sector's specific characteristics, which render an industrial strategy inappropriate. Accordingly, the management of healthcare supply chains involves elements linked to the management of relationships between stakeholders and the distribution of resources along the chain [69]. Providing health care that is both accessible and affordable is essential to ensuring the long-term viability of the industry. Therefore, people's wants and goals should go into the formulation of healthcare development policies that improve accessibility [80]. The capacity of healthcare institutions to enhance the quality of healthcare should be reflected in their social sustainability [81]. Due to the healthcare industry's particular characteristics, the supply chain management literature has been slow to focus on this kind of supply chain. More attention has to be paid to supply chain management concerns in the healthcare industry since it is a service supply chain. Efforts have been made to investigate how sustainability influences medical treatment. In recent years, researchers and policymakers have shifted their attention to healthcare innovations and the mitigation of healthcare's carbon footprint [82, 83]. The concepts of waste management and pollution are important to the consideration of environmental sustainability in the healthcare sector [84]. Furthermore, as healthcare organizations grow older, the long-term viability of care for both patients and staff becomes an essential foundation for cost-effectively providing high-quality treatment. Because of this, a coordinated effort is required for these groups to improve health outcomes [85]. Healthcare businesses may use an objective paradigm to analyze their environmental sustainability and compare outcomes across time to establish indicators of comparison [86]. Recognized as well as the work done to qualitatively emphasize the driving reasons for implementing sustainability in healthcare and to provide solutions to constantly strengthen sustainability compliance in healthcare [87], sustainability in healthcare supply chains has mostly addressed environmental stability and organizational stability, the latter of which is primarily concerned with financial difficulties.

9.5 Sustainability dimension and measurement

Many studies have been conducted to explore the different aspects of sustainability at the industrial, national, and global levels. Numerical models have been used in prior studies to illustrate the connection between supply networks and environmental sustainability [88]. The evaluation of green supply chains and their associated strategic concerns has been given equal consideration [89]. The current definitions of supply networks lack a comprehensive understanding of sustainability and fail to address the difficulties associated with assessing supply chain sustainability [90–92]. The major focus in analyzing the sustainability of supply chains has been on environmental concerns. This evaluation is often conducted using methodologies such as life cycle assessment (LCA) [93–96]. The concept of sustainability includes three dimensions components, often referred to as the triple bottom line: economic, social, and environmental.

The present studies only address the triple bottom line approach to sustainability, which seeks to expand beyond the conventional bottom line of a company to profits made socially, ecologically, and economically. Furthermore, the majority of the current information in this field examines supply networks in developed countries, and research on sustainable supply chains in low- and middle-income countries (LMICs) is limited due to the difficulty of collaboration between developing and industrialized countries. In LMICs, the shifting and unpredictable nature of business settings, combined with the absence of effective institutional frameworks, restrict the capacity of supply chains to acquire knowledge, promote innovation, and therefore impair the achievement of sustainability goals [97].

Several sustainability indicators have been proposed to assess the sustainability of supply networks, assisting stakeholders in formulating strategic choices. Optimization methods are used to optimize sustainability by using a linear benefit function that incorporates three components, namely the economic, social, and environmental elements of sustainability [98]. The measurement of sustainability performance in the supply chain may be facilitated by the use of sustainability metrics. The measures may be classified according to each component within the sustainability dimension. A metric may be derived from the literature analysis to assess the efficacy of an organization's sustainable supply chain. Table 9.1 presents the suggested metrics [99].

A triad-based analytical methodology was presented to examine the supply chain's sustainability [100]. However, sellers in the medical device business were evaluated across three criteria of sustainability using a fuzzy inference approach [101]. Several studies used a probabilistic model to evaluate the long-term sustainability of the supply chain. Additionally, a sustainability metric was developed that included several aspects of complex technological and economic efficiency [102, 103]. Certain organizations attempt to use the balanced scorecard methodology to assess the sustainable effects of their economic, social, and environmental endeavors. This technique enables managers to consider several performance indicators, rather than relying on just a single metric [104].

Table 9.1: The sustainability metrics.

S. no.	Dimension	Sustainability metrics
1.	Economical	Value added, output production, turnover, inputs utilized, profitability, competitiveness, market share, product durability, R&D efforts, company viability, financial stability, and benefit collaboration.
2.	Social	Employment, pay, productivity or labor-intensiveness, danger, noise, welfare, education, resources, personnel rotation rate, health and safety, human rights, legality assurance, and the security of raw materials and food.
3.	Environmental	Natural resources, solid waste management, air pollution and water pollution, transportation methods and distances, global warming, biodiversity conservation, acid deposition, landscape preservation, proper waste disposal methods, the ability to recycle products, restoration of forests and ecosystems, efforts to minimize deforestation, the conservation of primary forests, the use of genetically modified organisms (GMOs), the use of chemicals, pest control measures, and fertilizer usage.

Sustainability metric is implemented within the public health supply chains. The sustainability of a public health system relies on the collective efforts and contributions of its diverse participants operating at various levels. Each member plays a crucial role in enhancing the health outcomes of the supply chain. The Supply Chain Sustainability Index evaluates several aspects of sustainability throughout the supply chain, not only the triple bottom down method, using quantitative sub-indicators. The index, similar to previous activities, assesses the economic, social, and environmental impacts of supply networks. In addition, we have put out a proposal to examine the essential aspects of supply chains in the field of public health. The measurement of stakeholder outcomes includes the cooperative efforts of supply chains. The sustainability of health supply chains is dependent upon the individual contributions of each stakeholder and their respective roles in enhancing health outcomes. Since the health supply chain is responsible for improving health and saving lives, health results will have an impact on sustainability. Ensuring the quality of products and services as well as the quality of processes is a crucial need for the successful functioning of supply chains. Therefore, it is necessary to evaluate supply chains in terms of the several accreditations and certifications that are needed to ensure quality is maintained across different positions [105].

Governments, nongovernmental organizations, businesses, media outlets, research institutions, and any other parties that are engaged in decision-making on the long-term viability of an industry may be among the entities that take part in the process of assessing the supply chain's performance in terms of its contribution to environmental and social sustainability. The sustainability performance of the relevant parties may be examined by using industry-specific measures. Various approaches can be employed for conducting measurements, including LCA, analytical hierarchy process, multicriteria decision-making, nondominated sorting genetic algorithm, TOPSIS, behavior-based, supply chain operations reference, data envelopment analysis, structural equation model, fuzzy-based, and multiobjective optimization. After measuring the sustainability performance, the relevant stakeholders may implement a monitoring and standard-setting process to regulate the implementation of sustainable practices within an industry. The system comprises several components, such as the use of the complaint method, implementation of trademark protection measures, use of labels, posting of assessment results, regular involvement of third-party assessments, and the implementation of third-party certification and accreditation procedures. Figure 9.2 demonstrates a performance-measuring framework for the supply chain based on these procedures.

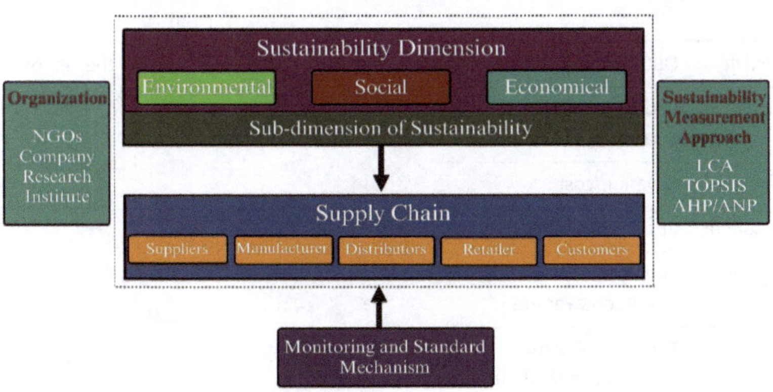

Figure 9.2: A framework for measuring sustainability in supply chains.

9.5.1 Economic output

The economic dimension of the supply chain is a complex concept that includes not only the profits and financial well-being of the individual chain members but also the larger economic benefits developed by the countries, regions, and local communities in which these members operate [106]. First, it evaluates the monetary performance, which reflects a company's ability to generate revenue and conduct its operations efficiently. This includes sales strategies, production costs, and market competition. Additionally, the examination of financial health is a concentrated evaluation of the firm's potential to maintain long-term viability and stability. This assessment includes the analysis of profitability, solvency, and investment capacity, all of which play a crucial role in maintaining development and enhancing competitiveness. Efficiency is a crucial factor in assessing the most effective distribution and exploitation of resources throughout the supply chain, with the primary objective of minimizing inefficiencies and maximizing production, all the while fostering sustainable practices. The comprehensive focus on the economic dimension acknowledges the interdependence between supply chain activities and the economic welfare of different stakeholders. It underscores the significance of responsible and efficient practices that not only enhance profitability but also make a positive contribution to the economic well-being of host regions and communities. This dimension, which is frequently divided into three essential characteristics, offers a comprehensive perspective on the economic viability and the effect of the operations that occur inside the supply chain. The economic component, as given in Table 9.2, is organized into three different categories.

Table 9.2: Economic output indicators.

Categories	Metric	Measurement		Economic subindex (Ei)
Monetary performance	Productivity	Output/input economic	Economic sustainability of supply chain function	Si will be the GM of all the six economic metrics
	Intensity measurement	Orders fulfilled as a percentage of total		
Financial health	Profitability	Revenue/cost		
	Value added	Price/profit ratio		
Efficiency	Supply days	Supply and demand every month on average		
	Product service delivery	The ratio of actual deliveries to total output and purchases		

9.5.1.1 Monetary performance

The concept of monetary performance pertains to the capacity of a corporation to produce money and effectively execute its operational activities. The evaluation of financial performance within the supply chain is a crucial metric that assesses the effectiveness of a company in producing money and managing its daily activities. This dimension comprises several crucial aspects. Firstly, the process entails evaluating the organization's capability to create income, which is contingent upon its proficiency in marketing and selling items or services within its target market. Efficient sales methods, competitive pricing, and efficient distribution channels are all significant factors that contribute to the success of a business. Additionally, the concept of operational efficiency is of significance, as it entails a concerted effort to minimize expenses associated with production, decrease the time required for the completion of tasks, and enhance the management of inventories. Efficient operations not only result in financial savings but also improve the organization's capacity to rapidly fulfill client expectations. Finally, the market position of a corporation has significant importance since it may provide several benefits, including enhanced revenue and better negotiating power within the supply chain, resulting from a considerable market share.

9.5.1.2 Financial health

It refers to the overall state of well-being and prolonged sustainability of a corporation about its financial resources. The economic aspect of financial well-being includes more than just the immediate generating of revenues and profit but also extends to the long-term sustainability and stability of a company. The concept comprises a variety of essential aspects that together contribute to the firm's ability to withstand and recover from challenges and interruptions. Profitability is a crucial aspect of financial well-being, as it signifies the organization's capacity to make earnings from its expenses [106]. The ability to generate sustainable profits allows a company to allocate resources toward reinvesting in its operations, facilitating expansion, and maintaining resilience during periods of economic decline. The evaluation of solvency is an important factor in determining the ability of a company to fulfill its financial commitments in the long run. This includes the assessment of debt levels, creditworthiness, and the ability of the company to meet its financial responsibilities. In addition, the evaluation of financial health includes the examination of the organization's capacity to strategically allocate resources toward critical areas such as research and development, innovation, and capital enhancements. These investments have the potential to improve the firm's competitive advantage and secure its sustainability in the long term within the supply chain ecosystem.

9.5.1.3 Efficiency

Efficiency is defined as the strategic allocation and effective use of resources to meet the needs of stakeholders while minimizing waste and inefficiency [107]. Efficiency plays a fundamental role within the economic aspect of supply chain management, as it applies to the strategic allocation and effective usage of resources to meet the needs of stakeholders, while avoiding wastage and inefficiencies. Efficiency includes several aspects of the supply chain, spanning the stages of manufacturing through the processes of distribution. The use of efficient resource allocation strategies guarantees the optimal utilization of inputs, including labor, materials, and energy, resulting in cost savings and enhanced production. When it comes to waste reduction, strategies such as lean manufacturing and just-in-time inventory management are often used to minimize surplus inventory and optimize operational efficiency. Furthermore, the enhancement of efficiency serves to advance sustainability objectives by mitigating the environmental impact associated with supply chain activities [106]. In general, the optimization of a supply chain not only improves economic efficiency but also supports the principles of environmental and social sustainability, thereby harmonizing with the larger goals of responsible corporate conduct.

9.5.2 Social output

The social output refers to the outcomes or results that arise from social interactions or activities within a particular environment. The social aspect of sustainability refers to human resources throughout the supply chain. The concept comprises a broad range of factors and efforts aimed at promoting justice, equality, and good social outcomes for the workforce, the communities in which supply chain activities take place, and the areas affected by these operations. Primarily, enhancing sustainability within the social dimension involves encouraging and maintaining business practices that promote the well-being of all stakeholders engaged in the process. This includes assuring fair salaries that represent the value of the employees' efforts, maintaining safe and healthy working conditions that safeguard both physical and emotional well-being, and providing opportunities for skill development, growth, and career promotion for employees.

Enhancing sustainability with the social dimension entails the establishment and perpetuation of corporate strategies that are equitable and advantageous to the workforce, communities, and regions affected by the supply chain. Table 9.3 displays the different sub-indicators of social output. They may be classified into two broad categories. A socially sustainable supply chain actively connects with and contributes to the communities in which it operates. The involvement could occur via the provision of assistance to local initiatives, allocation of resources toward community development projects, and cultivation of constructive rapport with inhabitants in the area. It makes an effort to be a responsible corporate citizen by honoring the cultural and social norms of the neighborhood and working toward leaving a good impact on the world. Furthermore, the social aspect covers the implementation of ethical sourcing methods, thereby extending the dedication to justice and ethical treatment to both suppliers and their employees. The maintenance of social sustainability within the supply chain necessitates the establishment of mechanisms to ensure that suppliers comply with ethical labor practices and environmental requirements. The social dimension of sustainability is a comprehensive approach that acknowledges the significant impact of supply chain activities on human lives and communities.

Table 9.3: Social output indicators.

Categories	Metric	Measurement		Social subindex (Si)
Workplace (internal)	Wage ratio	Top or bottom wages	Social sustainability of supply chain function	Si will be the GM of all the six social metrics
	Health delivery	Health expenditure/total expenditure		
	Gender equality	Female or male personnel		
	Capacity develops	Training or development budget/total expenditure		
Community (external)	Product recalls	Total sale/notification		
	Employment opportunities	Total jobs Offered/unemployment		

9.5.2.1 Workplace

The term "workplace" refers to the collective group of persons who are integral to the functioning of a business, including its employees at various levels – ranging from entry-level positions to managerial roles. The role of human resources in the workplace is crucial for the effective functioning of a supply chain since they bear the responsibility of strategizing, implementing, and overseeing a range of operational activities. The prioritization of empowering and safeguarding the well-being of these workers is of utmost importance to maintain a flourishing and productive workforce. To begin with, the concept of workplace empowerment involves the provision of workers with the necessary resources, expertise, and authorization to make choices that have a direct influence on their jobs. The variety of potential actions includes providing workers on the manufacturing floor with more autonomy as well as engaging employees in decision-making processes that have an impact on the efficiency of the supply chain. The use of empowerment strategies inside an organization cultivates a heightened perception of ownership, responsibility, and motivation among employees, resulting in a favorable impact on the overall productivity and resilience of the supply chain. Additionally, the welfare of workers is closely related to the efficient operation of the supply chain. The overall well-being, including both physical and mental health, of employees, is of utmost importance since it significantly influences their capacity to function at an optimum level. It is essential to acknowledge the importance of worker empowerment and well-being within the realm of supply chain management. A staff that is motivated, engaged, and in good health has the potential to improve operational efficiency, decrease employee turnover, and bolster the overall resilience of the supply chain. Organizations have a responsibility to allocate resources toward strategies and activities that emphasize internal human resources, recognizing their crucial role in achieving supply chain success.

9.5.2.2 Community

The term "community" includes people outside the supply chain who are impacted in some way, either directly or indirectly. The community consists of nearby inhabitants and companies located near the supply chain's physical sites, such as factories, warehouses, or distribution centers. The traffic, noise, and pollution created by supply chain activities as well as the job possibilities they provide, have direct effects on these people. For a healthy relationship and to avoid conflicts, it is critical that these communities are not negatively impacted by the supply chain's presence. Consumers who depend on supply chain-delivered goods and services are also considered part of the greater community. The quality, affordability, and availability of these products can significantly influence the well-being and satisfaction of consumers. Also, the environmental practices of the supply chain, such as efforts to be sustainable and reduce waste, can resonate with consumers who care about the environment.

This can affect their buying decisions and loyalty to a brand. Regulatory bodies, government organizations, and advocacy groups are other parts of the community. Their involvement is essential in defining the legal and ethical frameworks that govern the supply chain. Compliance with legislation, as well as alignment with community values and expectations, are essential for avoiding legal problems and preserving a favorable public image. Additionally, supply chain suppliers and coworkers are members of the extended community. The supply chain's performance and its ties with the community at large are all susceptible to the reputation of its participants. Sustainable and responsible supply chain management focuses on identifying stakeholders beyond the direct scope of the supply chain and responding to their needs. In addition to being morally required, investing in strong connections, limiting negative externalities, and making beneficial contributions to the greater community may improve the supply chain's future sustainability and profitability.

9.5.3 Environmental output

Improving environmental sustainability requires decreasing the supply chain's environmental impact. Most research has been done on one of the three elements of supply chain management, specifically its impact on the environment [106]. Enhancing environmental sustainability within the supply chain is an important aim, with the environmental aspect of supply chain management being the subject of numerous studies and analyses. This dimension comprises a variety of measures that are designed to decrease the environmental impact of supply networks. The reduction of carbon emissions, with a focus on transportation optimization, the adoption of low-emission cars, and the establishment of aggressive carbon reduction objectives is the basis of this initiative. Equally important is reducing waste, which leads to the acceptance of circular economy ideas, sustainable packing solutions, and less waste being generated overall. Collaboration and transparency with suppliers play an essential part in this process, as audits and collaborative activities contribute to ensuring compliance with environmental requirements and promoting advances in sustainability. Additionally, the implementation of transparency initiatives, such as comprehensive sustainability disclosures, promotes trust and responsibility among stakeholders, underscoring the utmost significance of addressing the environmental dimension of supply chain management in the current corporate climate. Enhancing environmental sustainability within supply chains is of great significance. This multidimensional aspect includes the decrease of ecological footprint via the mitigation of environmental consequences, including carbon emissions and waste reduction. Companies are engaging in the optimization of transportation routes, the use of low-emission cars, and the acceptance of circular economy ideas to decrease waste. The use of collaborative initiatives with suppliers, such as audits and transparency initiatives, continues to increase to ensure compliance with environmental regulations and ethical principles. Transparency is especially important in a world where people want complete information about efforts to be sustainable. The environmental aspect of supply chain management continues to be at the center of sustainable business practices, which reflects the urgent need to solve environmental concerns in today's globalized economy. As may be seen in Table 9.4, they are divided into multiple categories.

Table 9.4: Environmental output indicators.

Categories	**Metric**	**Measurement**		**Environmental subindex (ENi)**
Sources	Emission	Emission per unit/total unit	Environmental sustainability of supply chain function	ENi will be the GM of all the four social metrics
	Energy	Renewable energy/total energy used		
Disposal	Recycling	Recyclable weight this month/recyclable weight last month		
	Waste diversion	Waste diverted/total amount of waste		

9.5.3.1 Sources

Sources refer to the emissions resulting from localized effects and the energy consumption associated with supply chain functions. When considering the environmental effect of supply chain management, two major categories of emissions must be taken into account: those resulting from local impacts and those related to energy use throughout the different supply chain functions. Local impacts refer to the direct effects that a supply chain's activities have on the nearby area in which it operates. These effects include the pollution of air and water, disturbance of habitats, and the depletion of resources. These impacts may have serious ecological and socioeconomic effects, requiring rigorous monitoring and mitigation.

Consumption of energy is a significant contributor to emissions from supply chain activities. The carbon footprint of a supply chain is significantly influenced by the energy consumption associated with transportation, production, storage, and distribution activities. There is an increasing movement among businesses to prioritize energy efficiency initiatives, which include the adoption of renewable energy sources and the optimization of operational processes to mitigate energy-related emissions. To attain significant advancements in environmental sustainability, it is essential to effectively handle both of these types of emissions. This includes the adoption of strategies aimed at minimizing the adverse effects on the local environment, including the use of environmentally sustainable manufacturing techniques and the responsible purchase of materials. Additionally, efforts are made to optimize energy consumption throughout the whole supply chain to reduce carbon emissions. By recognizing and minimizing these two factors of environmental influence, corporations could work toward a supply chain that is more ecologically sustainable and socially responsible, yielding positive outcomes for both the natural environment and the advancement of humanity.

9.5.3.2 Disposal

The concept of disposal, as it relates to supply chain management, is a complete strategy aimed at effectively managing recycling and waste diversion initiatives throughout all stages of the supply chain. The process encompasses the strategic administration of materials and products that have reached the final stage of their lifespan to reduce their adverse impact on the environment. This involves not only the responsible disposal of waste products but also the implementation of sustainable measures such as recycling, reusing, or repurposing goods, wherever possible. Effective disposal strategies include waste reduction at the source, efficient collection and sifting processes, and the promotion of circular economy principles, in which products and materials are recycled or refurbished to extend their useful lifespan. By adopting these techniques across various supply chain functions, organizations have the potential to not only reduce the burden on landfills but also decrease the use of fresh resources and reduce the overall environmental impact of their operations, thereby promoting their sustainability objectives.

Disposal, in supply chain management, includes recycling and waste diversion measures that reduce environmental effects. This includes a thorough analysis of resource use, waste creation, and disposal systems at each level of the supply chain. Through careful data collection, analysis, and models, practitioners can find ways to use materials more efficiently, produce less waste, and improve recycling. The use of advanced methodologies such as life cycle assessment and carbon footprint analysis enable a systematic examination of the environmental impact related to supply chain operations. This scientific approach facilitates the quantification of the ecological footprint, hence assisting in the formulation and implementation of more sustainable policies. Furthermore, it is essential to include the concepts of circular economy, which prioritize the reuse and recycling of materials, resulting in a decreased dependence on new resources and a reduction in environmental impact. The scientific examination of waste management in supply chains highlights the need to implement environmentally friendly approaches that effectively reduce waste and promote a globally sustainable and environmentally conscious economy.

9.6 Challenges in social sustainability

The promotion of social sustainability, which aims to create more just and equal societies, faces several challenges. Many studies are being conducted to determine whether or not social problems can be solved within supply chains [107]. Some of these efforts are bringing attention to the challenges that are associated with sustainability in general as well as the sustainability of society, in particular. Limitations in economic resources and resource allocation restrict organizational initiatives to enhance community health [108–110]. Sustainability practices are also influenced by constraints arising from implemented laws, institutional rules, and guidelines. The presence of these limits has the potential to impact the moral judgment of workers, as they may encounter a personal problem whereby their own ideals conflict with the need to comply with corporate regulations [110, 111]. The presence of legislation alone may not be sufficient to ensure commitment to sustainability. In cases where there is a lack of government backing or inadequate law enforcement, the level of commitment to sustainability is predicted to be low [112, 113].

The study suggests accountability and social reporting as significant barriers to achieving social sustainability. Belal [114] conducted an analysis that indicated that social audit reports tend to place more emphasis on environmental issues as opposed to social concerns. According to the research conducted by Huq et al. [18], the presence of immoral behavior shown by certain suppliers, along with an organization's lack of awareness or disregard for these violations, contributes to the issue at hand. Additionally, Jiang [113] identified audit fraud as a significant factor contributing to such behaviors. In addition, the lack of effective coordination and communication among different stakeholders within a healthcare supply chain has a detrimental impact on the implementation of social sustainability [114]. Malik et al. [115], has emphasized that healthcare facilities that engage in infrequent or unstable ordering practices can face challenges related to cash flow for suppliers. Consequently, suppliers may prioritize sustaining their own business over socially sustainable actions that need long-term investments. Another problem raised by current studies is insufficient stakeholder participation and companies' unwillingness to incorporate input [114]. Comparably, the study conducted by Berman et al. [116] showed that companies tend to engage their stakeholders only in matters associated with organizational profitability while neglecting to include them in areas of social sustainability. The presence of organizational culture sometimes poses a significant challenge to the achievement of sustainability in healthcare practices, particularly when there is a disagreement between the organization's rules regarding behavior and the prevailing local culture [109,110]. A poor sustainability culture will make it more difficult to monitor activities, [18]. Furthermore, it is important to note that the organizational structure, business processes, and systems within a company can potentially hinder sustainability effort. This is particularly true if these elements prioritize the expansion of capacity and professionalism, rather than promoting long-term relationships with community partners who share a mutual responsibility to enhance the well-being of local citizens. The continuation of socioeconomic inequalities remains an incredible barrier; disparities in income, education, healthcare, and access to necessities hinder the accomplishment of social equality. The relationship between urbanization and globalization frequently leads to the degradation of established social frameworks and the displacement of people with limited opportunities, hence promoting social fragmentation and exclusion. There are still problems with labor rights, safety at the workplace, and fair salaries, and many people around the world are subjected to exploitative working conditions in a variety of industries. Furthermore, the growing digital gap, which is characterized by unequal access to information technology and the internet, worsens inequality. Moreover, climate change-induced migration, conflicts, and the erosion of cultural diversity pose significant threats to social stability and cohesion. Moreover, the absence of comprehensive and inclusive social policies, coupled with inadequate healthcare and education systems, keeps hindering progress toward social sustainability. Governments, corporations, civil society, and international organizations must work together to implement laws and practices that put social justice, inclusion, and the welfare of all people first to address these issues. Management and stakeholders may have a misunderstanding of sustainability's genuine meaning and purpose as well as a lack of understanding of their roles in advancing sustainability. The absence of support at the workplace and a lack of training and instruction on how to operate in a sustainable manner within businesses were cited as potential barriers to sustainability practices. Furthermore, it has been observed that management exhibits a deficiency in fundamental understanding of corporate social responsibility [113]. Consequently, it is important to implement initiatives aimed at enhancing consciousness regarding the significance of sustainability in wider terms, with a specific emphasis on social sustainability.

9.7 Conclusion

In conclusion, measuring social sustainability in the healthcare supply chain is an important and developing topic that has significant promise for furthering the triple bottom line of sustainability goals. This chapter has explored various dimensions of social sustainability, including worker well-being, ethical sourcing, community engagement, and equitable access to healthcare services – the assessment and quantification of social implications within healthcare supply chains to improve decision-making, promote transparency, and drive enhancements in the overall quality of care and ethical behavior. Furthermore, this chapter has explained the complexities and challenges related to the measurement of social sustainability, including issues such as the accessibility of data, involvement of stakeholders, and the formulation of appropriate metrics. Even with these problems, it is clear that a comprehensive approach to measuring social sustainability is necessary to make sure that healthcare supply chains not only provide services that are efficient and cost-effective but also place the well-being of people and communities in focus. Further research and stakeholder collaboration will be required to improve measurement methodologies, establish standards for the healthcare sector, and promote the incorporation of social sustainability principles into healthcare supply chain management practices in the future. This chapter emphasizes the importance of incorporating social sustainability into healthcare supply chain management, relating it with the overarching objectives of fostering a healthier, more equitable, and morally responsible healthcare ecosystem. This study is founded on the observations of senior managers in the healthcare industry who can precisely determine their stakeholders' perceptions. The objective of the study is to examine the social sustainability of a supply chain when considering elements that follow an exponential distribution. These factors can be utilized for measuring the sustainability of supply chains across various fields. The managers in issue engage with their clients regularly, enabling them to provide accurate accounts of their evaluations as well as those of their stakeholders regarding social sustainability. The study aims to investigate the barriers or factors that hinder the adoption and implementation of social sustainability practices within healthcare supply chains, taking into account the limited existing literature on this topic. The future sustainability of supply chains will be influenced by various additional problems that will require attention and resolution. Depending on the availability of data, future researchers may utilize the provided conceptual framework and approach to find more factors and develop an expanded repetition of the existing index. The preservation of the underlying structure of measurement allows for the attainment of such an objective. Even with these problems, the suggested index will help people think more about how supply chains in public health can be made more sustainable. Developing a comprehensive sustainability metric presents inherent challenges. However, the existing index has the potential to serve as a valuable instrument for individuals to assess their health status and make informed decisions to enhance their overall well-being and access to healthcare services. The study will serve as a fundamental basis for future research endeavors, facilitating the implementation of actual experimentation on the index within the context of public health supply chains.

References

[1] Monteiro NBR, da Silva EA, Moita Neto JM. Sustainable development goals in mining. Journal of Cleaner Production. 2019 Aug 10;228:509–20.

[2] Kapera I. Sustainable tourism development efforts by local governments in Poland. Sustainable Cities and Society. 2018;40(Apr):581–88.

[3] Everard M, Longhurst JWS. Reasserting the primacy of human needs to reclaim the 'lost half' of sustainable development. Science of the Total Environment. 2018 Apr 15;621:1243–54.

[4] Bird G, Rowlands D. The effect of IMF programmes on economic growth in low-income countries: An empirical analysis. Journal of Development Studies. 2017 Dec 2;53(12):2179–96.

[5] Wilcox J, Nasiri F, Bell S, Rahaman MS. Urban water reuse: A triple bottom line assessment framework and review. Sustainable Cities and Society. 2016 Nov 1;27:448–56.

[6] Dimond K, Webb A. Sustainable roof selection: Environmental and contextual factors to be considered in choosing a vegetated roof or rooftop solar photovoltaic system. Sustainable Cities and Society. 2017 Nov 1;35:241–49.

[7] Ehrgott M, Reimann F, Kaufmann L, Carter CR. Social sustainability in selecting emerging economy suppliers. Journal of Business Ethics. 2011 Jan;98:99–119.

[8] Pfeffer J. Building sustainable organizations: The human factor. Academy of Management Perspectives. 2010 Feb;24(1):34–45.

[9] Epstein MJ, Widener SK. Identification and use of sustainability performance measures in decision-making. Journal of Corporate Citizenship. 2010 Dec;1(40):43–73.

[10] Holden M. Urban policy engagement with social sustainability in metro Vancouver. Urban Studies. 2012 Feb;49(3):527–42.

[11] Isherwood J, Laybourn-Langton L. Health and sustainability. Future Healthcare Journal. 2018 Jun;5(2):84.

[12] Olsen IT. Sustainability of health care: A framework for analysis. Health Policy and Planning. 1998 Jan 1;13(3):287–95.

[13] Carter CR, Rogers DS. A framework of sustainable supply chain management: Moving toward new theory. International Journal of Physical Distribution & Logistics Management. 2008;38(5):360–87.

[14] Laguna J. Institutional politics, power constellations, and urban social sustainability: A comparative-historical analysis (Ph.D.). Available from ProQuest Dissertations & Theses Global. The Humanities and Social Sciences Collection. (1617974013); 2014.

[15] Drakakis-Smith D. Third world cities: Sustainable urban development, 1. Urban Studies. 1995 May;32(4–5):659–77.

[16] Ranjbari M, Shams Esfandabadi ZS, Zanetti MC, Scagnelli SD, Siebers PO, Aghbashlo M, et al. Three pillars of sustainability in the wake of COVID-19: A systematic review and future research agenda for sustainable development. Journal of Cleaner Production. 2021 May 15;297:126660.

[17] Pullman ME, Dillard J. Values-based supply chain management and emergent organizational structures. International Journal of Operations & Production Management. 2010 Jun 22;30(7):744–71.

[18] Anisul Huq F, Stevenson M, Zorzini M. Social sustainability in developing country suppliers: An exploratory study in the ready-made garments industry of Bangladesh. International Journal of Operations & Production Management. 2014 Apr 28;34(5):610–38.

[19] Holden M. Urban policy engagement with social sustainability in metro Vancouver. Urban Studies. 2012 Feb;49(3):527–42.

[20] Hoop ED, Loeber A, Essink D. Exploring, diversifying and debating sustainable health (care) approaches. Sustainability. 2022 Feb 1;14(3):1698.

[21] Marshall D, McCarthy L, McGrath P, Claudy M. Going above and beyond: How sustainability culture and entrepreneurial orientation drive social sustainability supply chain practice adoption. Supply Chain Management: An International Journal. 2015 Jun 8;20(4):434–54.

[22] Mani V, Gunasekaran A, Papadopoulos T, Hazen B, Dubey R. Supply chain social sustainability for developing nations: Evidence from India. Resources, Conservation and Recycling. 2016 Aug 1;111:42–52.

[23] Johnson E. Book Review: Walking the Talk: The Business Case for Sustainable Development. Charles Holliday, Stephan Schmidheiny and Philip Watts, Greenleaf Publishing, ISBN 1-874719-50-0;£ 21.95 (British pounds sterling),£ 18.95 to readers of EIA Review, 288 pages. Environmental Impact Assessment Review. 2003;23(1):131.

[24] Carter CR, Jennings MM. Logistics social responsibility: An integrative framework. Journal of Business Logistics. 2002 Mar;23(1):145–80.

[25] Santilli J, Vogenberg FR. Key strategic trends that impact healthcare decision-making and stakeholder roles in the new marketplace. American Health & Drug Benefits. 2015 Feb;8(1):15–20.

[26] Hussain M, Ajmal MM, Gunasekaran A, Khan M. Exploration of social sustainability in healthcare supply chain. Journal of Cleaner Production. 2018 Dec 1;203:977–89.

[27] Akamp M, Müller M. Supplier management in developing countries. Journal of Cleaner Production. 2013 Oct 1;56:54–62.

[28] Robèrt KH, Schmidt-Bleek B, De Larderel JA, Basile G, Jansen JL, Kuehr R, et al. Strategic sustainable development – Selection, design, and synergies of applied tools. Journal of Cleaner Production. 2002 Jun 1;10(3):197–214.

[29] Mossialos E, Oliver A. An overview of pharmaceutical policy in four countries: France, Germany, the Netherlands and the United Kingdom. The International Journal of Health Planning and Management. 2005 Oct;20(4):291–306.

[30] Nakamba CC, Chan PW, Sharmina M. How does social sustainability feature in studies of supply chain management? A review and research agenda. Supply Chain Management: An International Journal. 2017 Nov 13;22(6):522–41.

[31] Schaltegger S, Burritt R. Business cases and corporate engagement with sustainability: Differentiating ethical motivations. Journal of Business Ethics. 2018;147(2):241–59. http://www.jstor.org/stable/45022375

[32] Mueller M, Dos Santos VG, Seuring S. The contribution of environmental and social standards towards ensuring legitimacy in supply chain governance. Journal of Business Ethics. 2009 Nov;89(4):509–23.

[33] Kleindorfer PR, Singhal K, Van Wassenhove LN. Sustainable operations management. Production and Operations Management. 2005 Dec;14(4):482–92.

[34] Montalbán-Domingo L, García-Segura T, Sanz MA, Pellicer E. Social sustainability criteria in public-work procurement: An international perspective. Journal of Cleaner Production. 2018 Oct 10;198:1355–71.

[35] Shrivastava P. The role of corporations in achieving ecological sustainability. Academy Management Review. 1995 Oct 1;20(4):936–60.

[36] Starik M, Rands GP. Weaving an integrated web: Multilevel and multisystem perspectives of ecologically sustainable organizations. Academy Management Review. 1995;20(4):908–35.

[37] Seuring S, Müller M. From a literature review to a conceptual framework for sustainable supply chain management. Journal of Cleaner Production. 2008 Oct 1;16(15):1699–710.

[38] Labuschagne C, Brent AC, Van Erck RPG. Assessing the sustainability performances of industries. Journal of Cleaner Production. 2005 Mar 1;13(4):373–85.

[39] Maloni MJ, Brown ME. Corporate social responsibility in the supply chain: An application in the food industry. Journal of Business Ethics. 2006 Sep;68(1):35–52.

[40] Wilhelm PG. International validation of the corruption perceptions index: Implications for business ethics and entrepreneurship education. Journal of Business Ethics. 2002 Feb;35:177–89.

[41] Ghauri P, Hatani F, Zhang-Zhang Y, Rohlfer S, Bu M. International business and the SDGs: Current issues and future directions. International Business and Sustainable Development Goals. 2023 Jul;31(17):17–30.

[42] Horner, M., Price, A., Bebbington, J. and Emmanuel, R., (eds.) SUE-Mot Conference 2009: Second International Conference on Whole Life Urban Sustainability and its Assessment: conference proceedings. Loughborough: Loughborough University, 2009, pp. 865885.
[43] Khan M, Hussain M, Gunasekaran A, Ajmal MM, Helo PT. Motivators of social sustainability in healthcare supply chains in the UAE – Stakeholder perspective. Sustainable Production and Consumption. 2018 Apr 1;14:95–104.
[44] Konovsky MA, Pugh SD. Citizenship behavior and social exchange. Academy of Management Journal. 1994 Jun 1;37(3):656–69.
[45] Thin N. Social Progress and Sustainable Development. Bloomfield: Kumarian Press; 2002 Sep.
[46] Dempsey N, Bramley G, Power S, Brown C. The social dimension of sustainable development: Defining urban social sustainability. Sustainable Development. 2011 Sep;19(5):289–300.
[47] Vifell ÅC, Soneryd L. Organizing matters: How 'the social dimension' gets lost in sustainability projects. Sustainable Development. 2012 Jan;20(1):18–27.
[48] Wu Z, Pagell M. Balancing priorities: Decision-making in sustainable supply chain management. Journal of Operations Management. 2011 Sep 1;29(6):577–90.
[49] Friedman AL, Miles S. Stakeholders: Theory and Practice. United Kingdom: OUP Oxf; 2006 Apr 13.
[50] Ashby A, Leat M, Hudson-Smith M. Making connections: A review of supply chain management and sustainability literature. Supply Chain Management: An International Journal. 2012 Aug 3;17(5):497–516.
[51] Hussain M, Malik M, Al Neyadi HS. AHP framework to assist lean deployment in Abu Dhabi public healthcare delivery system. Business Process Management Journal. 2016 Jun 6;22(3):546–65.
[52] Linton JD, Klassen R, Jayaraman V. Sustainable supply chains: An introduction. Journal of Operations Management. 2007 Nov 1;25(6):1075–82.
[53] Horner M, Price A, Bebbington J, Emmanuel R. (eds.). SUE-Mot Conference 2009: Second International Conference on Whole Life Urban Sustainability and its Assessment: Conference Proceedings. Loughborough: Loughborough University, 2009; pp. 865885.
[54] Tate WL, Ellram LM, Kirchoff JF. Corporate social responsibility reports: A thematic analysis related to supply chain management. Journal of Supply Chain Management. 2010 Jan;46(1):19–44.
[55] Porter ME, Kramer MR. Strategy and society: The link between competitive advantage and corporate social responsibility. Harvard Business Review. 2006 Dec;84(12):78–92.
[56] Bai C, Sarkis J. Integrating sustainability into supplier selection with grey system and rough set methodologies. International Journal Of Production Economics. 2010 Mar 1;124(1):252–64.
[57] Lu RX, Lee PK, Cheng TC. Socially responsible supplier development: Construct development and measurement validation. International Journal of Production Economics. 2012 Nov 1;140(1):160–67.
[58] Miao Z, Cai S, Xu D. Exploring the antecedents of logistics social responsibility: A focus on Chinese firms. International Journal Of Production Economics. 2012 Nov 1;140(1):18–27.
[59] Klassen RD, Vereecke A. Social issues in supply chains: Capabilities link responsibility, risk (opportunity), and performance. International Journal of Production Economics. 2012 Nov 1;140(1):103–15.
[60] Mani V, Agarwal R, Gunasekaran A, Papadopoulos T, Dubey R, Childe SJ. Social sustainability in the supply chain: Construct development and measurement validation. Ecological Indicators. 2016 Dec 1;71:270–79.
[61] Mani V, Agrawal R, Sharma V. Social sustainability in the supply chain: Analysis of enablers. Management Research Review. 2015 Sep 21;38(9):1016–42.
[62] Sodhi MS, Tang CS. Corporate social sustainability in supply chains: A thematic analysis of the literature. The International Journal of Production Research. 2018 Jan 17;56(1–2):882–901.
[63] Mani V, Gunasekaran A, Delgado C. Enhancing supply chain performance through supplier social sustainability: An emerging economy perspective. International Journal of Production Economics. 2018 Jan 1;195:259–72.

[64] Lehtimäki H, Helén I, Snell K, Eriksson P, Montonen T. Sustainable value creation in the commercialisation of innovation: The case of Auria Biobank. International Journal of Entrepreneurship and Innovation Management. 2019;23(5):451–65.

[65] Hutchins MJ, Sutherland JW. An exploration of measures of social sustainability and their application to supply chain decisions. Journal of Cleaner Production. 2008 Oct 1;16(15):1688–98.

[66] Phillips RA. Stakeholder theory and a principle of fairness. Business Ethics Quarterly. 1997;7(1): 51–66. https://doi.org/10.2307/3857232

[67] Samuel C, Gonapa K, Chaudhary PK, Mishra A. Supply chain dynamics in healthcare services. International Journal of Health Care Quality Assurance. 2010;23(7):631–42.

[68] Aptel O, Pourjalali H. Improving activities and decreasing costs of logistics in hospitals: A comparison of US and French hospitals. International Journal of Accounting. 2001 Feb 1;36(1):65–90.

[69] De Vries J, Huijsman R. Supply chain management in health services: An overview. Supply Chain Management: An International Journal. 2011 May 3;16(3):159–65.

[70] Christoph F. Innovation for health: Success factors for the research-based pharmaceutical industry. Management for professionals. In: Franz C, Bieger T, Herrmann A (eds.), Evolving Business Models (pp. 93–112). Springer;2017. DOI:10.1007/978-3-319-48938-4_7

[71] Linner T, Solcanu G, van den Boom C, Lingegard H, Istamto T, Proctor G, Lu Y, Steebakkers J. Business model innovation for value and technology based preventive health care. Gerontechnology, Journal. 2017;16(3):200–06.

[72] Pencheon D. Making health care more sustainable: The case of the English NHS. Public Health. 2015 Oct 1;129(10):1335–43.

[73] Sagha Zadeh RS, Xuan X, Shepley MM. Sustainable healthcare design: Existing challenges and future directions for an environmental, economic, and social approach to sustainability. Facilities. 2016 Apr 4;34(5/6):264–88.

[74] Santilli J, Vogenberg FR. Key strategic trends that impact healthcare decision-making and stakeholder roles in the new marketplace. American Health & Drug Benefits. 2015;8(1):15–20.

[75] Rivard-Royer H, Landry S, Beaulieu M. Hybrid stockless: A case study: Lessons for health-care supply chain integration. International Journal of Operations & Production Management. 2002 Apr 1;22(4):412–24.

[76] Shih SC, Rivers PA, Hsu HS. Strategic information technology alliances for effective health-care supply chain management. Health Services Management Research. 2009 Aug;22(3):140–50.

[77] Langabeer J. The evolving role of supply chain management technology in healthcare. Journal of Healthcare Information Management: JHIM. 2005 Jan 1;19(2):27–33.

[78] Hussain M, Malik M. Prioritizing lean management practices in public and private hospitals. Journal of Health Organization and Management. 2016 May 16;30(3):457–74.

[79] Al Hammadi F, Hussain M. Sustainable organizational performance: A study of healthcare organizations in the United Arab Emirates. International Journal of Organizational Analysis. 2019;27(1):169–86.

[80] Jana A, Harata N. Provisioning health care infrastructure in communities: Empirical evidences from West Bengal, India. Socio-Economic Planning Sciences. 2016 Jun 1;54:37–46.

[81] Djukic A, Marić J. Towards socially sustainable healthcare facilities–the role of evidence-based design in regeneration of existing hospitals in Serbia. Procedia Environment Sciences. 2017 Jan 1;38:256–63.

[82] Price R, Sergelen O, Unursaikhan C. Improving surgical care in Mongolia: A model for sustainable development. World Journal of Surgery. 2013 Jul;37(7):1492–99.

[83] Rosenberg-Yunger ZR, Daar AS, Singer PA, Martin DK. Healthcare sustainability and the challenges of innovation to biopharmaceuticals in Canada. Health Policy. 2008 Sep 1;87(3):359–68.

[84] Berwick DM, Hackbarth AD. Eliminating waste in US health care. JAMA. 2012;307(14):1513–16.

[85] Saad SG. Integrated environmental management for hospitals. Indoor and Built Environment. 2003 Feb;12(1–2):93–98.
[86] Carnero MC. Assessment of environmental sustainability in health care organizations. Sustainability. 2015 Jun 29;7(7):8270–91.
[87] Marimuthu M, Paulose H. Emergence of sustainability-based approaches in healthcare: Expanding research and practice. Procedia – Social and Behavioral Sciences. 2016;224:554–61.
[88] de Brito MP, Dekker R, Flapper SDP. Reverse logistics: A review of case studies. In: B Fleischmann, A Klose (eds.), Distribution Logistics. Lecture Notes in Economics and Mathematical Systems, 544 (pp. 243–81). Berlin, Heidelberg: Springer; 2005.
[89] Sarkis J. A strategic decision framework for green supply chain management. Journal of Cleaner Production. 2003 Jun 1;11(4):397–409.
[90] Svensson G. Aspects of sustainable supply chain management (SSCM): Conceptual framework and empirical example. Supply Chain Management: An International Journal. 2007 Jun 26;12(4):262–66.
[91] Gunasekaran A, Patel C, McGaughey RE. A framework for supply chain performance measurement. International Journal Of Production Economics. 2004 Feb 18;87(3):333–47.
[92] Hervani AA, Helms MM, Sarkis J. Performance measurement for green supply chain management. Benchmarking: An International Journal. 2005 Aug 1;12(4):330–53.
[93] Khasreen MM, Banfill PF, Menzies GF. Life-cycle assessment and the environmental impact of buildings: A review. Sustainability. 2009 Sep 18;1(3):674–701.
[94] Lambert AJD. Life-cycle chain analysis, including recycling. In Sarkis J (ed.), Greener Manufacturing and Operations: From Design to Delivery and Back (pp. 36–55). Greenleaf; 2001.
[95] EPA U. Life Cycle Assessment: Principles and Practice. Cincinnati: National Risk Management Research Laboratory; 2006.
[96] McIntyre K, Smith H, Henham A, Pretlove J. Environmental performance indicators for integrated supply chains: The case of Xerox Ltd. Supply Chain Management: An International Journal. 1998 Sep 1;3(3):149–56.
[97] Silvestre BS. Sustainable supply chain management in emerging economies: Environmental turbulence, institutional voids and sustainability trajectories. International Journal Of Production Economics. 2015 Sep 1;167:156–69.
[98] Bhinge R, Moser R, Moser E, Lanza G, Dornfeld D. Sustainability optimization for global supply chain decision-making. Procedia CIRP. 2015 Jan 1;26:323–28.
[99] Lsaputri VH, Hisjam M, Sutopo W. A review on sustainable metrics for Sustainability Measurement in Supply Chain. IOP Conference Series: Materials Science and Engineering. 2020 Oct 1;943(1):012056
[100] Chardine-Baumann E, Botta-Genoulaz V. A framework for sustainable performance assessment of supply chain management practices. Computers & Industrial Engineering. 2014 Oct 1;76:138–47.
[101] Ghadimi P, Heavey C. Sustainable supplier selection in medical device industry: Toward sustainable manufacturing. Procedia CIRP. 2014 Jan 1;15:165–70.
[102] Ahi P, Searcy C. Assessing sustainability in the supply chain: A triple bottom line approach. Applied Mathematical Modelling. 2015 Jun 1;39(10–11):2882–96.
[103] Barbiroli G. New indicators for measuring the manifold aspects of technical and economic efficiency of production processes and technologies. Technovation. 1996 Jul 1;16(7):341–56.
[104] Epstein MJ, Roy MJ. Making the business case for sustainability: Linking social and environmental actions to financial performance. Journal of Corporate Citizenship. 2003 Mar 1;9:79–96.
[105] Subramanian L, Alexiou C, Nellis JJG, Steele P, Tolani F. Developing a sustainability index for public health supply chains. Sustain Futures. 2020 Jan 1;2:100019.
[106] Handfield R, Walton SV, Sroufe R, Melnyk SA. Applying environmental criteria to supplier assessment: A study in the application of the Analytical Hierarchy Process. European Journal of Operation Research. 2002 Aug 16;141(1):70–87.

[107] Mani V, Agarwal R, Gunasekaran A, Papadopoulos T, Dubey R, Childe SJ. Social sustainability in the supply chain: Construct development and measurement validation. Ecological Indicators. 2016 Dec 1;71:270–79.

[108] Pagell M, Wu Z, Wasserman ME. Thinking differently about purchasing portfolios: An assessment of sustainable sourcing. Journal of Supply Chain Management. 2010 Jan;46(1):57–73.

[109] Walker H, Jones N. Sustainable supply chain management across the UK private sector. Supply Chain Management. 2012 Jan 20;17(1):15–28.

[110] Dunphy JL. Healthcare professionals' perspectives on environmental sustainability. Nursing Ethics. 2014 Jun;21(4):414–25.

[111] Liu AMM, Fellows R, Tuuli MM. The role of corporate citizenship values in promoting corporate social performance: Towards a conceptual model and a research agenda. Construction Management and Economics. 2011 Feb 1;29(2):173–83.

[112] Park-Poaps H, Rees K. Stakeholder forces of socially responsible supply chain management orientation. Journal of Business Ethics. 2010 Mar;92(2):305–22.

[113] Jiang B. Implementing supplier codes of conduct in global supply chains: Process explanations from theoretic and empirical perspectives. Journal of Business Ethics. 2009 Mar;85:77–92.

[114] Belal AR. Stakeholder accountability or stakeholder management: A review of UK firms' social and ethical accounting, auditing and reporting (SEAAR) practices. Corporate Social Responsibility and Environmental Management. 2002 Mar;9(1):8–25.

[115] Malik MM, Abdallah S, Hussain M. Assessing supplier environmental performance: Applying analytical hierarchical process in the United Arab Emirates healthcare chain. Renewable and Sustainable Energy Reviews. 2016 Mar 1;55:1313–21.

[116] Berman SL, Wicks AC, Kotha S, Jones TM. Does stakeholder orientation matter? The relationship between stakeholder management models and firm financial performance. Academy of Management Journal. 1999 Oct 1;42(5):488–506.

Chapter 10
Importance of telemedicine on healthcare sustainability during pandemics

Abstract: People who live in rural places suffer from lack of health facilities, which could make it challenging for them to get specialist care under universal healthcare. The provision of essential healthcare services in places with limited access may be effectively accomplished by using novel technological advancements, such as telemedicine. This approach enhances the efficiency and effectiveness of service delivery procedures. Hospital visits can lead to significant expenses, especially in rural regions, mostly due to the associated costs of transportation. People prefer telemedicine during the COVID-19 pandemic when direct contact is risky because of the virus. Using telemedicine services, such as video conferencing or other virtual technologies, may lessen the need for in-person medical appointments. Therefore, telemedicine effectively reduces both the time burden on patients and healthcare providers, as well as the financial costs associated with treatment. The use of technological innovations within the healthcare industry, including telemedicine services, has shown to be beneficial for both the government and individuals under the unprecedented circumstances of the continuing COVID-19 epidemic. The chapter examines the significant potential of telemedicine in enhancing the sustainability of healthcare, particularly during the pandemic. Telemedicine, which uses digital technology to provide healthcare to remote patients, has become an essential tool for crisis-prepared healthcare systems. The chapter starts by presenting a comprehensive review of the many challenges that pandemics pose to healthcare systems. The problems include a notable surge in patient numbers, shortages of resources, and the risk for healthcare personnel to be exposed to the infection. The chapter explores the analysis of the role of telemedicine in enhancing healthcare accessibility, especially for those living in underprivileged or physically remote areas. It analyzes the potential financial benefits of telemedicine and its ability to improve patient outcomes by enabling timely interventions and reducing delays in treatment. The study examines the utilization of telemedicine in healthcare delivery to relatively remote villages that are situated at a considerable distance from urban hospitals. It aims to analyze and discuss the level to which telemedicine infrastructures have successfully implemented sustainability measures in delivering healthcare services to underserved regions, with particular focus on managerial perspectives. The primary objective of the study is to stimulate discussions about health policy on the use of emerging digital technologies as catalysts for the future sustainability of healthcare systems.

https://doi.org/10.1515/9783111436432-010

10.1 Introduction

Telemedicine and telehealth are terms used to refer to the utilization of healthcare information that is sent from one location to another using information and communication technology to enhance the health condition of the patient. The primary objective of the study is to examine teleconsultations, a specific kind of telemedicine, whereby healthcare practitioners who are physically located in different regions engage in consultations using information and communications technology. These teleconsultations aim to enhance the delivery of healthcare services by including additional value-added features [1–3]. Telemedicine refers to the exchange of health information and the provision of healthcare services using electronic communication and information technology, particularly when the individuals involved are situated in separate geographical places. According to WHO, telemedicine refers to the delivery of healthcare facilities by healthcare professionals using information communication technologies (ICT) in situations where distance plays a crucial role. These services include the exchange of valid information for purposes such as diagnosis, treatment, and prevention of diseases and injuries, as well as research, evaluation, and continuing education for healthcare providers. The ultimate goal of telemedicine is to advance the health of both individuals and communities [4]. Telemedicine has the potential to have a substantial impact on the healthcare sector in industrialized countries, with the following benefits:

- Real-time communication facilitates connection between patients and healthcare providers through the use of videoconferences, home-health monitoring devices, and telephone, allowing for communication to occur at any time and from any location.
- Store and forward involves the sharing of images, data, voice, video, and other medical information with physicians for the purpose of diagnosing patients who are located at a significant distance.
- Remote patient monitoring entails the continuous monitoring of patient health parameters and the collection of various health-related data while patients are in the comfort of their own homes.

The data is then submitted for monitoring, evaluation, and response by healthcare professionals. Additionally, telemedicine serves as a crucial solution for addressing the issue of limited healthcare accessibility among patients residing in rural areas. This phenomenon is particularly true within the marine domain since telemedicine facilitates the provision of healthcare services and guidance to those working at sea [5–9]. It is predicted that the use of telemedicine in poor nations will lower the load on healthcare facilities, decrease patient suffering, eliminate the need for transportation, eliminate patient nervousness in hospitals, reduce expenses, and improve service quality [10]. However, the deployment of telemedicine poses significant challenges in developing nations, as shown by research indicating that over 90% of telemedicine initiatives in these regions have been discontinued or unsuccessful [10, 11]. Telehealth can enhance the efficacy, organization, and accessibility of healthcare. The current state of study in this field is characterized by its early stages; however, it is progressively advancing. Telemedicine has improved the lives of people with heart disease by lowering their risk of death and hospitalization, while also improving their quality of life. People should seek out a diagnostic or rehabilitation program for a variety of important reasons. This can give people the peace of mind that they are getting the best care. This approach is aimed to reduce hospital visits and mitigate the widespread transmission of the virus [12–15]. E-health services can provide the necessary healthcare support and services by using electronic tools, in conjunction with ICT platforms [16–18]. By reducing unnecessary hospital visits, it lowers physical contact between patients, doctors, and healthcare personnel [12, 18]. Digital healthcare services include several components such as mobile health, medical information management, and telemedicine facilities [19]. Even in underdeveloped countries, interest in telemedicine was low before the COVID-19 epidemic. Hence, the adoption of services at a greater scale was hindered by the poor infrastructure, technology, and psychological preparation of both patients and medical personnel [14].

Telemedicine allows patients to conveniently obtain medical care while ensuring the safety of both the doctor and the person receiving treatment. This may suggest that individuals are not required to take leave from their professional responsibilities or make childcare arrangements. Visiting a medical facility involves the potential risk of contracting infections due to proximity to other individuals. This poses a significant danger, particularly for those with chronic medical conditions or compromised immune systems. It serves as a preventive measure against the potential acquisition of an infection inside the medical facility. Telemedicine service companies may potentially see reduced overhead costs. Clinicians may discover that the integration of telemedicine into their practice may serve as a supplementary source of revenue, enabling them to provide treatment to a larger number of patients. When healthcare providers conduct virtual consultations with patients, they are not subjected to the potential transmission of illnesses that the patient may carry. Patient satisfaction with their physician may be influenced by factors such as the convenience of accessing healthcare services, minimizing waiting times for treatment, and reducing the risk of nosocomial infections [20, 21].

The implementation of telemedicine has enhanced healthcare providers' ability to deliver care to a larger number of individuals without the need for physical presence. While the initial adoption of video conferencing in telehealth led to the emergence of numerous service providers, the subsequent advancements in telemedicine technologies are poised to offer significantly greater benefits. During a clinical encounter, healthcare professionals have the capability to employ natural language processing techniques to automatically generate patient notes. During emergency operations, remote specialists provide their expertise. The latest developments in telemedicine technologies encompass the integration of artificial intelligence (AI) systems, which aim to enhance the efficiency and effectiveness of physicians' practices. This technology enables patients to stay informed through the use of wearables and other remote patient-tracking resources. Additionally, it utilizes robots to administer specialized treatment to previously inaccessible regions [22–24]. Telemedicine is promising for patients in remote regions. Countries with few and insufficient healthcare facilities have the greatest impact. To maintain accurate medical records, patients and physicians need adequate hardware and software security. Some clinics provide online video conferencing for medical visits. When in-person visits are unnecessary, these appointments allow patients to continue treatment with their regular doctor. A web-based doctor or nurse practitioner visits are another interactive appointment type [25–27].

10.2 Telemedicine in healthcare

Telemedicine, defined as the utilization of electronic information and communication technologies to facilitate healthcare services when there is physical separation between the involved parties, is widely recognized as a means to enhance healthcare delivery and patient outcomes. This is especially true in regions and communities where healthcare facilities are limited or not accessible [28–31]. Telemedicine is a healthcare service that utilizes telecommunication and electronic information technology. It describes the whole set of products created to help people and their doctors or healthcare professionals. The technology has a wide variety of applications, including online conversations with patients, remote control functionality, telehealth nursing, as well as remote physical and psychiatric rehabilitation. The use of this technology facilitates improved decision-making in healthcare, enhances the quality and effectiveness of emergency services, accelerates the diagnostic process, and generates cost savings for healthcare providers and patients via the optimization of clinical processes and the reduction of hospital-related travel expenses [32, 33]. Telemedicine has emerged as a very favorable option for addressing mental health concerns. The strategy effectively addresses some factors that hinder patients from accessing such essential modes of therapy [34–36]. Telemedicine is a dynamic and progressive healthcare service that involves a wide range of functions, such as diagnosis, treatment, prevention, research, evaluation, and healthcare education [37]. The primary objective of this initiative is to provide a range of services, including consultation, diagnostic advice, and monitoring, which facilitate remote connectivity for healthcare professionals and patients [38]. Remotely connecting primary care or emergency department doctors to medical experts is a B2B model (through electronic information and telecommunications) [39].

This was replaced during the epidemic by person-to-person consultations using a B2C approach, or, in this instance, doctor-to-patient. Effective communication and interaction between doctors and patients plays a crucial role in the delivery of healthcare services. Similarly, the physical environment, such as the space in which healthcare is provided, may also have a significant impact on the healthcare system [40–42]. The use of telemedicine has the potential to provide remedies for healthcare institutions that possess inadequate infrastructure, restricted mobility, insufficient funds, and a shortage of personnel [43]. The terms telemedicine and telehealth are sometimes used interchangeably in scholarly discourse. According to the United States Department of Health and Human Services, telemedicine is defined as the provision of clinical services from a distance. Additionally, telehealth includes nonclinical activities such as training, administrative meetings, and continuing medical education [44]. Telemedicine covers all facets of healthcare services, as defined by the World Health Organization [37]. The American Telemedicine Association uses the terms telemedicine and telehealth interchangeably, but it acknowledges that telehealth encompasses a wider range of applications [45]. Another closely related word is eHealth, which is a broader concept that includes telemedicine, digital healthcare records, and other forms of healthcare information technology [46].

Telemedicine refers to online consultations, virtual conferences, remote surgeries, live monitoring, medical test collaborations, continuous patient data streams, and home monitoring. Although the interviews were exploratory and broad concerning the meaning of telemedicine, our empirical analysis focused on the patient-clinician relationship. The use of telemedicine has resulted in enhanced accessibility to healthcare facilities that provide high-quality services. Patients will now get healthcare treatments that are more personalized to their individual needs. In addition, individuals can conveniently connect with superior healthcare professionals via the use of video application software. This allows for remote consultations, enabling clinicians to effectively collaborate, save data, manage reports, and capitalize on one another's specialized expertise through enhanced networking technologies. This enhancement contributes to the enhancement of medical practice by reducing the time physicians allocate to rural assignments, enabling them to allocate more time and attention to patient care. Telemedicine has the capacity to facilitate the practice of private healthcare professionals, hence augmenting the overall patient experience. The use of electronic files in healthcare settings has many benefits, including improved patient experience and enhanced efficiency for doctors. By reducing the need for patients to wait in lengthy lineups, the adoption of electronic files enables a more streamlined and easy process. Additionally, clinicians can access patient information more easily, leading to reduced total wait times and enhanced productivity. In addition, the use of remote visits enables physicians to allocate less time to individual patients, hence enabling them to attend to a larger number of patients [47, 48]. Telemedicine strategies enhance healthcare accessibility in remote regions, hence generating value for patients. Several telemedicine job experiences have proven to improve patient outcomes and access, as well as reduce cost [49–51].

One significant benefit of telemedicine compared to traditional healthcare is the possibility of enhanced accessibility to medical services for those who face isolation, such as prisoners, individuals working at sea, or those in conflict zones [52]. Several researchers have effectively shown the favorable effects of telemedicine approaches on healthcare provision in geographically isolated regions, including both timely detection and subsequent monitoring. Rising healthcare expenditures and the demand for improved treatment are prompting additional organizations to consider the advantages of telemedicine. There is a need for enhanced communication between doctors and remote patients, as well as increased use of healthcare facilities. Telemedicine also facilitates improved connection, leading to a reduction in hospital readmissions and increased patient adherence to prescribed treatment regimens. The greater interaction that comes with telemedicine also helps doctors talk to each other. Telemedicine may be used by healthcare professionals as a means to establish support networks, facilitating the exchange of expertise and ultimately enhancing the delivery of healthcare services. Telemedicine refers to the provision of medical services through online platforms, mostly via video conferencing technology. This technology offers several benefits for both patients and healthcare practitioners. Despite the presence of technological challenges and supporters, telemedicine has the potential to complement and improve the overall experience of patients [53–55].

10.3 Telemedicine's potential and characteristics in a healthcare sustainability management system

The approach of telemedicine and its related services has been firmly established and shown to be beneficial for society. Figure 10.1 illustrates the many characteristics and facilities provided by the telemedicine concept, particularly within the healthcare sector. This technique offers several benefits such as chronic health management, prescription compliance, remote services, and treatment for essential and severe situations. These advantages eventually contribute to the support of the healthcare and medical care sectors. Furthermore, a collection of tele-wearables can facilitate the recovery of patients while simultaneously providing them with real-time updates about their health state in a distinctive manner [56–57]. Telemedicine, being an essential component of modern healthcare management systems, provides an extensive variety of capabilities and features that effectively improve patient care and optimize healthcare procedures. Telemedicine plays a significant part in providing remote consultations, thereby allowing patients to seek healthcare services from the convenience of their residences. This aspect has particular significance for those facing mobility limitations or living in places with limited access to adequate healthcare resources. Additionally, it minimizes the burden on healthcare facilities by decreasing overcrowding in waiting rooms and emergency departments.

Telemedicine is a technologically advanced achievement that has been widely regarded as a disruptive development in the healthcare industry. To serve patients who are located far away, telemedicine utilizes several electronic communication methods, including teleconferencing, image-sharing, and remote patient observation. In addition, physicians may use automation as a means to provide high-quality medical care to their patients. There is a need for the organization to enhance its information technology support systems and acquire proficiency in a novel approach to file management. An example of this is when primary care doctors use virtual appointments to seek guidance from specialists on any concerns they may have regarding a particular sickness or treatment. The expert is provided with various materials for inspection, including exam

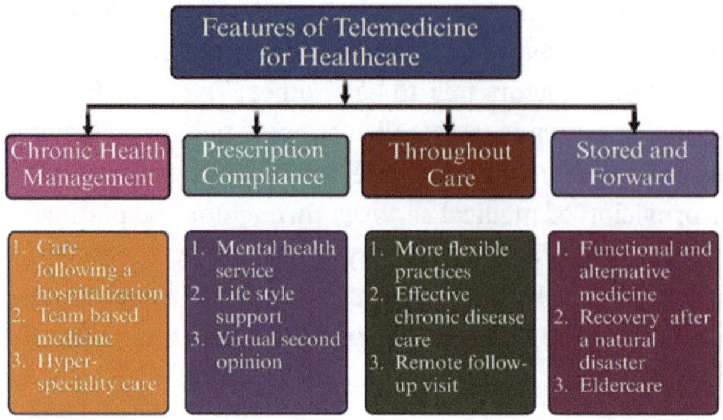

Figure 10.1: Various telemedicine features and capabilities for the healthcare industry.

reports, historical records, medical results, X-rays, and other photographs. The expert can respond electronically and arrange a virtual appointment with the doctor. Virtual consultations have the potential to eliminate the need for excessive referrals to experts that require physical presence, decrease the duration of waiting periods for professional input, and remove the requirement for wasteful travel. Telemedicine approaches are particularly advantageous in situations when a physician can visually assess the patient, accurately diagnose a medical condition, and document the experience [58–61]. Telemedicine allows healthcare practitioners to collaborate, discuss, and share medical data in real time. This facilitates a comprehensive and coordinated approach to patient care, thereby enhancing diagnostic precision and treatment planning. Using personal devices and remote sensors, telemedicine also makes it possible to keep an eye on patients with long-term conditions. This lets doctors find health problems early and take care of them right away. Moreover, the integration of telemedicine with electronic health records (EHRs) ensures that patient data is stored securely and is readily accessible to authorized healthcare professionals, thereby enhancing the overall continuity of care. Telemedicine in healthcare management systems improves access to care, encourages efficient communication, data sharing, monitoring, and record integration, and leads to better patient outcomes and a more patient-centered approach.

An electronic personal health record system monitors and preserves health records. Anyone can use a web-enabled device like a PC or smartphone anytime. In an emergency, a personal health report will immediately offer emergency responders crucial information, such as diagnoses, prescriptions, preferred substances, and the doctor's contact information. All of a patient's medical records may now be kept in one secure area because of the applications made possible by this technology. Recovery programs should set goals to improve patient behavior, made easier by mobile health technology. This tool helps patients track calories, vital signs, exercise, and medication dosages and timetables. Telemedicine may be used to expand patient self-management, beyond diabetes, to other chronic conditions including hypertension and digestive problems [62–64]. When integrated with AI, it has the potential to enhance its effectiveness. The use of this technology has the potential to improve routine procedures, resulting in a decrease in the burden on doctors and an enhancement in their job satisfaction. The program effectively transmits data in a prompt and superior way to guarantee the smooth running of appointments. As a result, medical professionals would be able to conduct more thorough examinations and reviews of their patients. Patients are advised to consult the doctor's available time slots as well as their approaching appointments with the option to postpone them. Healthcare analytics often serves as a means to engage with the collected data and see it via a user interface. Store and forward methods lead to better time and cost utilization over time. Telemonitoring is a healthcare practice that utilizes a range of electronic technologies to immediately communicate patient data to an analytical interface belonging to a healthcare professional [65, 66].

The use of this technology has shown its significance and need in many roles within the telehealth and medical sectors. It is worth noting that this technology is significantly transforming the fields of surgery, medical training, and education. After receiving approval from the doctor, the patient needs to receive confirmation of the planned appointment. The profiles of individuals may include internal information, such as reviews, modifications, and essential hospital warnings. Physicians can arrange meetings by using schedules. The incorporation of scheduling and rescheduling functionalities is a prevalent characteristic seen in contemporary telemedicine systems. While receiving notification of an appointment, doctors possess the ability to retrieve the patient's medical record and any other pertinent information, thereby facilitating a comprehensive consultation and accurate diagnosis [67, 68]. Virtual reality (VR) technologies have enhanced telemedicine communication by creating a more immersive experience. By using VR, surgeons and their staff can view 3D monitors throughout surgeries. Physicians may perform surgeries on patients hundreds of kilometers away via video conferencing. Through video discussions, global medical teams may coordinate and address urgent problems. A telehealth platform may employ VR to mimic patient data and a graphical interface with video conferencing to improve doctor–patient communication. This strategy assists in rural or remote areas and is crucial for delivering local healthcare to people. In rural places, this boosts faith in local healthcare systems.

This technique increases financing for distant patients' local health care. Additionally, this technology enables patients to stay with relatives [69–71]. Adopting digital health technology platforms has been shown to improve the resilience of healthcare teams [72, 73]. Martínez-Caro et al. [74] observed that educating healthcare personnel is crucial for implementing e-health services effectively utilizing ICT. Sims [75] and Lu et al. [76] suggested creating a sustainable healthcare expert community using virtual platforms, and gamification to engage doctors.

Digital health services improve efficiency, enhance service quality, reduce response time, and improve post-discharge schedules, resulting in patient, clinician, and healthcare satisfaction [76–80]. Several studies highlight the significance of healthcare information technology in collecting, securing, and analyzing large amounts of data for e-health and telemedicine services [81–83] Telemedicine is the use of information and communication technology for secure consultation, analysis, and transfer of patients' healthcare data through video, still images, audio, words, and other methods for the diagnosis of disease. Healthcare institutions and governments have equal responsibilities for advancing telemedicine across the world [19, 84, 85]. An innovative study was conducted by Cobelli [86] that uncovered both the benefits and limitations of contactless medical care delivery methods like telemedicine. To better understand what influences community pharmacists in Italy to use or not use telemedicine during COVID-19, Cobelli et al. [86] recently performed research. Patients' opinions on teleconsultations were investigated by Baudier et al. [87] for the COVID-19 conference. Kaplan [88] provided a framework for addressing the ethical, social, technological, privacy, and legal concerns raised by the use of telemedicine services during the COVID-19 pandemic. Rezaei et al. [89] studied ethical difficulties in telemedicine adoption, focusing on privacy, security, and autonomy. Various factors have contributed to a lack of interest in telemedicine services, particularly in developing economies. COVID-19 threw up several challenges, including those related to technology, privacy, clinician familiarity, and patient perception [16, 84, 86, 87, 90]. These included clinician's intention and willingness [18, 84, 91], patient loss of autonomy, fear, cost of technology [92], lack of technology, technical skill of staff, age and education level [93], and patient's lack of trust and acceptance [94,95].

10.4 The challenges associated with attaining sustainability

The initial implementation rates of telemedicine projects that were implemented in the 1990s were found to be unsatisfactorily low [96–98]. Since then, several studies have demonstrated the effectiveness and efficiency of various clinical applications in teleconsultations [30, 99, 100]. Additionally, significant financial support has been provided to telemedicine projects aimed at addressing the healthcare needs of medically underserved regions, particularly rural areas in America [101, 102]. Furthermore, there have been notable advancements in the capabilities, usability, and affordability of telemedicine technology [103]. Studies have also provided evidence of the successful implementation of telemedicine projects [104–106], and multiple telemedicine initiatives have been effectively executed [99, 107]. Telemedicine has shown considerable advantages in the field of healthcare; yet, it also poses several sustainability-related challenges. One of the foremost considerations relates to the possibility of heightened electronic waste as telehealth devices and infrastructure continuously advance. If not managed effectively, the rapid rate at which technology advances may have negative impacts on the environment. Moreover, the environmental implications of telemedicine operations, including energy use in data centers, servers, and devices, might be significant. The prevention of these difficulties requires the essential aspect of ensuring that telemedicine platforms are both energy-efficient and reliant on renewable sources of electricity. The viability of telemedicine comes into doubt due to concerns about data security and privacy. The need for strong cybersecurity protocols in safeguarding patient data contributes to the increased energy and resource requirements of telehealth systems. Maintaining an appropriate balance between the protection of patient data and the reduction of environmental impact caused by security measures is of great significance. It is important to recognize that the digital divide has the potential to increase inequalities in healthcare access and outcomes. Inequalities in healthcare access may arise due to the unequal availability of high-speed internet and requisite telemedicine devices, placing underprivileged communities at a potential disadvantage.

The task of effectively addressing this inequality sustainably, while also ensuring fair and equal access to technology and communication, is a considerable challenge. Healthcare systems must implement eco-friendly practices, make investments in renewable energy sources, prioritize responsible technology disposal, and work toward inclusive telehealth solutions that take into account the needs of all patients while minimizing environmental impact, to address these sustainability challenges. However, it is noteworthy that the usage rates of telemedicine have remained low [107]. Additionally, only a limited number of telemedicine projects have been successfully maintained over time, despite the constant demand for the services they provide [104, 106, 108]. Sustainability refers to the capability of a telehealth service to maintain its operations sustainably, either by integrating telehealth activities into an established clinical environment or by achieving a significant level of independent telehealth engagement. This concept is recognized as a multifaceted and complicated phenomenon, often regarded as the ultimate goal of telehealth. Previous studies have investigated the sustainability of telemedicine services, with a particular example being the meso-level model that incorporates many factors [109]. This model is grounded in the normalization process theory [110, 111]. The primary determinant was the acceptability of clinicians. Additional criteria that were considered in this study were the availability of clinicians in the workforce, presence of suitable technology, presence of telehealth champions, favorable opinions held by healthcare professionals regarding telehealth, establishment of excellent relationships between healthcare providers, demand for telehealth services by clinicians, and the allocation of resources [109]. Numerous variables influence and emphasize the significance of healthcare providers' involvement in assessing the sustainability of teleconsultation projects. This implies that to get a deeper comprehension of the sustainability of telemedicine services, it is advisable to concentrate on teleconsultation initiatives that have been effectively managed, and analyze them from the perspectives of the healthcare professionals involved. Regarding the acceptability of medical professionals, there are different variations in the features of teleconsultation session objectives, activities, and procedures between ongoing and inactive initiatives. The availability of the clinician workforce is a crucial determinant.

10.5 The use of telemedicine to reduce healthcare's carbon footprint

The integration of healthcare has not been a prominent aspect of climate protection efforts. Given that the primary emphasis of healthcare is on treating individual health concerns, this may be understood, at least at first appearance. However, it is important to note that occupational, public, and global health are different fields within the healthcare sector that prioritize the significance of environmental health. Collectively, they have effectively increased public consciousness about the significance of environmental variables in the development of cancer, autoimmunity, and several other health disorders [112]. There is growing data addressing the carbon footprint of the healthcare industry and how it might be reduced by further integrating telemedicine into modern healthcare [113]. The carbon emissions associated with conventional medical care are significant. The emissions originating directly from hospitals and healthcare-owned vehicles constitute 17% of the global carbon footprint of the medical sector. An additional 12% of the emissions may be attributed to indirect sources, including acquired energy sources such as electric power, steam, air conditioning, and heaters. The majority of emissions, namely 71%, originate from the healthcare supply chain. These emissions are generally a result of the manufacture, transportation, and disposal of various goods and services within the industry. Examples of such goods and services include medicines, chemical compounds, agricultural and food goods, healthcare products, medical equipment, and instrumentation [114]. Overall, the healthcare sector is responsible for emitting 2 gigatons (GT) of CO_2 equivalents, which corresponds to about 4.4% of the global greenhouse gas emissions [115–117]. There is little support for advocating a decrease in this level of consumption, given the potential risks to human health and well-being. Nevertheless, empirical data indicates that there exists a potential adverse correlation between the efficacy of healthcare services and the carbon footprint of the medical sector [118]. There exists a positive correlation between the level of complexity of a medical facility and its carbon footprint and net emissions. The association between extended periods of in-hospital treatment and the constant transporting of patients as well as medical supplies to healthcare facilities has been connected with elevated levels of carbon emissions. Hence, the use of telemedicine as a means of delivering healthcare services remotely might potentially serve as a viable alternative, facilitating environmentally sustainable and patient-centered healthcare practices. While telemedicine is unable to fully substitute the interaction between doctors and patients in traditional face-to-face encounters, it does have the potential to enhance healthcare by providing ongoing, readily available, personalized treatment, all of which are crucial aspects of patient-centered medical services [119].

Telemedicine involves the provision of healthcare services through separate means, facilitated by telecommunications technology. The comprehensive characterization of telemedicine acknowledges that this expanding discipline comprises a range of different products and services. Telemedicine is used as a means of preventing and monitoring health conditions, serving as a substitute for in-person consultations. However, it is important to recognize that there is potential for remote execution of surgical procedures in the future [120]. The current capabilities of telesurgery are restricted in terms of patient assessment and education. However, the integration of robots and wireless networks has the potential to facilitate surgical procedures at local hospitals, reducing the need for patient and resource transportation to specialist facilities [121]. There are apprehensions about the efficacy of remote healthcare delivery due to the ongoing learning process for both patients and clinicians. Nevertheless, a significant number of physicians and patients have accumulated years of experience using telehealth. One example of a country that has implemented an e-health strategy is Ghana, which began this approach in 2010 [122]. The COVID-19 epidemic has served as a catalyst, prompting doctors and patients globally to introduce themselves to teleconsultations and achieve their advantages, all the while adhering to social distancing measures. Teleconsultations may be a viable option for people in distant places or near healthcare institutions in the post-COVID-19 age [123]. The variation in carbon emissions reduction may vary across different countries or populations, particularly among those who need regular in-person consultations or supplementary laboratory and imaging examinations.

While telemedicine has made significant progress, there is still space for development to reduce patient and environmental burdens. The spread of COVID-19 has already helped achieve this goal [124]. However, there is potential for further advancements in the integration of novel technologies into the organization of healthcare services [125]. The integration of telemedicine and AI has the potential to contribute to a reduction in the carbon footprint of the healthcare industry [126]. Over the last several years, AI systems have advanced to the point where they can reliably and accurately monitor patients with chronic diseases or undertake emergency evaluation (triage) of patients needing medical treatment [127]. Previous research has examined the uses for inflammatory bowel disorders and acute abdominal discomfort [128]. Additional testing is essential, especially for health disorders that have not previously been analyzed using AI. Applying existing information in acute and chronic care might reduce healthcare-related CO_2 emissions by reducing standby staff and avoiding needless emergency department visits. The AI–telemedicine synergy is anticipated to contribute to the reduction of carbon emissions from an environmental perspective [120, 124].

There are several strategies for reducing in-hospital treatment, but there is a significant deficiency in the infrastructure for in-home care, particularly in low- and middle-income countries [129]. The provision of this kind of care is often characterized by its informal nature, which has the potential to impose significant burden on those who provide it [130]. Patients transitioning from the hospital to a home setting should be provided with many opportunities for virtual or in-person consultations. It will enable them to experience a welcoming and secure atmosphere, free from nosocomial diseases and other risks associated with hospitalization. According to Rosen et al. [129], increasing the use of telemedicine for home consultations might result in a reduction in patients' visits to hospitals, hence contributing to a decrease in the environmental impact of healthcare [130, 131]. A sustainability approach to healthcare has the capacity to fundamentally transform medical and environmental policies. In contrast to several other political or financial transformations, safeguarding the environment does not need major investments. However, the reduction of carbon emissions and the expansion of telemedicine exhibit concurrent developments. Indeed, the adoption of pragmatism is important. To reduce healthcare-related carbon emissions, telemedicine applications need study, awareness, education, and advocacy initiatives to provide tangible effects.

10.6 Challenges to telemedicine implementation in healthcare

Figure 10.2 demonstrates the many barriers encountered during the effective adoption of telemedicine to provide optimal treatment and assistance to healthcare facilities and patients. When considering the implementation of telehealth techniques in the field of healthcare and its connected areas, it is important to address the common challenges that are often encountered. The aforementioned matters must remain free of any potential improper disclosure of sensitive information, criminal activity, misuse, and inaccurate solutions. The presence of any of these difficulties may lead to discouragement among individuals or complicate matters about health concerns [132–134]. The use of telemedicine services and other telehealth technology monitoring techniques has the potential to help patients and healthcare providers in effectively managing chronic health conditions, including diabetes and asthma. Healthcare providers may also monitor patients at their homes after their release from the hospital or throughout their recovery from an injury. The mishandling of patients is a significant cause of concern for clinicians. Despite the progress achieved in the field of medicine facilitating the integration of technology, there have been some occurrences of device outages.

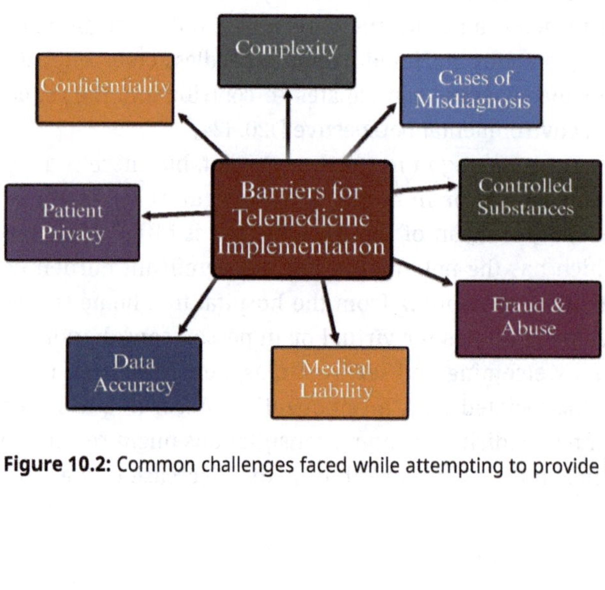

Figure 10.2: Common challenges faced while attempting to provide healthcare through telemedicine.

Healthcare systems contemplating the use of telemedicine technology engage in consultations with industry stakeholders. Practitioners who are interested in integrating telemedicine into their clinics have a wide range of practical alternatives available to them, facilitating the process of integration [135, 136]. The telemedicine approach facilitates the sharing of patient information between doctors situated in different countries. In contemporary healthcare practices, primary care practitioners can remotely share patient history and medical information with doctors, eliminating the need for physical proximity between the two parties. Systems can transmit data over significant distances across different systems, hence facilitating the exchange of information among practitioners. This enables one practitioner to stay informed about the actions and progress made by another practitioner. Consequently, there is a reduction in the duplication of monitoring efforts and a decrease in instances of insufficient drug management. Physicians can gather and share information with their patients via the use of patient portals. Medical gadgets have the capability of transmitting crucial physiological indicators and other information to healthcare professionals, enabling them to implement necessary modifications to the treatment regimen. Patients are now able to transmit biometric data to their healthcare providers via wearable gadgets or remote monitoring tools, such as pulse monitors or blood pressure cuffs. The system can retrieve patient information via a dashboard or clinical decision support system. It may then gather and present data, providing a near-real-time representation of the patient's state [137–139].

Telemedicine facilitates the communication of medical information between patients and healthcare providers over significant geographical distances, hence revolutionizing the healthcare landscape. Real-time interchange of healthcare information, including imaging photos, blood tests, and other relevant data, facilitates effective patient assessment. The use of telemedicine, when patients first meet with a distant physician via video chat, presents a potential solution to address the issue of overcrowding in emergency departments. Healthcare networks that have successfully integrated telemedicine technology can attest to the considerable investment of time and money involved. To reap the advantages, practice administrators, clinicians, physicians, and other relevant individuals need to possess a comprehensive understanding of the device's operational procedures. Although the initial cost of implementing telemedicine might be high, healthcare facilities can potentially get a favorable return on investment in the long run as a result of increased patient volume and reduced staffing requirements. The technology enables clinicians to assess a patient's medical history, conduct clinical exams, and perform other tasks [140–142]. Currently, there are telemedicine alternatives that allow patients the opportunity to get medical care from the comfort of their residences.

The use of telemedicine offers significant advantages to those who want medical attention but lack the financial resources or ability to travel extensive distances or endure prolonged waiting periods. By just transmitting the film, a doctor at another institution can quickly assess high-risk patients. This prevents the costly and time-consuming practice of transferring newborns to a separate facility. Patients who previously had limited access to medical facilities will now be able to consult with specialists without having to leave the comfort of their own homes. People with infectious conditions should avoid exposing themselves in crowded waiting rooms to reduce disease transmission. People frequently schedule consultations during their lunch breaks or after their regular workday has ended [143–145]. Telemedicine introduces a new degree of technology-assisted therapy by linking patients with their caretakers, providing access to doctors and specialists, and encouraging both patients and providers to avoid high-risk locations. Various telemedicine strategies may be used for the treatment of individuals suffering from drug misuse. In general, the expense associated with medical care is reduced, leading to financial savings. The cost reductions would become more evident as technology progresses. Telemedicine is very beneficial for both doctors and patients in terms of treatment and diagnosis. It has the potential to function as a very effective support system. This particular telemedicine technology is advantageous for healthcare professionals in facilitating the provision of suitable medicines and treatment for patients. Telemedicine has the potential to provide cost savings for both patients and healthcare providers in comparison to traditional care methods. To provide appropriate medical care, healthcare providers need to have on-site telemedicine services [146, 147]. The use of telemedicine facilitates continuous patient care. This will facilitate the practice of appropriate self-care and contribute to the accurate identification of medical conditions. AI-enabled telemedicine systems can independently analyze patient data and provide prompt assistance to doctors in adapting to emerging technology. The use of telemedicine technology will enable healthcare providers to effectively attend to a larger number of patients. The duration of visits may be shortened, and the physical dimensions of hospitals can be decreased. The provision of immediate telemedicine services can reduce the frequency of visits to emergency departments. The technology facilitates the collaboration of professionals in remote healthcare facilities. These exchanges provide live streaming sessions with faraway professionals, enabling on-screen consultations.

Telemedicine refers to the process of transmitting medical information from one site to another remotely. This practice is undertaken to enhance an individual's physical fitness. Securing an appointment with a primary care physician or a specialized medical practitioner might present difficulties. The remote network is now engaged in the collection and transmission of data to other healthcare organizations for analysis. This technology enables mental health practitioners to increase their patient load while ensuring a high level of medical treatment. This leads to increased profitability and enhanced time management. Parents would not need to leave their sick children at home to go to the doctor anymore [148–150]. Telemonitoring refers to the use of technological devices to transmit medical information from individuals located in different geographical regions. Doctors and primary care providers may use electronic devices to keep tabs on their patients using this form of monitoring. Telemonitoring systems facilitate the provision of healthcare services to patients in the comfort of their own homes. This intervention reduces the frequency of hospital visits and enables individuals to have more autonomy in managing their medical care. Patients who live in remote areas and do not have access to telemedicine's cost-effective medical services are unable to go to hospitals or other medical facilities. Interacting with other healthcare professionals regularly may be a significantly time-consuming aspect of their everyday responsibilities. Radiologists can acquire high-quality images and receive feedback remotely via telemedicine. The need for the supplier and the photos to be at the precise place is no longer necessary, resulting in improved operational efficiency. Telemedicine is often used in the field of mental health, making it one of the most prevalent specialties. This approach enables treatment sessions to be conducted remotely, without the restriction of physical distance [151–153].

10.7 Telemedicine in healthcare: an important application

Telemedicine provides access to numerous treatment options, such as primary care assessments, psychotherapy, and physical therapy, among others. It employs wireless technologies such as laptops and smartphones to administer treatment. Telemedicine utilizes video conferencing the majority of the time. However, some services offer treatment via email or phone. Often patients use telemedicine, in conjunction with their primary care physician. This technology is useful when a patient is unable to travel to a healthcare facility. It enables practices to work in extended or weekend hours without incurring the expense of having their office open. This also makes practices more appealing to the increasing number of patients who demand telecommuting as a primary care provider option [154–156]. The utilization of this approach has shown to be both economically advantageous and straightforward in assisting individuals afflicted with severe illnesses to effectively manage their condition, actively participate in their treatment regimen, and mitigate the progression of complications. Table 10.1 presents a comprehensive overview of the significant benefits of telemedicine within the healthcare sector. A multitude of pioneering technologies are employed within the field of engineering. Healthcare and its related disciplines have been extensively studied and discussed in the chapter [157–160]. These resources are beneficial in addressing a range of issues about design, manufacturing, and the establishment of a sustainable environment [161–163]. With the help of telemedicine, doctors can operate on a patient in one location while receiving advice from specialists in another.

This technology offers significant advantages in remote or geographically challenging regions where access to experts is limited. It is a convenient tool for conducting remote visits while saving time and money. Significant advancements in the field of telemedicine have been made possible by the proliferation of the internet in recent decades. With the development of this new intelligent technology capable of high-quality video streaming, primary and specialized care can now be provided to patients in the comfort of their own homes, workplaces, or assisted living facilities, saving them time and money. It is widely claimed that telemedicine has been in existence, in various forms, since the development of the telephone. It is possible to transmit visual content as a substitute for numerical information through the mobile device. Telemedicine is widely recognized as a crucial element in the delivery of healthcare services across different nations. The technology exhibits a significant range of practical uses, enabling patients to schedule subsequent appointments through video communication, enhancing adherence to posttreatment visit protocols, and alleviating the burden on both patients and healthcare providers [164–166].

Current mobile health applications facilitate telemedicine by integrating an interactive clinical interface into their software. Examples of noncritical events include the treatment of patients with minor diseases, exchange of data related to investigations, and the presentation of imaging findings. Patients have the choice to purchase medications and obtain prescriptions through a specialized mobile application. The establishment of a strong and effective collaboration with payment gateways may be essential. Analyses and the transmission of information occur in real-time. Telemedicine systems are frequently integrated, facilitating seamless collaboration between patients and physicians through a unified application for communication and data exchange. Direct patient data can be readily collected and subsequently transmitted to the relevant healthcare professional. The provided evidence could potentially be a transcript of a scheduled appointment or a test report that has been stored within a designated folder in an electronic health record system. Furthermore, this technology enables medical professionals to analyze and make sense of the gathered data [215–217].

The utilization of confidential medical records is prevalent in the field of telemedicine. The utilization of this technology involves the acquisition, retention, and dissemination of data, thereby facilitating the consideration of crucial factors within this domain. Telemedicine is an online platform that facilitates remote communication with healthcare professionals, enabling individuals to engage in virtual consultations with doctors [218–220]. Mobile applications function as a means of establishing a link between the mobile device and the internal infrastructure of the hospital. Specialized modules can provide alerts, research findings, and visual representations of data to support clinical decision-making. The aforementioned program utilizes a remote-control device to provide care for individuals afflicted with severe and chronic medical conditions. Every individual who is admitted to the hospital and remains under medical supervision during a surgical procedure occupies a valuable hospital bed. The concept of home health refers to the provision of medical and healthcare services to individuals in the comfort of their own homes. Telemedicine encompasses the collection of vital signs, the provision of video conferencing services, and the ability to generate alerts to a nurse's station [221–224].

Telemedicine utilizes emerging networking systems and application software to facilitate communication and file management. It also utilizes advanced devices, such as computers and mobile devices, to enable users to remotely access and control healthcare facilities. This allows individuals to conveniently view their medical reports, schedule appointments, request prescription refills, make payments, and communicate with healthcare providers through an online patient portal. A progressive

Table 10.1: Significant benefits of telemedicine within the healthcare sector.

S. no.	Application	Description	References
1.	Telehealth	Their range of applications has been greatly increased by technological development and medical innovation. The concept of telehealth may also be utilized by clinicians, investigators, lab staff, and physicians with the introduction of the technology in the healthcare sector. Hospital systems, medical groups, and suppliers must integrate telemedicine into their medical services at present. Furthermore, numerous organizations express a desire to engage in collaboration with a telemedicine provider to facilitate the process and implementation. Telemedicine is executed through the utilization of a telemedicine apparatus, comprising a computer and portable medical instruments. Physicians additionally employ high-resolution imaging cameras to provide precise diagnostic photographs to specialists.	[167–170]
2.	Remote therapy	The concept of remote therapy refers to the provision of medical or therapeutic services to individuals, without the need for in-person contact. Virtual remote treatment has been identified as an effective method for preventing hospitalization, as it can be conducted remotely. Patients who desire to communicate with a healthcare professional possessing extensive expertise in a particular ailment can now do so without the need to travel long distances, thanks to the availability of video visits. The training and skills possessed by each specialist can be optimally utilized. Healthcare can exhibit competitive dynamics, as primary care and specialist physicians engage in collaboration to optimize patient health outcomes. Telemedicine enables healthcare providers to effectively communicate with patients and collaborate through a secure remote connection.	[171–174]
3.	Medical treatment for students	The utilization of telemedicine offers significant potential in addressing the healthcare needs of our increasingly elderly demographic. When a child gets sick at school, they can see the school nurse or their parents can pick them up and take them to an urgent care center. This is usually inconvenient and, maybe, not even necessary. Innovative educational institutions have the potential to establish partnerships with medical professionals to facilitate virtual tours from the confines of the classroom. The provider has the authority to determine the necessary course of action and offer parents appropriate guidance or reassurance. It has also been demonstrated that having on-call doctors available 24/7 and on holidays will help keep patients who reside in assisted living facilities, out of the hospital.	[175–178]

(continued)

Table 10.1 (continued)

S. no.	Application	Description	References
4.	Assist handicapped patients	Telemedicine plays a crucial role in enhancing the accessibility of healthcare services for those with impairments. Furthermore, it enhances accessibility for many demographics, including the elderly, culturally challenged individuals, and those who are imprisoned. Telemedicine has the potential to assist in the management and care of many medical diseases. The effectiveness of medical care is enhanced when a person seeks assistance from a qualified healthcare professional who provides comprehensive information about their symptoms. According to studies, there is evidence to suggest that patients who use telemedicine have reduced hospitalization durations, resulting in cost savings. Furthermore, reduction in travel time may result in decreased secondary expenses, such as fuel costs.	[179–182]
5.	Improve the overall functioning of the healthcare system	Telehealth includes a diverse range of technological instruments and infrastructure used to deliver healthcare services to patients and improve the overall efficiency of the healthcare system. Telemedicine includes a wider range of digital healthcare services. Furthermore, the use of telehealth expands beyond healthcare delivery and includes other nonclinical activities, including but not limited to provider staffing, administrative conferences, and continuing healthcare education. People who are unable to go to a doctor's office may still get medical attention via telemedicine. It is also applicable for subsequent visits, management of chronic illnesses, consultation with experts, medication management, and several other healthcare resources that may be remotely accessible via secure video and audio connections.	[183–185]
6.	Beneficial for illnesses without lab testing	Telemedicine is an effective approach for managing disorders that do not need laboratory investigations or a physical examination. Additionally, this technology offers ongoing therapeutic interventions, such as psychotherapy, to ensure continuous treatment. A patient who lives a significant distance away from a primary care center may still be able to get improved treatment as a result of the COVID-19 epidemic. If a patient is unable to go to their appointment on their own, doctors may be more inclined to treat a wider variety of ailments. A physician, for instance, may utilize telemedicine to prescribe medications to a patient remotely if the patient is suffering from an illness.	[186–189]

7.	Enhancing patient coordination	Telemedicine can enhance patient coordination and improve the provision of healthcare services. Fragmented care may lead to several issues such as treatment shortages, excessive use of medical care, insufficient medication utilization, and not needed or overlapping care. The patient can promptly remunerate the modest fees associated with telehealth services. This service may be conveniently supplied to patients, thereby enhancing their coordination with the healthcare provider.	[190–192]
8.	Virtual doctor's consultation	In the current era of social distancing, a significant number of individuals eagerly anticipate engaging in virtual consultations with healthcare professionals. The use of telemedicine has resulted in enhancements in medication management, improvements in patient's quality of life, and reduction in healthcare expenses. The expansion of telemedicine applications has become a primary objective for healthcare practitioners aiming to offer healthcare services to patients via online and distant means. This technology facilitates the development of telemedicine applications that are appropriate for both medical practitioners and patients. Furthermore, it has been shown that patients perceive a system to be reliable and easy to use when it has a strong authentication method with low specific requirements.	[193–196]
9.	Reduce patient travel time	The implementation of telehealth services would result in a significant reduction in the available appointment duration. Patients will no longer be required to spend significant amount of time traveling to or waiting inside a medical facility. In the present scenario, individuals can schedule appointments using an Internet platform and then use an application to guide them to their designated meeting location. Further, it is significant that healthcare has become ever more accessible in terms of affordability. Patients can engage in remote consultations with physicians, ensuring their safety and convenience in the confines of their residences. Additionally, it might potentially assist anyone requiring urgent medical care. Physicians who want immediate access to information rely on an internal database containing preestablished queries. Various application programming interfaces and facilities may also provide efficient and direct access.	[197–200]

(continued)

Table 10.1 (continued)

S. no.	Application	Description	References
10.	Cost-effectiveness	In contrast to traditional methods of healthcare management, telemedicine offers a more economically efficient alternative. This assertion has validity since it eliminates the need for doctors and patients to engage in frequent travel to facilitate doctor–patient interactions. Furthermore, video consultations often involve lower costs compared to in-person consultations, resulting in savings on expenses related to travel and the establishment of medical facilities. In addition, the incorporation of online booking and video conferencing into healthcare solutions has the potential to provide cost savings. This is especially advantageous in regions where individuals may have difficulties accessing healthcare facilities for medical treatment. Telemedicine is also contributing to the expansion of the medical industry's boundaries. The accessibility of research, scientific advancements, and expansion in this particular discipline has significantly increased. The use of video conferencing technology has enhanced the educational experience and training of medical students, making it more authentic and practical.	[201–203]
11.	Enhanced service provision for healthcare professionals and patients	The field of telemedicine has seen significant expansion, resulting in several advantages for both healthcare providers and those receiving medical care. The primary need for healthcare providers is to enhance patient services, a goal that may be accomplished via the implementation of telemedicine systems. These technologies not only facilitate the improvement of doctors' private practices but also enable them to improve and perfect their services. Thus, the development of medical video conferencing technologies in medical institutions is vital for enhancing the field of medicine. The absence of a barrier is a well-recognized factor in the healthcare industry that may be addressed by the use of telemedicine technologies. The reason for this is the severity of the worldwide scarcity of proficient healthcare practitioners. The present deficiency of medical professionals may be effectively addressed by the use of this technological solution.	[204–207]

12.	Electronic health monitoring	Telemedicine systems that include digital health monitoring capabilities, with the integration of artificial intelligence and machine learning, have the potential to enhance the precision of diagnosis and treatment recommendations. The use of video conferencing and machine learning technologies, in conjunction with the analysis of individual patient cases, will allow the enhancement of the algorithm, hence improving the accuracy of diagnosis for each patient. Telemedicine refers to the practice of using networking technologies to establish a connection between doctors located in one geographical region and patients situated in a different one. Appointment, review, and photographic transcript software provide a viable substitute for face-to-face meetings. In the context of a pandemic, the use of digital platforms enables healthcare providers and individuals seeking medical assistance to ensure their safety, while facilitating the expansion of medical services.	[208–210]
13.	Protect patients against infectious diseases	Telemedicine makes it possible for everyone to live in a more secure society. Patients experiencing symptoms of influenza or the common cold must obtain guidance from their healthcare provider before entering the workplace, to prevent the transmission of infectious agents. When offering assistance and monitoring the progress of patients, healthcare personnel are effectively protected against the transmission of infectious illnesses. The comprehensive solution integrates sophisticated network security features with regulated access rights management. This measure may enhance the security of patient information sent via telemedicine equipment. Additionally, it serves as a means to decrease the frequency of trips to the emergency room.	[211–214]

educational institution collaborates with medical professionals to conduct remote visits inside the classroom setting. The healthcare professional can assess the gravity of the circumstances and provide parents with guidance or comfort. The use of telemedicine technology has shown its effectiveness in preventing hospitalization among individuals residing in assisted living homes. Telemedicine is used by on-call doctors to conduct remote visits to determine the need for hospitalization at clinics that are situated at a distance. A variety of telehealth alternatives are now emerging to facilitate the management of healthcare and the receipt of medical treatments. Telemedicine is increasingly being utilized in various medical fields, including dermatology, mental health, medicine, and cardiology, to provide enhanced treatment to populations that have limited access to doctors and clinics. It is widely believed that telemedicine offers a promising avenue for significantly reducing the costs associated with managing health conditions such as asthma and diabetes, which require continuous monitoring of a patient's condition. In the realm of recovery, telemedicine technology addresses these challenges through a range of interventions, such as conducting cognitive or psychiatric assessments, facilitating group counseling sessions, providing patients with internet services, and remotely monitoring their progress. For instance, this technology can assist individuals in recuperating from lung disease by enabling them to undergo video therapy sessions and receive supervised physical rehabilitation at a local satellite site. Notably, cardiovascular disease remains the leading cause of mortality worldwide; however, many adverse events can be prevented through the use of drug therapies or modifications to one's social and lifestyle habits.

10.8 Telemedicine's future in healthcare sustainability

The potential of telemedicine in the healthcare sector to improve sustainability is very promising across multiple areas. Initially, it can reduce the carbon emissions related to conventional healthcare delivery approaches. Telemedicine has the potential to reduce emissions resulting from transportation, by reducing the need for patients to travel to healthcare facilities for regular checkups and nonurgent consultations. Telemedicine facilitates the effective allocation of resources. By leveraging the capability to do remote consultations and diagnostics, healthcare providers may enhance the efficient distribution of staff and equipment, resulting in a reduction in excessive use and wastage. This might potentially result in the implementation of more sustainable healthcare practices via the conservation of resources and reduction of operating expenses. Additionally, the use of telemedicine has the potential to enhance patient participation and education, playing a crucial role in encouraging healthy lives and preventative care. Patients who possess a sense of empowerment and are equipped with comprehensive knowledge about their health tend to have a greater propensity for making choices that are conducive to long-term sustainability.

Consequently, this inclination toward sustainable decision-making serves to alleviate the strain placed on the healthcare system over an extended period. Telemedicine, combining telemonitoring and wearables, allows proactive chronic disease management. It has the potential to reduce acute symptoms, hospitalizations, and their corresponding environmental consequences using remotely monitoring patient data and acting as deemed necessary. The future of telemedicine in the healthcare sector is in line with the sustainability objectives as it effectively mitigates travel-related emissions, optimizes resource allocation, fosters patient participation and education, and facilitates proactive healthcare management. Telemedicine, as technology progresses and becomes increasingly integrated with healthcare systems on a global scale, has the potential to make a substantial contribution toward the establishment of a healthcare ecosystem that is both sustainable and efficient.

Patients will be able to arrange an appointment with the doctor of their choosing in the future after completing a fast and simple registration procedure. Patients will be able to upload their medical histories, verification papers, medical reports, and previous prescriptions rather than typing them out. The patient interface is an effective instrument that enables healthcare professionals to generate a timely and prioritized treatment plan. The system presents a comprehensive overview of the patient's medical and personal documentation, aiding the physician in making informed decisions. The local healthcare infrastructure can efficiently deploy and provide a wide range of medical services, including both urgent and nonurgent treatment.

This approach enables doctors to prioritize in-person attention on high-demand and complex situations, while delegating the management of low-level cases that may be effectively handled remotely. In the future, it is anticipated that emergency departments will use video consultations as a means of triaging their patients, resulting in the expedited release of those with nonemergency conditions. Consequently, there has been a reduction in emergency diversion and an enhancement in the well-being of patients. Numerous healthcare organizations are increasingly embracing telemedicine application software as a strategic measure to proactively plan for future healthcare demands. The platform facilitates the connection between cardiologists and people seeking medical treatment at the rural healthcare center. Additionally, these advancements will reduce the duration of subsequent visits and serve as a valuable adjunct to the existing system. Remote patient management has emerged as a prominent factor in the realm of virtual healthcare, presenting substantial prospects for several enterprises.

10.9 Conclusion

To facilitate long-term behavior modification, telemedicine connects doctors and patients. It is a huge help to the people working in medical offices. This allows staff to devote their time and energy to other valuable activities, such as patient care. With the option of doing virtual visits, doctors may see patients, while also possibly aiding other clinics that have been hit. As a result, the distance between a doctor and a patient is less of a barrier when it comes to making a diagnosis, receiving treatment, or preventing illness. People in remote places, where receiving appropriate medical care is difficult, may benefit from the most comprehensive telemedicine program. This technology has been proven to improve healthcare facilities in recent years by facilitating communication over large geographic distances. It makes it less difficult for people in disadvantaged regions to make and keep appointments. Those who are physically challenged have easier access to medical advice and medications. In general, interactions between telemedicine and healthcare facilities provide an exciting subject for healthcare professionals and researchers. However, the existence of a research deficit in this area may lead to a doubt about the future of the healthcare industry, as we now understand and experience it. Telemedicine catalyzes novel research about the field of built environment, as well as the development of guidelines to support the use of this growing technology. Moreover, the use of telemedicine technology and associated procedures has the potential to significantly influence the planning and design of healthcare facilities. Moreover, telemedicine can disrupt the planning and design of healthcare facilities, as well as restructure them to provide a more sustainable supply of healthcare that takes into account sociodemographic factors and reduces carbon emissions. The use of telemedicine can affect the advancement of healthcare systems and determine the future direction of healthcare facilities in a manner that is characterized by enhanced environmental sustainability, equal access, and social sustainability.

References

[1] Roine R, Ohinmaa A, Hailey D. Assessing telemedicine: A systematic review of the literature. CMAJ. 2001 Sep 18;165(6):765–71. Roine R, Ohinmaa A, Hailey D. Assessing telemedicine: a systematic review of the literature. CMAJ. 2001 Sep 18;165(6):765–71. Roine R, Ohinmaa A, Hailey D. Assessing telemedicine: a systematic review of the literature. CMAJ. 2001 Sep 18;165(6):765–71.

[2] Chin TL. Telemedicine. Secrets to success. Health Data Management. 1998 Jul;6(7):76–79.

[3] Brantley D, Laney-Cummings K, Spivack R. Innovation, Demand and Investment in Telehealth. February 2004. US Department of Commerce. Office of Technology Policy (OTP). Accessed December. 2009;14.

[4] WHO A. Health telematics policy in support of WHO's Health-For-All strategy for global health development. Report of the WHO group consultation on health telematics. 1997,11:11–16.

[5] Cxakır E, Arslan O Turkish Telemedical Assistance Service: Last four years of activity. International Maritime Health. 2018; 69:184–91.

[6] Mahdi SS, Amenta F. Eighty years of CIRM. A journey of commitment and dedication in providing maritime medical assistance. International Maritime Health. 2016; 67:187–95.

[7] Nittari G, Pirillo I, Amenta F et al. The right to medical assistance for seafarers. Ethical and practical consequences of the introduction of telemedicine to improve healthcare on board ships. Marine Policy, Elsevier. 2019;106: 103525.

[8] Nittari G, Khuman R, Baldoni S, Pallotta G, Battineni G, Sirignano A, Amenta F, Ricci G. Telemedicine practice: Review of the current ethical and legal challenges. Telemedicine and e-Health. 2020 Dec 1;26(12):1427–37.

[9] Sagaro GG, Amenta F Past, present, and future perspectives of telemedical assistance at sea: A systematic review. International Maritime Health. 2020;71: 97–104.

[10] Laouyane A. Telemedicine and developing countries. Journal of Telemedicine and Telecare. 1998 Mar;4(2_suppl):1–88.

[11] Alaboudi A, Atkins A, Sharp B, et al. Barriers and challenges in adopting Saudi telemedicine network: The perceptions of decision-makers of healthcare facilities in Saudi Arabia. Journal of Infection and Public Health. 2016; 9:725–33.

[12] Hollander JE, Carr BG. Virtually perfect? Telemedicine for COVID-19. The New England Journal of Medicine. 2020;382:1679–81

[13] Ferenczi GG, Mahmood AN, Bergmann RK. Telemedicine pre and post covid-19: Lessons for commercialisation based on previous use cases. Journal of the International Society for Telemedicine and eHealth. 2020 Dec 14;8:e8–1.

[14] Bhaskar S, Bradley S, Chattu VK, Adisesh A, Nurtazina A, Kyrykbayeva S, Sakhamuri S, Moguilner S, Pandya S, Schroeder S, Banach M. Telemedicine as the new outpatient clinic gone digital: Position paper from the pandemic health system REsilience PROGRAM (REPROGRAM) international consortium (Part 2). Frontiers in Public Health. 2020 Sep 7;8:410.

[15] Wang X, Zhang Z, Yang L, Zhao J. Price and capacity decisions in a telemedicine service system under government subsidy policy. The International Journal of Production Research. 2020:1–14.

[16] Armfield NR, Bradford M, Bradford NK. The clinical use of Skype – For which patients, with which problems and in which settings? A snapshot review of the literature. International Journal of Medical Informatics. 2015;84(10):737–42.

[17] Lolich L, Ricco I, Deusdad B, Timonen V. Embracing technology? Health and social care professionals" attitudes to the deployment of e-health initiatives in elder care services in Catalonia and Ireland. Technological Forecasting and Social Change. 2019;147:63–71.

[18] Portnoy J, Waller M, Elliott T. Telemedicine in the era of COVID-19. Journal of Allergy & Clinical Immunology in Practice. 2020;8(5):1489–91.

[19] Adenuga KI, Iahad NA, Miskon S. Towards reinforcing telemedicine adoption amongst clinicians in Nigeria. International Journal of Medical Informatics. 2017;104:84–96.

[20] Sarhan F. Telemedicine in healthcare. 1: Exploring its uses, benefits and disadvantages. Nursing Times. 2009 Oct 1;105(42):10–13.

[21] Moghadas A, Jamshidi M, Shaderam M. Telemedicine in the healthcare system, 2008. In: World Automation Congress (pp. 1–6). IEEE; 2008 Sep.

[22] Chunara R, Zhao Y, Chen J, Lawrence K, Testa PA, Nov O, Mann DM. Telemedicine and healthcare disparities: A cohort study in a large healthcare system in New York City during COVID-19. Journal of the American Medical Informatics Association. 2021 Jan;28(1):33–41.

[23] Flumignan CD, Rocha AP, Pinto AC, Milby KM, Batista MR, Atallah AN, Saconato H. What do Cochrane systematic reviews say about telemedicine for healthcare? São Paulo Medical Journal. 2019 Apr;137 (2):184–92.

[24] Kaspar BJ. Legislating for a new age in medicine: Defining the telemedicine standard of care to improve healthcare in Iowa. Iowa Law Review. 2013;99:839.

[25] Rockwell KL, Gilroy AS. Incorporating telemedicine as part of COVID-19 outbreak response systems. The American Journal of Managed Care. 2020 Apr 1;26(4):147–48.

[26] Bashshur R, Shannon G, Krupinski E, Grigsby J. The taxonomy of telemedicine. Telemedicine and e-Health. 2011 Jul 1;17(6):484–94.

[27] Funderskov KF, Boe Danbjørg D, Jess M, Munk L, Olsen Zwisler AD, Dieperink KB. Telemedicine in specialised palliative care: Healthcare professionals and their perspectives on video consultations – A qualitative study. Journal of Clinical Nursing. 2019 Nov;28(21–22):3966–76.
[28] Mechanic D. Rethinking medical professionalism: The role of information technology and practice innovations. Milbank Quarterly. 2008;86:327–58.
[29] Graves BA. Telehealth for communities: Toward eliminating rural health disparities. Online Journal of Rural Nursing and Health Care. 2012;10:4–6.
[30] Doarn CR, Pruitt S, Jacobs J, Harris Y, Bott DM, Riley W, Lamer C, Oliver AL. Federal efforts to define and advance telehealth – A work in progress. Telemedicine and e-Health. 2014 May 1;20(5):409–18.
[31] Palen T, Bodily M. PS3-28: Telemedicine specialty consultation in a medically underserved community. Clinical Medicine and Research. 2010;8:200–01
[32] Weinstein RS, Lopez AM, Joseph BA, Erps KA, Holcomb M, Barker GP, Krupinski EA. Telemedicine, telehealth, and mobile health applications that work: Opportunities and barriers. American Journal of Medicine. 2014 Mar 1;127(3):183–87.
[33] Parimbelli E, Bottalico B, Losiouk E, Tomasi M, Santosuosso A, Lanzola G, Quaglini S, Bellazzi R. Trusting telemedicine: A discussion on risks, safety, legal implications and liability of involved stakeholders. International Journal of Medical Informatics. 2018 Apr 1;112:90–98.
[34] Wilson LS, Maeder AJ. Recent directions in telemedicine: Review of trends in research and practice. Healthcare Informatics Research. 2015 Oct;21(4):213.
[35] Hajesmaeel-Gohari S, Bahaadinbeigy K. The most used questionnaires for evaluating telemedicine services. BMC Medical Informatics and Decision Making. 2021 Dec;21(1):1–1.
[36] Lupton D, Maslen S. Telemedicine and the senses: A review. Sociology of Health and Illness. 2017 Nov;39(8):1557–71.
[37] Sun C, Chrysikou E, Savvopoulou E, Hernandez-Garcia E, Fatah Gen. Schieck A. Healthcare built environment and telemedicine practice for social and environmental sustainability. Sustainability. 2023 Feb 2;15(3):2697.
[38] Shaw DK. Overview of telehealth and its application to cardiopulmonary physical therapy. Cardiopulmonary Physical Therapy Journal. 2009;20:13–18.
[39] Wu X, Hąbek P. Trends in corporate social responsibility reporting. The case of Chinese listed companies. Sustainability. 2021 Aug 3;13(15):8640.
[40] Lyon BL. Stress, Coping, and Health. Handbook of Stress, Coping and Health: Implications for Nursing Research, Theory, and Practice (pp. 3–23). 2000.
[41] Golden BR, Martin RL. Aligning the stars: Using systems thinking to (re) design Canadian healthcare. Healthcare Quarterly (Toronto, Ont.). 2004 Jan 1;7(4):34–42.
[42] Chrysikou E. Why we need new architectural and design paradigms to meet the needs of vulnerable people. Palgrave Communications. 2018;4:116.
[43] Mashima PA, Doarn RC. Overview of telehealth activities in speech-language pathology. Telemedicine E-Health. 2008;14:1101–17.
[44] Kovac M. HIPAA and telehealth: protecting health information in a digital world. Journal of Intellectual Freedom & Privacy. 2021 Jul 28;6(2):6–9.
[45] Turvey C, Coleman M, Dennison O, Drude K, Goldenson M, Hirsch P, Jueneman R, Kramer GM, Luxton DD, Maheu MM, Malik TS. ATA practice guidelines for video-based online mental health services. Telemedicine and e-Health. 2013 Sep 1;19(9):722–30.
[46] Kluge Eike-Henner W Electronic Health Record: Ethical Considerations (pp. 105–33). Academic Imprint: San Diego, CA, USA; 2020.
[47] Hutchings R. The impact of Covid-19 on the use of digital technology in the NHS. Nuffield Trust. 2020 Aug 27;27:2002–08.
[48] Lacey Bryant S, Bridgen R, Hopkins E, McLaren C, Stewart D. NHS knowledge and library services in England in the digital age. Health Information & Libraries Journal. 2022 Dec;39(4):385–91.

[49] Burri H, Heidbüchel H, Jung W, Brugada P. Remote monitoring: A cost or an investment? Europace 2011;13:ii44–ii48.

[50] Calò L, Gargaro A, De Ruvo E, Palozzi G, Sciarra L, Rebecchi M, Guarracini F, Fagagnini A, Piroli E, Lioy E. Economic impact of remote monitoring on ordinary follow-up of implantable cardioverter defibrillators as compared with conventional in-hospital visits. A single-center prospective and randomized study. Journal of Interventional Cardiac Electrophysiology. 2013;37:69–78.

[51] Hasan A, Paul V. Telemonitoring in chronic heart failure. European Heart Journal. 2011;32:1457–64.

[52] Norton SA, Burdick AE, Phillips CM, Berman B. Teledermatology and underserved populations. Archives of Dermatology. 1997;133:197–200.

[53] Albahri AS, Alwan JK, Taha ZK, Ismail SF, Hamid RA, Zaidan AA, Albahri OS, Zaidan BB, Alamoodi AH, Alsalem MA. IoT-based telemedicine for disease prevention and health promotion: State-of-the-Art. Journal of Network and Computer Applications. 2021 Jan 1;173:102873.

[54] Bashshur RL, Shannon GW, Krupinski EA, Grigsby J, Kvedar JC, Weinstein RS, Sanders JH, Rheuban KS, Nesbitt TS, Alverson DC, Merrell RC. National telemedicine initiatives: Essential to healthcare reform. Telemedicine and e-Health. 2009 Jul 1;15(6):600–10.

[55] Manchanda S. Telemedicine–getting care to patients closer to home. American Journal of Respiratory & Critical Care Medicine. 2020 Jun 15;201(12):P26–7

[56] El-Shafai W, Khallaf F, El-Rabaie ES, El-Samie FE. Robust medical image encryption based on DNA-chaos cryptosystem for secure telemedicine and healthcare applications. Journal of Ambient Intelligence and Humanized Computing. 2021 Oct;12:9007–35.

[57] Gareev I, Gallyametdinov A, Beylerli O, Valitov E, Alyshov A, Pavlov V, Izmailov A, Zhao S. The opportunities and challenges of telemedicine during COVID-19 pandemic. Frontiers in Bioscience-Elite. 2021 Dec 20;13(2):291–98.

[58] Mars M. Telemedicine and advances in urban and rural healthcare delivery in Africa. Progress in Cardiovascular Diseases. 2013 Nov 1;56(3):326–35.

[59] Chau PY, Hu PJ. Investigating healthcare professionals' decisions to accept telemedicine technology: An empirical test of competing theories. Information and Management. 2002 Jan 1;39(4):297–311.

[60] Heinzelmann PJ, Lugn NE, Kvedar JC. Telemedicine in the future, Journal of Telemedicine and Telecare. 2005 Dec;11(8):384–90.

[61] Kohnke A, Cole ML, Bush R. Incorporating UTAUT predictors for understanding home care patients' and clinician's acceptance of healthcare telemedicine equipment. International Journal of Innovation and Technology Management. 2014 Jul;9(2):29–41.

[62] Bajowala SS, Milosch J, Bansal C. Telemedicine pays billing and coding update. Current Allergy and Asthma Reports. 2020 Oct;20(10):1–9.

[63] Lin JC, Kavousi Y, Sullivan B, Stevens C. Analysis of outpatient telemedicine reimbursement in an integrated healthcare system. Annals of Vascular Surgery. 65 (2020 May 1) 100–06.

[64] Javaid M, Haleem A, Singh RP, Suman R. Significance of Quality 4.0 towards comprehensive enhancement in manufacturing sector. Sensors International. 2021 Jan 1;2:100109.

[65] Ning AY, Cabrera CI, D'Anza B. Telemedicine in otolaryngology: A systematic review of image quality, diagnostic concordance, and patient and provider satisfaction. Annals of Otology, Rhinology & Laryngology. 2021 Feb;130(2):195–204.

[66] Salehahmadi Z, Hajialiasghari F. Telemedicine in Iran: Chances and challenges. World Journal of Plastic Surgery. 2013 Jan;2(1):18.

[67] Von Wangenheim A, De Souza Nobre Lf, Tognoli H, Nassar SM, Ho K. User satisfaction with asynchronous telemedicine: A study of users of Santa Catarina's system of telemedicine and telehealth. Telemedicine and e-Health. 2012 Jun 1;18(5):339–46.

[68] Ayatollahi H, Mirani N, Nazari F, Razavi N. Iranian healthcare professionals' perspectives about factors influencing the use of telemedicine in diabetes management. World Journal of Diabetes. 2018 Jun 6;9(6):92.

[69] Kadir MA. Role of telemedicine in healthcare during COVID-19 pandemic in developing countries. Telehealth and Medicine Today. 2020 Apr 30;5(2).
[70] Whitten PS, Mair FS, Haycox A, May CR, Williams TL, Hellmich S. Systematic review of cost-effectiveness studies of telemedicine interventions. BMJ. 2002 Jun 15;324(7351):1434–37.
[71] Hooshmand M, Yao K. Challenges facing children with special healthcare needs and their families: Telemedicine as a bridge to care. Telemedicine and e-Health. 2017 Jan 1;23(1):18–24.
[72] Papa A, Mital M, Pisano P, Del Giudice M. E-health and well-being monitoring using smart healthcare devices: An empirical investigation. Technological Forecasting and Social Change. 2020;153:119–226.
[73] Rubbio I, Bruccoleri M, Pietrosi A, Ragonese B. Digital health technology enhances resilient behaviour: Evidence from the ward. International Journal of Operations & Production Management. 2020;40:34–67.
[74] Martínez-Caro E, Cegarra-Navarro JG, García-p´erez A, Fait M. Healthcare service evolution towards the Internet of Things: An end-user perspective. Technological Forecasting and Social Change. 2018;136:268–76.
[75] Sims JM. Communities of practice: Telemedicine and online medical communities. Technological Forecasting and Social Change. 2018;126:53–63.
[76] Lu W, Hou H, Ma R, Chen H, Zhang R, Cui F, Zhang Q, Gao Y, Wang X, Bu C, Zhao J, Zhai Y. Influencing factors of patient satisfaction in teleconsultation: A cross-sectional study. Technological Forecasting and Social Change. 2021;168:120775.
[77] Laurenza E, Quintano M, Schiavone F, Vrontis D. The effect of digital technologies adoption in the healthcare industry: A case-based analysis. Business Process Management Journal. 2018;24:1124–44
[78] Liu X, Hu M, Helm JE, Lavieri MS, Skolarus TA. Missed opportunities in preventing hospital readmissions: Redesigning post-discharge checkup policies. Production and Operations Management. 2018;27:2226–50.
[79] Miao R, Wu Q, Wang Z, Zhang X, Song Y, Zhang H, Sun Q, Jiang Z. Factors that influence users' adoption intention of mobile health: A structural equation modeling approach. International Journal of Production Research. 2017;55:5801–15.
[80] Senot C, Chandrasekaran A, Ward PT. Collaboration between service professionals during the delivery of health care: Evidence from a multiple-case study in US hospitals. Journal of Operations Management. 2016;42:67–79.
[81] Vesselkov A, Hämmäinen H, Töyli J. Technology and value network evolution in telehealth. Technological Forecasting and Social Change. 2018;134;207–22.
[82] Villalba-Mora E, C I, V FL, M I. Adoption of health information technologies by physicians for clinical practice: The Andalusian case. International Journal of Medical Informatics. 2015;84:477–85.
[83] Wang Y, Kung L, Byrd TA. Big data analytics: Understanding its capabilities and potential benefits for healthcare organizations. Technological Forecasting and Social Change. 2018;126:3–13.
[84] Al-Samarraie H, Ghazal S, Alzahrani AI, Moody L. Telemedicine in Middle Eastern Countries: Progress, barriers, and policy recommendations. International Journal of Medical Informatics. 2020;141:104232
[85] Lee SM, Lee D. Opportunities and challenges for contactless healthcare services in the post-COVID-19 Era. Technological Forecasting and Social Change. 2021;167:120712.
[86] Cobelli N, Cassia F, Burro R. Factors affecting the choices of adoption/nonadoption of future technologies during coronavirus pandemic. Technological Forecasting and Social Change. 2021;169, 120814.
[87] Baudier P, Kondrateva G, Ammi C, Chang V, Schiavone F. Patients' perceptions of teleconsultation during COVID-19: A cross-national study. Technological Forecasting and Social Change. 2021;163:120510.

[88] Kaplan B. Revisiting health information technology ethical, legal, and social issues and evaluation: Telehealth/telemedicine and COVID-19. International Journal of Medical Informatics. 2020 Nov 1;143:104239.
[89] Rezaei M, Jafari-Sadeghi V, Cao D, Mahdiraji HA. Key indicators of ethical challenges in digital healthcare: A combined Delphi exploration and confirmative factor analysis approach with evidence from Khorasan province in Iran. Technological Forecasting and Social Change. 2021 Jun 1;167:120724.
[90] Rho MJ, Choi I, Young, Lee J. Predictive factors of telemedicine service acceptance and behavioral intention of physicians. International Journal of Medical Informatics. 2014;83:559–71.
[91] Alajlani M, Clarke M. Effect of culture on acceptance of telemedicine in Middle Eastern countries: Case study of Jordan and Syria. Telemedicine and e-Health. 2013 Apr 1;19(4):305–11.
[92] Scott Kruse C, Karem P, Shifflett K, Vegi L, Ravi K, Brooks M. Evaluating barriers to adopting telemedicine worldwide: A systematic review. Journal of Telemedicine and Telecare. 2018;24:4–12.
[93] Jansen-Kosterink S, Dekker-van Weering M, Van Velsen L. Patient acceptance of a telemedicine service for rehabilitation care: A focus group study. International Journal of Medical Informatics. 2019;125:22–29.
[94] van Velsen L, Tabak M, Hermens H. Measuring patient trust in telemedicine services: Development of a survey instrument and its validation for an anticoagulation web-service. International Journal of Medical Informatics. 2017;97:52–58.
[95] Doarn CR. Telehealth: A very useful tool that enables and improves patient access. The Journal of the American Board of Family Medicine. 2016 Jul 1;29(4):430–31.
[96] Mach AL, Kinzer J. Legislative Actions to Modify the Affordable Care Act in the 111th-115th Congresses. Washington, DC: Congressional Research Service; 2018, 27.
[97] Paul DL, Pearlson KE, McDaniel RR. Assessing technological barriers to telemedicine: Technology-management implications. IEEE Transactions on Engineering Management. 1999 Aug;46(3):279–88.
[98] Hailey D, Roine R, Ohinmaa A. Systematic review of evidence for the benefits of telemedicine. Journal of Telemedicine and Telecare. 2002;8:1–7.
[99] Marcin JP, Nesbitt TS, Cole SL. et al. Changes in diagnosis, treatment, and clinical improvement among patients receiving telemedicine consultations. Telemedicine Journal and E-health. 2005;11:36–43.
[100] Kruger LG. Broadband Loan and Grant Programs in the USDA's Rural Utilities Service. Congressional Research Service. Washington, DC: Library of Congress; 2013.
[101] Lobley D. The economics of telemedicine. Journal of Telemedicine and Telecare. 1997 Sep 1;3(3):117–25.
[102] Doolittle GC, Spaulding AON, Williams AR. The decreasing cost of telemedicine and telehealth. Telemed E-Health. 2011;17:671–75.
[103] Broens T, Vollenbroek-Hutten M, Hermens H. et al. Determinants of successful telemedicine implementations: A literature study. Journal of Telemedicine and Telecare. 2007;13:303–09.
[104] Doarn CR. The last challenges and barriers to the development of telemedicine programs. Studies in Health Technology and Informatics. 2008;131:45.
[105] Jennett P, Yeo M, Pauls M. et al. Organizational readiness for telemedicine: Implications for success and failure. Journal of Telemedicine and Telecare. 2003;9:27–30.
[106] Grigsby B, Brega AG, Bennett RE. et al. The slow pace of interactive video telemedicine adoption: The perspective of telemedicine program administrators on physician participation. Telemedicine Journal and E-health. 2007;13:645–56
[107] Hartley D. Rural health disparities, population health, and rural culture. American Journal of Public Health. 2004;94:1675–78.
[108] Wade VA, Eliott JA, Hiller JE. Clinician acceptance is the key factor for sustainable telehealth services. Qualitative Health Research. 2014;24(5):682–92.

[109] May C. A rational model for assessing and evaluating complex interventions in health care. BMC Health Services Research. 2006;6:86.
[110] May CR, Mair F, Finch T. et al. Development of a theory of implementation and integration: Normalization process theory. Implementation Science. 2009;4:29.
[111] Trakoli A. Raising awareness of the health effects of environmental exposures. BMJ. 2019;365:l1878.
[112] Malik A, Lenzen M, McAlister S, McGain F. The carbon footprint of Australian health care. Lancet Planet Health. 2018;2(1):e27–35.
[113] Purohit A, Smith J, Hibble A. Does telemedicine reduce the carbon footprint of healthcare? A systematic review. Future Healthcare Journal. 2021 Mar;8(1):e85.
[114] Lenzen M, Malik A, Li M. et al. The environmental footprint of health care: A global assessment. Lancet Planet Health. 2020;4(7):e271–9.
[115] Goshua A, Gomez J, Erny B, Burke M, Luby S, Sokolow S, LaBeaud AD, Auerbach P, Gisondi MA, Nadeau K. Addressing climate change and its effects on human health: A call to action for medical schools. Academic Medicine. 2021 Mar 1;96(3):324–28.
[116] Nidhi A. Reducing the carbon footprint of clinical trials: Implementing sustainable practices in clinical research. Nature Environment & Pollution Technology. 2023 Dec 1;22(4).
[117] Bi P, Hansen A. Carbon emissions and public health: An inverse association? Lancet Planet Health 2018;2(1):e8–9.
[118] Woodward A, Smith KR, Campbell-Lendrum D. et al. Climate change and health: On the latest IPCC report. Lancet. 2014;383(9924):1185–89.
[119] Yellowlees PM, Chorba K, Burke Parish M, Wynn-Jones H, Nafiz N. Telemedicine can make healthcare greener. Telemedicine Journal e-Health. 2010;16(2):229–32.
[120] Choi PJ, Oskouian RJ, Tubbs RS. Telesurgery: Past, present, and future. Cureus. 2018;10(5), e2716–e2716.
[121] Afarikumah E. Electronic health in Ghana: Current status and future prospects. Online Journal of Public Health Informatics. 2014;5(3), 230–230.
[122] Gillman-Wells CC, Sankar TK, Vadodaria S. COVID-19 reducing the risks: Telemedicine is the new norm for surgical consultations and communications. Aesthetic Plastic Surgery. 2021 Feb;45:343–48.
[123] Holmner A, Ebi KL, Lazuardi L, Nilsson M. Carbon footprint of telemedicine solutions–unexplored opportunity for reducing carbon emissions in the health sector. PLoS One. 2014;9(9):e105040.
[124] Alotaibi YK, Federico F. The impact of health information technology on patient safety. Saudi Medical Journal. 2017;38(12):1173–80.
[125] Tsagkaris C, Hoian AV, Ahmad S, Essar MY, Campbell LW, Grobusch L, Angelopoulos T, Kalaitzidis K. Using telemedicine for a lower carbon footprint in healthcare: A twofold tale of healing. The Journal of Climate Change and Health. 2021 Mar 1;1:100006.
[126] Jeddi Z, Bohr A. Chapter 9 Remote patient monitoring using artificial intelligence. In: Bohr A, Memarzadeh K, (eds.). Artificial Intelligence in Healthcare (pp. 203–34). Academic Press; 2020.
[127] Farahmand S, Shabestari O, Pakrah M, Hossein-Nejad H, Arbab M, Bagheri-Hariri S. Artificial intelligence-based triage for patients with acute abdominal pain in emergency department; a diagnostic accuracy study. Advanced Journal of Emergency Medicine. 2017;1(1):e5.
[128] Swedberg L, Chiriac EH, Törnkvist L, Hylander I. From risky to safer home care: Health care assistants striving to overcome a lack of training, supervision, and support. International Journal of Qualitative Studies on Health and Well-being. 2013;8(1):20758
[129] Rosen JM, Kun L, Mosher RE. et al. Cybercare 2.0: Meeting the challenge of the global burden of disease in 2030. Health Technol (Berl). 2016;6(1):35–51
[130] Ploeg J, Valaitis RK, Cleghorn L. et al. Perceptions of older adults in Ontario, Canada on the implementation and impact of a primary care programme, health teams advancing patient

experience: Strengthening Quality (Health TAPESTRY): A descriptive qualitative study. BMJ Open [Internet]. 2019;9(6), e026257.

[131] Acharya RV, Rai JJ. Evaluation of patient and doctor perception toward the use of telemedicine in Apollo TeleHealth Services, India. Journal of Family Medicine and Primary Care. 2016 Oct;5(4):798–803.

[132] Shamim-Uzzaman QA, Bae CJ, Ehsan Z, Setty AR, Devine M, Dhankikar S. et al. The use of telemedicine for the diagnosis and treatment of sleep disorders: An American academy of sleep medicine update. Journal of Clinical Sleep Medicine. 2021 Feb 18:jcsm-9194.

[133] Ohinmaa A, Hailey D, Roine R. Elements for assessment of telemedicine applications. International Journal of Technology Assessment in Health Care. 2001 Apr 1;17(2):190–202.

[134] Zanaboni P, Knarvik U, Wootton R. Adoption of routine telemedicine in Norway: The current picture. Global Health Action Glob Health Action. 2014 Dec 1;7(1):22801.

[135] Wootton R, Bahaadinbeigy K, Hailey D. Estimating travel reduction associated with the use of telemedicine by patients and healthcare professionals: Proposal for quantitative synthesis in a systematic review. BMC Health Services Research. 2011 Dec;11(1):185.

[136] Xue Y, Liang H, Mbarika V, Hauser R, Schwager P, Kassa Getahun MK. Investigating the resistance to telemedicine in Ethiopia. International Journal of Medical Informatics. 2015 Aug 1;84(8):537–47.

[137] Khoong EC, Butler BA, Mesina O, Su G, DeFries TB, Nijagal M. et al. Patient interest in and barriers to telemedicine video visits in a multilingual urban safety-net system. Journal of the American Medical Informatics Association. 2021 Feb;28(2):349–53.

[138] De Toledo P, Jiménez S, Del Pozo F, Roca J, Alonso A, Hernandez C. Telemedicine experience for chronic care in COPD. IEEE Transactions on Biomedical Engineering. 2006 Jul 5;10(3):567–73.

[139] Lin CF. Mobile telemedicine: A survey study. Journal of Medical Systems [Internet]. 2012 Apr;36(2):511–20.

[140] Whitten P. Telemedicine: Communication technologies that revolutionise healthcare services. Generations. 2006 Jul 1;30(2):20–24.

[141] Hu PH. Evaluating telemedicine systems success: A revised model. In: 36th Annual Hawaii International Conference on System Sciences, 2003. Proceedings of the 2003 Jan 6 (pp. 8–).

[142] Acheampong F, Vimarlund V. Business models for telemedicine services: A literature review. Health System. 2015 Nov 1;4(3):189–203.

[143] Kim YS. Telemedicine in the USA with focus on clinical applications and issues. Yonsei Medical Journal. 2004 Oct 1;45(5):761–75.

[144] Klaassen B, Van Beijnum Bj, Hermens HJ. Usability in telemedicine systems – A literature survey. International Journal of Medical Informatics. 2016 Sep 1;93:57–69.

[145] Fox KC, Somes GW, Waters TM. Timeliness and access to healthcare services via telemedicine for adolescents in state correctional facilities. The Journal of Adolescent Health: Official Publication of the Society for Adolescent Medicine. 2007 Aug 1;41(2):161–67.

[146] Coldebella B, Armfield NR, Bambling M, Hansen J, Edirippulige S. The use of telemedicine for delivering healthcare to bariatric surgery patients: A literature review. Journal of Telemedicine and Telecare. 2018 Dec;24(10):651–60.

[147] Xiao Y, Shen X, Sun BO, Cai L. Security and privacy in RFID and applications in telemedicine. IEEE Communications Magazine. 2006 May 15;44(4):64–72.

[148] Hersh WR, Helfand M, Wallace J, Kraemer D, Patterson P, Shapiro S. et al. Clinical outcomes resulting from telemedicine interventions: A systematic review. BMC Medical Informatics and Decision Making. 2001 Dec;1(1):5.

[149] Al-Sofiani ME, Alyusuf EY, Alharthi S, Alguwaihes AM, Al-Khalifah R, Alfadda A. Rapid implementation of a diabetes telemedicine clinic during the coronavirus disease 2019 outbreak: Our protocol, experience, and satisfaction reports in Saudi Arabia. Journal of Diabetes Science and Technology. 2021 Mar;15(2):329–38.

[150] Ateriya N, Saraf A, Meshram VP, Setia P. Telemedicine and virtual consultation: The Indian perspective. The National Medical Journal of India. 2018 Jul 1;31(4):215–18.
[151] Leite H, Hodgkinson IR, Gruber T. New development: 'Healing at a distance' – Telemedicine and COVID-19. Public Money and Management. 2020 Aug 17;40(6):483–85.
[152] Constantinides P, Barrett M. Negotiating ICT development and use: The case of a telemedicine system in the healthcare region of Crete. Information and Organization. 2006 Jan 1;16(1):27–55.
[153] Vasquez-Cevallos LA, Bobokova J, González-Granda PV, Iniesta JM, Gómez EJ, Hernando ME. Design and technical validation of a telemedicine service for rural healthcare in Ecuador. Telemedicine Journal and E-health. 2018 Jul 1;24(7):544–51.
[154] Kerleau M, Pelletier-Fleury N. Restructuring of the healthcare system and the diffusion of telemedicine. The European Journal of Health Economics. 2002 Sep 1;3(3):207–14.
[155] Ishfaq R, Raja U. Bridging the healthcare access divide: A strategic planning model for rural telemedicine network. Science Journal. 2015 Aug;46(4):755–90.
[156] Javaid M, Khan IH, Vaishya R, Singh RP, Vaish A. Data analytics applications for COVID-19 pandemic. Current Medicine Research and Practice. 2021 Mar 1;11(2):105.
[157] Haleem A, Javaid M, Suman R, Singh RP. 3D printing applications for radiology: An overview. Indian Journal of Radiology and Imaging. 2021 Jan;31(1):10–17.
[158] Haleem A, Javaid M, Rab S. Impact of additive manufacturing in different areas of Industry 4.0. International Journal of Logistics Systems and Management. 2020;37(2):239–51.
[159] Suman R, Javaid M, Choudhary SK, Haleem A, Singh RP, Nandan D. et al. Impact of COVID-19 pandemic on particulate matter (PM) concentration and harmful gaseous components on Indian metros. Sustainable Operations and Computers. 2021 Jan 1;2:1–.
[160] Gupta P, Haleem A, Javaid M. Designing of a carburettor body for ethanol blended fuel by using CFD analysis tool and 3D scanning technology. 2019;78:466–72.
[161] Javaid M, Babu S, Rab S, Vaishya R, Haleem A. Tribological review of medical implants manufactured by additive manufacturing. In Tribology and Sustainability. 379–95.
[162] Larsen SB, Sørensen NS, Petersen MG, Kjeldsen GF. Towards a shared service centre for telemedicine: Telemedicine in Denmark, and a possible way forward. Health Informatics Journal. 2016 Dec;22(4):815–27.
[163] Ekeland K, Jensen AG, Rasmussen LK, Pedersen J, Bowes CD, Flottorp A. et al. A model for assessment of telemedicine applications: Mast. International Journal of Technology Assessment in Health Care. 2012 Jan;28(1):44–51.
[164] DeSilva S, Vaidya SS. The application of telemedicine to pediatric obesity: Lessons from the past decade. Telemedicine and E-health. 2021 Feb 1;27(2):159–66.
[165] Philips R, Seim N, Matrka L, Locklear B, Moberly AC, Inman G M et al. Cost savings associated with an outpatient otolaryngology telemedicine clinic. Laryngoscope Investig Otolaryngol. 2019 Apr;4(2):234–40.
[166] Majerowicz A, Tracy S, Majerowicz S. Tracy, telemedicine: Bridging gaps in healthcare delivery. Journal AHIMA. 2010 May;81(5):52–53.
[167] Barbosa W, Waddell KZ, Myers E, Dorsey T, ER. Improving access to care: Telemedicine across medical domains. Annual Review of Public Health. 2021 Apr 1;42:463–81.
[168] Baldwin LP, Eldabi MC, Jones T, RW. Telemedicine and its role in improving communication in healthcare, Logist. Information Management. 2002 Oct 1.
[169] B, Razavi-Khorasani N, Moghadam AD, Farokhi E, Rezaei N. COVID-19 and telemedicine: Immediate action required for maintaining healthcare providers well-being. Journal of Clinical Virology. 2020 May 1;126:104345.
[170] Adewale OS. An internet-based telemedicine system in Nigeria. International Journal of Information Management. 2004 Jun 1;24(3):221–34.

[171] G, Rabbito C, Di Lazzaro V, Amantini A, Grippo A, Carrai R, Pasqui R. et al. The Italian technical/administrative recommendations for telemedicine in clinical neurophysiology. Neurological Sciences. 2021 May;42(5):1923–31.
[172] Jnr BA. Use of telemedicine and virtual care for remote treatment in response to COVID-19 pandemic. Journal of Medical Systems [Internet]. 2020 Jul;44(7):1–9.
[173] Mathur P, Lalchandani SS, Mehta A. Evolving role of telemedicine in health care delivery in India. Primary Health Care. 2017;7(260):2167–1079.
[174] Hojabri R, Borousan E, Manafi M. Impact of using telemedicine on knowledge management in healthcare organizations: A case study. African Journal of Business Management. 2012 Feb 1;6(4):1604–13.
[175] Krupinski EA, Weinstein RS. Telemedicine is an academic center – The Arizona telemedicine program. Telemedicine and E-health. 2013 May 1;19(5):349–56.
[176] Haleem A, Javaid M, Singh RP, Suman R. Significant roles of 4D printing using smart materials in the field of manufacturing. Advanced Industrial and Engineering Polymer Research. 2021 Oct 1;4(4):301–11.
[177] Iasbech PA, Lavarda RA. Strategy and practices: A qualitative study of a Brazilian public healthcare system of telemedicine. International Journal of Public Sector Management. 2018 Apr 9.
[178] Javaid A, Singh M, Suman RP, Haleem R. Quality 4.0 technologies to enhance traditional Chinese medicine for overcoming healthcare challenges during COVID19, Digital Chinese Medicine. 2021 Jun 1;4(2):71–80.
[179] Ng HS, Tan HS, Wong CM, Wong CC. Wireless technologies for telemedicine. BT Technology Journal. 2006 Apr 1;24(2):130–37.
[180] Al-Qirim A-QN. Championing telemedicine adoption and utilisation in healthcare organisations in New Zealand. International Journal of Medical Informatics. 2007 Jan 1;76(1):42–54.
[181] Luciano E, Mahmood MA, Mansouri Rad P. Telemedicine adoption issues in the United States and Brazil: Perception of healthcare professionals. Health Informatics Journal. 2020 Dec;26(4):2344–61.
[182] McKelvey EF, Hitt SS, Smith WC, Azam MV, Lowery GA, Lowery CL. The use of telemedicine in obstetrics: A review of the literature. Obstetrical & Gynecological Survey. 2011 Mar 1;66(3):170–78.
[183] Palozzi G, Schettini I, A C. Enhancing the sustainable goal of access to healthcare: Findings from a literature review on telemedicine employment in rural areas. Sustainability. 2020 Jan;12(8):3318.
[184] De Moraes ER, Cirenza C, Lopes RD, Carvalho AC, Guimaraes PO, Rodrigues AA, De Paola AAE. et al. Prevalence of atrial fibrillation and stroke risk assessment based on telemedicine screening tools in a primary healthcare setting. European Journal of Internal Medicine. 2019 Sep 1;67:36–41.
[185] Ghassemi A, Sumon M, Amini K, Hood SB, Sikka C. Lack of telemedicine training in academic medicine: Are we preparing the next generation? Telemedicine and e-Health. Pourmand. 2021 Jan 1;27(1):62–67.
[186] Nirupama VG, Taneja Y, Chellaiyan N, A. Telemedicine in India: Where do we stand? Journal of Family Medicine and Primary Care. 2019 Jun;8(6):1872–76.
[187] Tasneem I, Ariz A, Bharti D, Haleem A, Javaid M, Bahl S. 3D printing technology and its significant applications in the context of healthcare education. Journal of Industrial Integration and Management. 2021 May 25;1–8.
[188] Memon M, Memon N, Misbahuddin MS. In: The Role of Service-Oriented Architecture in Telemedicine Healthcare System International Conference On Complex, Intelligent and Software Intensive Systems (pp. 208–14). IEEE Publications; 2009.
[189] Drake C, Lian T, Cameron B, Medynskaya K, Bosworth HB, Shah K. Understanding telemedicine's 'new normal': Variations in telemedicine use by specialty line and patient demographics. Telemedicine and E-health. 2021 Mar 25.
[190] Brown EM. The Ontario telemedicine network: A case report. Telemedicine Journal and E-health. 2013 May 1;19(5):373–76.

[191] Pandian PS. An overview of telemedicine technologies for healthcare applications. International Journal of Biomedical and Clinical Engineering. 2016 Jul 1;5(2):29–52.
[192] Vitacca S, Paletta M, Giordano L, Balbi B A. Telemedicine: A new frontier for effective healthcare services. Scalvini Monaldi Archives for Chest Disease. 2004 Dec 30;61(4).
[193] Ito J, Edirippulige S, Aono T, Armfield NR, Armfield NR. The use of telemedicine for delivering healthcare in Japan: Systematic review of literature published in Japanese and English languages. Journal of Telemedicine and Telecare. 2017 Dec;23(10):828–34.
[194] Greene G, Lightsey NE, Crawford IVHM, Striano AM, Simpson BM, Schoenfeld AK. Telemedicine use in orthopaedic surgery varies by race, ethnicity, primary language, and insurance status. Clinical Orthopaedics and Related Research. 2021 May 10:10–97.
[195] A.h, Sapci AH, Sapci HA. Digital continuous healthcare and disruptive medical technologies: M-Health and telemedicine skills training for data-driven healthcare. Journal of Telemedicine and Telecare. 2019 Dec;25(10):623–35.
[196] Ryu S. History of telemedicine: Evolution, context, and transformation. Healthc Inform Research. 2010 Mar;16(1):65.
[197] O'Shea J, Berger R, Samra C, Van Durme D, Van Durme D. Telemedicine in education: Bridging the gap. Education for Health (Abingdon). 2015 Jan 1;28(1):64–67.
[198] Sim R, Lee SWH. Patient preference and satisfaction with the use of telemedicine for glycemic control in patients with type 2 diabetes: A review. Patient Prefer Adherence. 2021;15:283–98.
[199] Sharma R, Fleischut P, Barchi D. Telemedicine and its transformation of emergency care: A case study of one of the largest US integrated healthcare delivery systems. International Journal of Emergency Medicine. 2017 Dec;10(1):1–4.
[200] Le LB, Viramontes HKR, Meneses KG, Dong TS, Saab S, Saab S Patient satisfaction and healthcare utilisation using telemedicine in liver transplant recipients. Digestive Diseases & Sciences. 2019 May;64(5):1150–57.
[201] L.r, Demaerschalk BM, Schwamm LH, Adeoye OM, Audebert HJ, Fanale CV, Hess DC. et al. Telemedicine quality and outcomes in stroke: A scientific statement for healthcare professionals from the American Heart Association/American Stroke Association. Stroke. 2017 Jan;48(1):e3–e25.
[202] Jennett P, Watanabe M. Healthcare and telemedicine: Ongoing and evolving challenges. Dis Manag Health Outcomes. 2006 Dec;14(1):9–13.
[203] Khan SA, Omar H. Teledentistry in practice: Literature review. Telemedicine and E-health. 2013 Jul 1;19(7):565–67.
[204] Fernández CE, Maturana CA, Coloma SI, Carrasco-Labra A, Giacaman RA. Teledentistry and mHealth for promotion and prevention of oral health: A systematic review and meta-analysis. Journal of Dental Research. 2021 Aug;100(9):914–27.
[205] S, Sood V, Mbarika S, Jugoo R, Dookhy CR, Doarn NP, Merrell RC. et al. What is telemedicine? A collection of 104 peer-reviewed perspectives and theoretical underpinnings. Telemedicine Journal and E-health. 2007 Oct 1;13(5):573–90.
[206] D.l, Saxena A, Rubens M, Valle CA, Williams LM, Veledar E, Gidel LT. et al. Impact of telemedicine on mortality, length of stay, and cost among patients in progressive care units: Experience from a large healthcare system. Critical Care Medicine. 2018 May;46(5):728–35.
[207] Schwalb P, Klecun E. The role of contradictions and norms in the design and use of telemedicine: Healthcare professionals' perspective. AIS Transactions on Human-Computer Interaction. 2019;11(3):117–35.
[208] Abdellatif MM, Mohamed W. Telemedicine: An IoT-based remote healthcare system. International Journal of Online and Biomedical Engineering. 2020 Jun 1;16(6).
[209] Jin Z, Chen Y. Telemedicine in the cloud era: Prospects and challenges. IEEE Pervasive Computing. 2015 Feb 4;14(1):54–61.

[210] Dasgupta A, Deb S. Telemedicine: A new horizon in public health in India. Indian Journal of Community Medicine: Official Publication of Indian Association of Preventive & Social Medicine. 2008 Jan;33(1):3.

[211] Djamasbi S, Fruhling A, Loiacono E. The influence of affect, attitude and usefulness in the acceptance of telemedicine systems. The Journal of Information Technology Theory and Application. 2009 Jun 22;10(1):5.

[212] C.P. Telemedicine and e-health. Waegemann mHealth Next Gener Telemed? 2010 Jan 1;16(1):23–26.

[213] Pilosof E, Hinings NP, Zimlichman B, Oborn E. Inst logics innov times crisis telemed digit 'PPE', Information Organisation. 2021 Mar 1;31(1):100340.

[214] Chen ET. Considerations of telemedicine in the delivery of modern healthcare. Academy of Journal Management. 2017 Sep 1;17(3):20–28.

[215] Bashshur RL, Smith GWS, Alverson BR, Antoniotti DC, Barsan N, Bashshur WG. et al. The empirical foundations of telemedicine interventions for chronic disease management. Telemedicine Journal and E-health. 2014 Sep 1;20(9):769–800.

[216] Justice EO. E-healthcare/telemedicine readiness assessment of some selected states in Western Nigeria. International Journal of Engineering & Technology. 2012 Feb;2(2):195–201.

[217] Sohn S, Helms TM, Pelleter JT, Müller A, Kröttinger AI, Schöffski O. Costs and benefits of personalized healthcare for patients with chronic heart failure in the care and education program "Telemedicine for the Heart". Telemedicine and e-Health. 2012 Apr 1;18(3):198–204.

[218] Mariani AW, Pêgo-Fernandes PM. Telemedicine: A technological revolution. São Paulo Medical Journal. 2012;130(5):277–78.

[219] Williams OE, Elghenzai S, Subbe C, Wyatt JC, Williams J The use of telemedicine to enhance secondary care: Some lessons from the front line. Future Healthcare Journal. 2017 Jun;4(2):109.

[220] Haleem M, Singh A, Suman RP. Industrial perspectives of 3D scanning: Features, roles and it's analytical applications. Sensors International 2021 Jun 30. Javaid. 100114.

[221] Purohit B, Vernekar PR, Shetti NP, Chandra P. Biosensor nanoengineering: Design, operation, and implementation for biomolecular analysis. Sensors International. 2020 Jan 1;1:100040.

[222] Malhotra BD, Chaubey A. Biosensors for clinical diagnostics industry. Sensors and Actuators, B: Chemical. 2003 Jun 1;91(1–3):117–27.

[223] Chandra P. Miniaturized label-free smartphone assisted electrochemical sensing approach for personalised COVID-19 diagnosis, sensors. International. 2020;1:100019.

Index

https://doi.org/10.1515/9783111436432-011

De Gruyter Series
Smart Computing Applications

ISSN 2700-6239
e-ISSN 2700-6247

Sustainability in Healthcare. mHealth AI and Robotics
Rishabha Malviya, Sonali Sundram and Babita Gupta, 2024
ISBN 978-3-11-143635-7, e-ISBN (PDF) 978-3-11-143643-2, e-ISBN (EPUB) 978-3-11-143658-6

Digital Transformation in Healthcare 5.0. Metaverse, Nanorobots and Machine Learning
Edited by Rishabha Malviya, Sonali Sundram, Rajesh Kumar Dhanaraj and Seifedine Kadry, 2024
ISBN 978-3-11-139738-2, e-ISBN (PDF) 978-3-11-139854-9, e-ISBN (EPUB) 978-3-11-139911-9

Digital Transformation in Healthcare 5.0. IoT, AI and Digital Twin
Edited by Rishabha Malviya, Sonali Sundram, Rajesh Kumar Dhanaraj and Seifedine Kadry, 2024
ISBN 978-3-11-132646-7, e-ISBN (PDF) 978-3-11-132785-3, e-ISBN (EPUB) 978-3-11-132786-0

Digital Twins. Internet of Things, Machine Learning, and Smart Manufacturing
Yogini Borole, Pradnya Borkar, Roshani Raut, Vijaya Parag Balpande and Prasenjit Chatterjee, 2023
ISBN 978-3-11-077878-6, e-ISBN (PDF) 978-3-11-077886-1, e-ISBN (EPUB) 978-3-11-077896-0

Deep Learning for Cognitive Computing Systems
Edited by M. G. Sumithra, Rajesh Kumar Dhanaraj, Celestine Iwendi, Anto Merline Manoharan, 2023
ISBN 978-3-11-075050-8, e-ISBN (PDF) 978-3-11-075058-4, e-ISBN (EPUB) 978-3-11-075061-4

Cloud Analytics for Industry 4.0
Edited by Prasenjit Chatterjee, Dilbagh Panchal, Dragan Pamucar and Sharfaraz Ashemkhani Zolfani, 2022
ISBN 978-3-11-077149-7, e-ISBN (PDF) 978-3-11-077157-2, e-ISBN (EPUB) 978-3-11-077166-4

Advances in Industry 4.0
Concepts and Applications
Edited by M. Niranjanamurthy, Sheng-Lung Peng, E. Naresh, S. R. Jayasimha, Valentina Emilia Balas, 2022
ISBN 978-3-11-072536-0, e-ISBN (PDF) 978-3-11-072549-0, e-ISBN (EPUB) 978-3-11-072553-7

Soft Computing and Optimization Techniques for Sustainable Agriculture
Debesh Mishra, Suchismita Satapathy, Prasenjit Chatterjee, 2022
ISBN 978-3-11-074495-8, e-ISBN (PDF) 978-3-11-074536-8, e-ISBN (EPUB) 978-3-11-074537-5

Knowledge Engineering for Modern Information Systems
Methods, Models and Tools
Edited by Anand Sharma, Sandeep Kautish, Prateek Agrawal, Vishu Madaan, Charu Gupta, Saurav Nanda, 2021
ISBN 978-3-11-071316-9, e-ISBN (PDF) 978-3-11-071363-3, e-ISBN (EPUB) 978-3-11-071369-5

www.degruyter.com

Knowledge Management and Web 3.0
Next Generation Business Models
Edited by Sandeep Kautish, Deepmala Singh, ZdzIslaw Polkowski, Alka Mayura, Mary Jeyanthi, 2021
ISBN 978-3-11-072264-2, e-ISBN (PDF) 978-3-11-072278-9, e-ISBN (EPUB) 978-3-11-072293-2

Cloud Security
Techniques and Applications
Edited by Sirisha Potluri, Katta Subba Rao, Sachi Nandan Mohanty, 2021
ISBN 978-3-11-073750-9, e-ISBN (PDF) 978-3-11-073257-3, e-ISBN (EPUB) 978-3-11-073270-2